D0203151

APPLIED MECHANICAL VIBRATIONS

McGraw-Hill Series in Mechanical Engineering

Consulting Editor

Jack P. Holman, *Southern Methodist University*

Barron: *Cryogenic Systems*
Eckert: *Introduction to Heat and Mass Transfer*
Eckert and Drake: *Analysis of Heat and Mass Transfer*
Eckert and Drake: *Heat and Mass Transfer*
Ham, Crane, and Rogers: *Mechanics of Machinery*
Hartenberg and Denavit: *Kinematic Synthesis of Linkages*
Hinze: *Turbulence*
Hutton: *Applied Mechanical Vibrations*
Jacobsen and Ayre: *Engineering Vibrations*
Juvinall: *Engineering Considerations of Stress, Strain, and Strength*
Kays and Crawford: *Convective Heat and Mass Transfer*
Lichty: *Combustion Engine Processes*
Martin: *Kinematics and Dynamics of Machines*
Phelan: *Dynamics of Machinery*
Phelan: *Fundamentals of Mechanical Design*
Pierce: *Acoustics: An Introduction to Its Physical Principles and Applications*
Raven: *Automatic Control Engineering*
Schenck: *Theories of Engineering Experimentation*
Schlichting: *Boundary-Layer Theory*
Shigley: *Dynamic Analysis of Machines*
Shigley: *Kinematic Analysis of Mechanisms*
Shigley: *Mechanical Engineering Design*
Shigley: *Simulation of Mechanical Systems*
Shigley and Uicker: *Theory of Machines and Mechanisms*
Stoecker: *Refrigeration and Air Conditioning*

APPLIED MECHANICAL VIBRATIONS

David V. Hutton, Ph.D.

College of Engineering
Clemson University

McGraw-Hill Book Company

New York St. Louis San Francisco Auckland Bogotá Hamburg
Johannesburg London Madrid Mexico Montreal New Delhi
Panama Paris São Paulo Singapore Sydney Tokyo Toronto

APPLIED MECHANICAL VIBRATIONS

1234567890 DODO 89876543210

This book was set in Times Roman by Science Typographers, Inc.
The editors were Frank J. Cerra and J. W. Maisel;
the production supervisor was John Mancia.
The drawings were done by ECL Art Associates, Inc.
The cover was designed by Robin Hessel.
R. R. Donnelley & Sons Company was printer and binder.

Library of Congress Catalog in Publicaton Data

Hutton, David V
Applied mechanical vibrations.

(McGraw-Hill series in mechanical engineering)
Bibliography: p.
Includes index.
1. Vibration. I. Title.
TA355.H87 620.3 80-11808
ISBN 0-07-031549-3

To my parents

CONTENTS

PREFACE

This text is an outgrowth of my lecture notes on vibrations, which were used as the basis for a mechanical vibrations course at Virginia Polytechnic Institute and State University. As such, it represents an integration of my teaching, research, and industrial experience in the field of vibration analysis. The primary objective is to present the basic theory of vibrations in an orderly manner and to include the practical implications of the theory through discussion, examples, and student exercises. Mathematical developments are treated in detail to foster a basic understanding. However, once a result is established, subsequent problems of a similar nature are solved by analogy to avoid unnecessary repetition. This text is not meant to cover the entire field of vibrations. Rather, it covers the basic areas important to the field and necessary for successful study of specialized advanced topics.

A knowledge of statics, dynamics, strength of materials, and calculus through differential equations is prerequisite to successful use of this text. The text can be used for a two-course sequence in vibrations beginning at the junior and senior levels. The second course may be a senior elective or a dual course open to seniors and graduate students.

Chapter 1 is a brief review of newtonian mechanics of one- and two-dimensional motion. I have found such a review to be essential, since as much as a year may have elapsed since the student's formal course in dynamics. In adherence to the trend toward metrication, both the International (SI) and U.S. Customary Systems of units are discussed. Text examples and problems based on each system are included in approximately equal numbers throughout.

Chapter 2 discusses the basic methods of solving linear differential equations with constant coefficients. This material is intended as a review for engineering students. Having taught vibrations at the senior level in engineering technology, I have found that a thorough coverage of this material is necessary for that purpose.

Chapter 3 addresses the basic theory of vibrations through simple harmonic oscillations. Emphasis is on the relationship between natural frequency and the physical parameters of the system. Energy methods are introduced both to provide knowledge of this approximate technique and to illustrate the effects of previous assumptions of massless springs.

Chapter 4 presents the response of undamped single-degree-of-freedom systems to harmonic forcing functions. Resonance and beating are discussed. This discussion of undamped forced vibrations is treated separately, so as to introduce one new concept at a time.

Chapter 5 introduces the concept of damping in mechanical systems, including viscous, structural, and Coulomb damping. Free and forced vibrations of viscously damped systems are treated in detail, and the concept of equivalent viscous damping is discussed. A brief discussion of self-excited vibrations is included.

Chapter 6 deals with field measurement and interpretation of vibration data for rotating equipment. The objective is to expose the student to identification of a vibration source through frequency analysis, utilizing the previous theory. A detailed discussion of dynamic-balancing procedures not often found in vibration texts is included. A new method of two-plane balancing calculations is introduced.

Chapter 7 is concerned with the methods of analyzing system response to nonharmonic forcing functions. Methods discussed include Fourier series, the convolution integral, Laplace transform, and numerical solution.

Chapter 8 introduces the mathematics of two-degrees-of-freedom systems. Matrix methods are introduced to make this chapter a foundation for the systems with multiple degrees of freedom to be studied later.

Chapter 9 is a brief course on the fundamentals and applications of electronic analog computers. No previous knowledge is assumed, and applications are presented for one and two degrees of freedom.

Chapter 10 is concerned with the vibration of systems with multiple degrees of freedom. The emphasis is on matrix formulation for digital-computer solution. FORTRAN programs for matrix operations are discussed, and example programs are included.

Chapter 11 is an introduction to the theory of vibrations of continuous systems. Exact solutions to classical problems are discussed, as are Fourier analysis and numerical techniques.

The material of Chapters 1 through 7 is expected to form the basis for a first course in vibrations. Chapters 8 through 11 may then be used for a second course. Several options are available, however, depending upon course objectives. Where suitable equipment is available, Chapters 6 and 9 provide a basis for practical laboratory exercises. Chapter 8 can be used in a first course in place of Chapter 7 without loss of continuity. Chapters 8, 10, and 11, with parallel emphasis on student use of the digital computer, can be used as an intermediate or graduate-level course in many programs.

I am indebted to many colleagues and former students whose criticisms and suggestions at various stages have contributed to the development of this text. Special thanks are due to Mrs. Vicky Trump, who typed most of the original draft, and Mrs. Jean Tulli, who typed the final manuscript. Finally, I thank my family for patience and encouragement during the many hours of preparation of this text.

David V. Hutton

APPLIED MECHANICAL VIBRATIONS

CHAPTER

ONE

INTRODUCTION

Owing to their inherent flexibility, the component parts of structures and machinery are capable of relative motion when subjected to internal or external force systems. If such motion is oscillatory in nature, the motion is referred to as *vibration*. Although applications exist where vibration performs a useful function, vibration is, in general, undesirable as it tends to accelerate the wearing of parts such as bearings and gears, create excessive noise, and transmit undesirable forces and/or motion to other equipment.

1.1 NEWTON'S LAWS OF MOTION

Vibration is a dynamic process, and, as such, the study of vibratory motion is based upon the fundamental laws of motion as set forth by Sir Isaac Newton (1642–1727). Newton's basic laws may be stated as follows:

First law If the resultant force on a particle is zero, the particle will remain at rest if initially at rest or will continue to move in a straight line with constant speed if initially in motion.

Second law If the resultant force on a particle is not zero, the particle will undergo an acceleration having a magnitude directly proportional to the magnitude of the resultant force and having the same line of action and sense as the resultant force.

Third law The mutual forces exerted by two particles on each other are equal in magnitude, opposite in sense, and have the same line of action.

Newton's laws of motion apply to particles, a particle being defined as an element of mass having physical dimensions so small as to occupy only a single point in space. In a very rigorous sense, this represents a sizable restriction on the applicability of Newton's laws of motion to engineering systems. However, the principles governing the dynamic behavior of larger, finite-sized bodies are extensions or adaptations of the basic laws of particle motion. Subsequent sections of this chapter will deal with the principles of rigid body motion.

1.2 PARTICLE MOTION; NEWTON'S SECOND LAW

In mathematical form, Newton's second law may be written as

$$\mathbf{R} = m\mathbf{a} \tag{1.1}$$

where $\mathbf{R}$ is the resultant external force acting on a particle, m is the mass of the particle, and $\mathbf{a}$ is the absolute acceleration of the particle. Since force and acceleration are vector quantities, Newton's second law is a vector equation; care must be exercised in treating the quantities properly in both magnitude and direction.

Consider the plane motion of a particle of mass m as shown in Fig. 1.1, where a rectangular coordinate system has been added for convenience. The particle is acted upon by forces $\mathbf{F}_1$, $\mathbf{F}_2$, and $\mathbf{F}_3$ having a resultant $\mathbf{R}$ with a direction defined by the angle θ from the x axis as shown. According to Newton's second law, the acceleration of the particle will have the same direction as the resultant force, as shown in the figure.

To simplify the mathematics, it is convenient to replace the vector form of Newton's law with equivalent scalar equations written with reference to the coordinate axes. This change is based upon the property that a vector quantity may be replaced by an equivalent set of vectors if the resultant of the set is the same as the original vector. Such equivalent vectors are known as the *components* of the original vector. It is particularly convenient to work with vector components which are mutually perpendicular as shown in Fig. 1.2, where the resultant force and the acceleration of the particle have each been resolved into a pair of vector components having directions parallel to the respective axes. Using

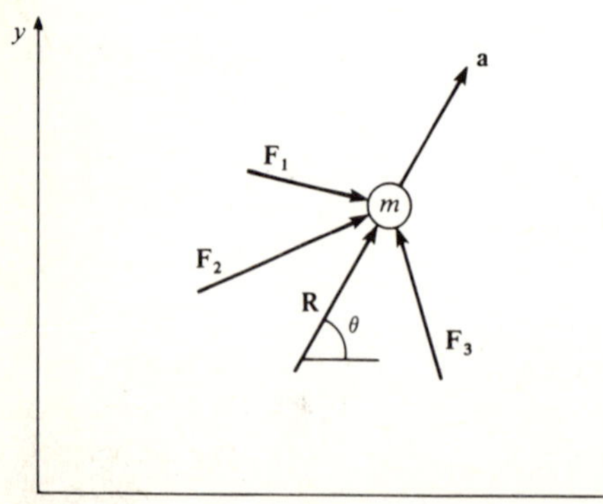

Figure 1.1

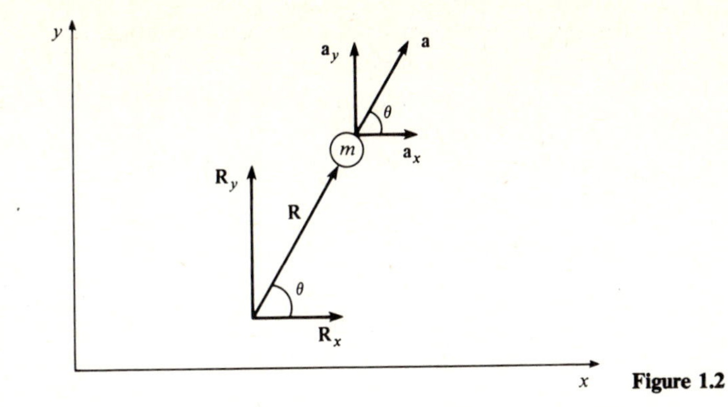

Figure 1.2

right-triangle trigonometry, we have $R_x = R\cos\theta$, $R_y = R\sin\theta$, $a_x = a\cos\theta$, and $a_y = a\sin\theta$. Note that it is not necessary to use vector notation with the components, since the subscript indicates that the direction of the component is parallel to either the x or y axis, and the algebraic sign of the component indicates whether the sense is in the positive or negative coordinate direction.

By using the rectangular components, Newton's second law for the motion of the particle can be written as the two scalar equations

$$R_x = ma_x \tag{1.2}$$

and

$$R_y = ma_y \tag{1.3}$$

The magnitudes of the resultant force and its components are related by

$$R = \sqrt{R_x^2 + R_y^2} \tag{1.4}$$

and similarly, for acceleration,

$$a = \sqrt{a_x^2 + a_y^2} \tag{1.5}$$

It may also be noted that

$$\tan\theta = \frac{R_y}{R_x} = \frac{a_y}{a_x} \tag{1.6}$$

may be used to obtain the orientation of the resultant force and acceleration if the components are known.

1.3 PLANE MOTION OF RIGID BODIES

To relate the motions of finite-sized bodies to the force systems which cause those motions, Newton's second law of motion must be supplemented by additional equations relating to the rotational motion of the body. Figure 1.3 depicts a rigid body undergoing general plane motion under the action of external forces $\mathbf{F}_1$ and $\mathbf{F}_2$. Point G locates the mass center of the body. The body

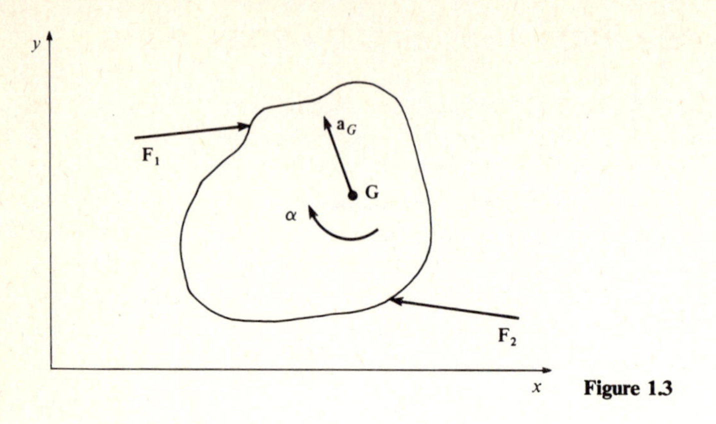

Figure 1.3

is assumed to be unconstrained so that it is free to translate and rotate in its plane of motion.

The motion of the mass center G is governed by Newton's second law as

$$\Sigma \mathbf{F} = m\mathbf{a}_G \tag{1.7}$$

where $\mathbf{a}_G$ is the absolute linear acceleration of the mass center, and m is the total mass of the body. In other words, the motion of the mass center is the same as that of a particle into which is concentrated the total mass of the body and to which the same external forces are applied. By the method of the previous section, the equation governing the motion of the mass center can also be written in component form as

$$\Sigma F_x = ma_{Gx} \tag{1.8}$$

and

$$\Sigma F_y = ma_{Gy} \tag{1.9}$$

In general, a body executing unconstrained plane motion will undergo rotation as well as translation. Whereas translation of the body is described by Eq. (1.7), a separate relation must be utilized to determine the rotational motion. As shown in essentially any undergraduate dynamics text,[1] the rotation of the body is governed by

$$\Sigma \mathbf{M}_G = I_G \boldsymbol{\alpha} \tag{1.10}$$

where $\Sigma \mathbf{M}_G$ is the net moment of all external forces about the mass center, I_G is the mass moment of inertia of the body about its mass center, and $\boldsymbol{\alpha}$ is the angular acceleration of the body. In plane motion, rotation is either clockwise or counterclockwise so that the vector equation (1.10) may be replaced with a scalar equation by adopting an appropriate sign convention.

[1] See F. P. Beer and E. R. Johnston, Jr., *Mechanics for Engineers*, vol. 2: *Dynamics*, 3d ed., McGraw-Hill, New York, 1976, chap. 16.

1.4 ROTATION OF A RIGID BODY ABOUT A FIXED POINT

Of more practical significance than general plane motion is the case in which a rigid body is constrained so that it must rotate about an axis through a fixed point. Figure 1.4*a* shows a rigid body subjected to external forces $\mathbf{F}_1$ and $\mathbf{F}_2$ and constrained to rotate about a fixed axis perpendicular to the plane of motion and passing through point C. Owing to the constraint at C, every point in the body moves in a circular path centered at C; in this case, the body is said to be in *pure rotation*. Figure 1.4*b* shows the free-body diagram, including the unknown reaction force $\mathbf{R}_C$ arising from the constraint. Applying the equations of the preceding section, we have

$$\Sigma F = \mathbf{F}_1 + \mathbf{F}_2 + \mathbf{R}_C = m\mathbf{a}_G \tag{1.11}$$

and

$$\Sigma \mathbf{M}_G = I_G \boldsymbol{\alpha} \tag{1.12}$$

which state that the original force system is equivalent to a force system composed of a single force $m\mathbf{a}_G$ and a single moment $I_G\boldsymbol{\alpha}$ acting at the center of mass as shown in Fig. 1.4*c*.

Utilizing the fact that a vector quantity can be replaced by an equivalent set of components, we resolve the equivalent force $m\mathbf{a}_G$ into the two components

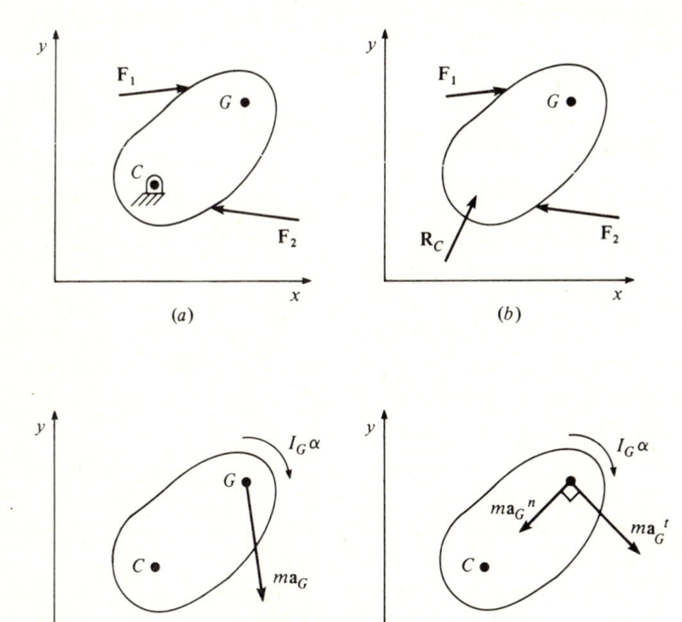

Figure 1.4

shown in Fig. 1.4*d*. The first component ma_G^n is directed from G toward the center of curvature C. The component a_G^n is called the *normal acceleration* of the mass center since its direction is normal, or perpendicular, to the path of motion of the mass center. The second component ma_G^t is in the direction of the tangent to the path of motion of G and is thus perpendicular to ma_G^n. The component a_G^t is the tangential acceleration of the mass center. Denoting by R_{CG} the distance from the axis of rotation to the mass center and taking the moment of the equivalent force system about point C give

$$\Sigma M_C = I_G\alpha + R_{CG}ma_G^t \tag{1.13}$$

Substituting the kinematic relation $a_G^t = R_{CG}\alpha$ into Eq. (1.13) gives

$$\Sigma M_C = (I_G + mR_{CG}^2)\alpha \tag{1.14}$$

as the equation governing the rotation of a rigid body about a fixed point. By application of the parallel-axis theorem (see App. A.2), the quantity in parentheses is recognized as the mass moment of inertia of the body about[2] point C. Thus, for rigid body rotation about a fixed point C, the equation governing the motion may be written as

$$\Sigma \mathbf{M}_C = I_C\boldsymbol{\alpha} \tag{1.15}$$

where $\Sigma \mathbf{M}_C$ is the resultant moment of all external forces about C, I_C is the mass moment of inertia of the body about point C, and $\boldsymbol{\alpha}$ is the absolute angular acceleration of the body.

1.5 VIBRATION AS A DYNAMIC PROCESS

The study of vibrations involves the solution of dynamics problems in which bodies are subjected to forces which may vary with time and with the displacement of the body itself. In mechanical equipment, forces which vary with time are present due to the motion and speed characteristics of the equipment, whereas forces which vary with displacement are due to the elastic behavior of the machine components.

For a rigid body subjected to forces which are constant or vary only with time, the equations of motion discussed in the previous sections may be used directly to determine the acceleration of the body. Once the acceleration is known, as either a constant or a function of time, successive integrations with respect to time will yield expressions for the velocity and displacement of the body.

In the case of a rigid body which moves under the action of one or more forces which vary with the displacement of the body, the equations of motion cannot be integrated directly to obtain the velocity and displacement. As an example, consider a problem in which a particle is constrained to move in a straight line and is acted upon by a force which is directly proportional to the

[2] Rigorously, about an axis perpendicular to the plane of motion and passing through point C.

displacement but which acts in a direction such as to oppose the motion. Denoting by k the proportionality constant for the force and recalling that acceleration is the second derivative of displacement with respect to time, one may write Newton's second law for the motion of the particle as

$$-kx = m\frac{d^2x}{dt^2} \tag{1.16}$$

and

$$m\frac{d^2x}{dt^2} + kx = 0 \tag{1.17}$$

Thus, for this problem, the equation of motion is a second-order differential equation which must be solved to determine the displacement of the body at any time. It is characteristic of vibrating systems that the equations of motion are differential equations. The methods of solution of the types of differential equations most commonly encountered in the study of mechanical vibrations are discussed in the next chapter.

1.6 SYSTEMS OF UNITS

Newton's second law of motion states that the acceleration of a particle is directly proportional to the resultant force which acts on the particle. If the units used to describe the physical quantities are chosen properly, the proportionality constant is the mass of the particle, and we can then write $\mathbf{F} = m\mathbf{a}$. This equality involves the four units of force, mass, length, and time. Three of the four units may be chosen arbitrarily, but the fourth unit must be chosen such that equality holds. The units so chosen are said to form a *consistent* system of units.

In this text, we shall use two systems of consistent units which are in common use in the United States. These are known as United States customary units and the International System of Units (SI units).[3]

United States Customary Units

This is the primary system used by engineers in the United States for many years. The base units which are chosen arbitrarily are the units of force, length, and time. These units are the *pound* (lb), the *foot* (ft), and the *second* (s), respectively. The foot is subdivided into 12 *inches* (in). Each of these units is precisely defined in terms of standards maintained by the National Bureau of Standards in Washington, D.C. Of particular note is the fact that the pound is defined as the weight (force of gravitational attraction) of a platinum standard at sea level and at the latitude of 45°. Thus, the United States customary system is a *gravitational* system rather than an absolute system.

[3] Systéme International d'Unites.

The unit of mass is chosen so that $\mathbf{F} = m\mathbf{a}$ is satisfied and is known as a *slug*. Substituting units gives

$$1 \text{ lb} = (1 \text{ slug})(1 \text{ ft/s}^2)$$

from which

$$1 \text{ slug} = 1 \text{ lb} \cdot \text{s}^2/\text{ft}$$

In words, one slug is the mass which will be accelerated at one foot per second per second by a force of one pound. Where it is convenient to express length in inches, mass is described in pound seconds squared per inch to form a consistent system.

In this system, a quantity of material is most often described by its weight in pounds. In dynamics it is necessary to work with mass, so a relation must be established between the two. Since a freely falling body is subjected only to the gravitational force (its weight) and acquires an acceleration equal to the acceleration of gravity g, we have

$$m = \frac{W}{g} \tag{1.18}$$

where g is expressed as 32.2 ft/s^2 or 386.4 in/s^2 in the vicinity of the earth's surface.

SI Units

In the International System of Units, the arbitrarily chosen units are those of mass, length, and time. These basic units are the *kilogram* (kg), the *meter* (m), and the *second* (s). The meter is often subdivided into 100 *centimeters* (cm) and 1000 *millimeters* (mm). These units are precisely defined in terms of standards maintained by the International Bureau of Weights and Measures near Paris, France.

The unit of force, the *newton* (N), is defined as that force which will impart an acceleration of one meter per second per second to a mass of one kilogram. Thus we have

$$1 \text{ N} = (1 \text{ kg})(1 \text{ m/s}^2) = 1 \text{ kg} \cdot \text{m/s}^2$$

Since the units of the SI system are independent of gravity, it is said to form an *absolute* system of units. If, however, we wish to express the weight of a body in SI units, we rewrite Eq. (1.18) as

$$W = mg \tag{1.19}$$

where $g = 9.81 \text{ m/s}^2$ to obtain the weight in newtons. This shows that the weight of a 1-kg mass is 9.81 N.

Conversion from United States customary units to SI units can be accomplished with the following conversion factors:

Force:	1 lb = 4.448 N
Mass:	1 slug = 14.59 kg
Length:	1 ft = 0.3048 m

CHAPTER
TWO

SECOND-ORDER, ORDINARY DIFFERENTIAL EQUATIONS WITH CONSTANT COEFFICIENTS

As previously discussed, Newton's second law for the motion of a particle and its counterpart for rigid-body rotation are second-order differential equations. In the mathematical study of vibrating systems, differential equations of the form

$$\frac{d^2x}{dt^2} + a_1\frac{dx}{dt} + a_2x = f(t) \tag{2.1}$$

are frequently encountered. Here, $x = x(t)$ represents the displacement of the system as a function of time, a_1 and a_2 are constant coefficients, and $f(t)$ is an external forcing function. The physical interpretation of each term will be examined in detail later. At present, we shall consider Eq. (2.1) in a purely mathematical sense and examine the various solutions which may exist.

2.1 HOMOGENEOUS EQUATION, $f(t) = 0$

If $f(t) = 0$, Eq. (2.1) becomes

$$\ddot{x} + a_1\dot{x} + a_2x = 0 \tag{2.2}$$

where a shorthand notation has been adopted in which the dots denote differentiation with respect to time. A differential equation such as Eq. (2.2), in which all terms contain the unknown function or its derivatives, is known as a *homogeneous* differential equation.

The solution to Eq. (2.2) is a function $x(t)$ which satisfies the equation for all values of time. This means that the function $x(t)$ must be such that its second derivative plus a constant multiple of its first derivative plus a constant multiple of the function itself must be identically zero independent of time. The implication is that the function $x(t)$ must be such that its form is unchanged by the process of differentiation. From elementary differential calculus, it is known that the linear exponential functions satisfy this criterion since their derivatives are also exponential functions. Thus, it seems reasonable to assume a solution to Eq. (2.2) of the form

$$x(t) = Ce^{st} \tag{2.3}$$

where C and s are constants which are to be determined. The derivatives of the assumed solution are

$$\dot{x}(t) = Cse^{st} \tag{2.4}$$

and

$$\ddot{x}(t) = Cs^2e^{st} \tag{2.5}$$

Substitution of the assumed solution and its derivatives into $\ddot{x} + a_1\dot{x} + a_2x = 0$ results in

$$Cs^2e^{st} + a_1Cse^{st} + a_2Ce^{st} = 0 \tag{2.6}$$

or

$$(s^2 + a_1s + a_2)Ce^{st} = 0 \tag{2.7}$$

If the assumed solution is the true solution to the differential equation, then Eq. (2.7) must be satisfied for all values of time. Since the exponential e^{st} is nonzero for all time, the only possible solutions to Eq. (2.7) are $C = 0$ and $s^2 + a_1s + a_2 = 0$. If $C = 0$, the assumed solution is itself identically zero, and we have obtained what is referred to as the *trivial* solution, since no worthwhile information is obtained. By this process of elimination, the solution of the differential equation is found to depend upon the algebraic equation

$$s^2 + a_1s + a_2 = 0 \tag{2.8}$$

Equation (2.8) is known as the *auxiliary* equation for the original differential equation (2.2).

Application of the quadratic formula gives two values,

$$s_1 = \frac{-a_1 + \sqrt{a_1^2 - 4a_2}}{2} \tag{2.9}$$

and

$$s_2 = \frac{-a_1 - \sqrt{a_1^2 - 4a_2}}{2} \tag{2.10}$$

which satisfy Eq. (2.8). The results of this procedure are two independent solutions

$$x_1(t) = C_1e^{s_1t} \tag{2.11}$$

and

$$x_2(t) = C_2e^{s_2t} \tag{2.12}$$

where C_1 and C_2 are arbitrary constants. The complete solution is then the sum of the two independent solutions[1] and is given by

$$x(t) = C_1 e^{s_1 t} + C_2 e^{s_2 t} \tag{2.13}$$

provided s_1 and s_2 are not equal. The case of equal roots will be discussed in detail later.

The character of the solution given by Eq. (2.13) depends directly upon the value of the term $a_1^2 - 4a_2$ appearing in Eqs. (2.9) and (2.10). Three cases are of interest and will be considered in turn.

Case 1: $a_1^2 > 4a_2$

In this case, $\sqrt{a_1^2 - 4a_2}$ is real and the roots s_1 and s_2 of the auxiliary equation are real and distinct. The character of the solution does not change, so its form remains

$$x(t) = C_1 e^{s_1 t} + C_2 e^{s_2 t} \tag{2.14}$$

The roots s_1 and s_2 may be both positive, both negative, or one positive and one negative. Thus the solution given by Eq. (2.14) will exhibit either exponential decay or exponential growth with time.

Example 2.1 Find the complete solution to the differential equation

$$\ddot{x} - 5\dot{x} + 6x = 0$$

SOLUTION Under the assumption that $x(t) = Ce^{st}$, differentiation and substitution in the differential equation give

$$Cs^2 e^{st} - 5Cse^{st} + 6Ce^{st} = 0$$

which is the same as

$$(s^2 - 5s + 6)Ce^{st} = 0$$

The auxiliary equation is then

$$s^2 - 5s + 6 = 0$$

which factors into

$$(s - 2)(s - 3) = 0$$

The roots are then $s_1 = 2$ and $s_2 = 3$, so that the complete solution is

$$x(t) = C_1 e^{2t} + C_2 e^{3t}$$

[1] The sum of the independent solutions to a *linear* differential equation is also a solution to the differential equation. A linear differential equation is one which contains no products of the solution function and/or its derivatives.

Case 2: $a_1^2 < 4a_2$

In this instance, the quantity $a_1^2 - 4a_2$ is negative so that $\sqrt{a_1^2 - 4a_2}$ is a complex, or imaginary, number which may be written as

$$\sqrt{a_1^2 - 4a_2} = \sqrt{(-1)(4a_2 - a_1^2)} = i\sqrt{4a_2 - a_1^2} \tag{2.15}$$

where $i = \sqrt{-1}$ = imaginary unit. The roots of the auxiliary equation are then

$$s_1 = \frac{-a_1 + i\sqrt{4a_2 - a_1^2}}{2} = a_1' + ia_2' \tag{2.16}$$

and

$$s_2 = \frac{-a_1 - i\sqrt{4a_2 - a_1^2}}{2} = a_1' - ia_2' \tag{2.17}$$

where $a_1' = -a_1/2$ and $a_2' = \sqrt{4a_2 - a_1^2}\,/2$. The solution is then

$$x(t) = C_1 e^{(a_1' + ia_2')t} + C_2 e^{(a_1' - ia_2')t} \tag{2.18}$$

By utilizing the properties of exponential multiplication, the solution can be rewritten as

$$x(t) = e^{a_1't}\left(C_1 e^{ia_2't} + C_2 e^{-ia_2't}\right) \tag{2.19}$$

which is still not a workable form, owing to the appearance of the complex quantity i. At this point, we shall make use of two general relations, known as *Euler's formulas*, to convert the complex exponentials to equivalent trigonometric functions. These relations are

$$e^{i\theta} = \cos\theta + i\sin\theta \tag{2.20}$$

and

$$e^{-i\theta} = \cos\theta - i\sin\theta \tag{2.21}$$

where θ is a real function. In the problem at hand, we have $\theta = a_2't$, so we can write

$$e^{ia_2't} = \cos a_2't + i\sin a_2't \tag{2.22}$$

and

$$e^{-ia_2't} = \cos a_2't - i\sin a_2't \tag{2.23}$$

Substitution of Eqs. (2.22) and (2.23) into Eq. (2.19) yields

$$x(t) = e^{a_1't}(C_1\cos a_2't + iC_1\sin a_2't + C_2\cos a_2't - iC_2\sin a_2't) \tag{2.24}$$

or

$$x(t) = e^{a_1't}\left[(C_1 + C_2)\cos a_2't + i(C_1 - C_2)\sin a_2't\right] \tag{2.25}$$

Finally, we define two new constants A and B as

$$A = C_1 + C_2 \tag{2.26}$$

and

$$B = i(C_1 - C_2) \tag{2.27}$$

to obtain the desired form of the solution as

$$x(t) = e^{a_1't}(A\cos a_2't + B\sin a_2't) \tag{2.28}$$

It may appear that Eq. (2.27) is used to remove the complex nature of the solution artificially, and this is true in general. However, for differential equations having the form of Eq. (2.2) and corresponding to physical systems, it can be shown that C_1 and C_2 are complex conjugates,[2] and A and B are therefore real numbers.

The solution for case 2 as given by Eq. (2.28) is the product of an exponential function which grows or decays with time and a harmonic function which oscillates with time. The physical interpretation of this solution in terms of vibrating systems will be given in detail in Chap. 5.

Example 2.2 Find the solution of the differential equation

$$4\ddot{x} - 8\dot{x} + 7x = 0$$

SOLUTION To apply the procedure developed thus far, we rewrite the given differential equation as

$$\ddot{x} - 2\dot{x} + \tfrac{7}{4}x = 0$$

from which $a_1 = -2$, $a_2 = \frac{7}{4}$, and $a_1^2 - 4a_2 = -3 < 0$ so that the method of case 2 is applicable. Then

$$a_1' = \frac{-a_1}{2} = 1 \qquad a_2' = \frac{\sqrt{4a_2 - a_1^2}}{2} = \frac{\sqrt{3}}{2}$$

and substitution for a_1' and a_2' into Eq. (2.28) gives the solution as

$$x(t) = e^t\left(A\cos\frac{\sqrt{3}}{2}t + B\sin\frac{\sqrt{3}}{2}t\right)$$

Case 3: $a_1 = 4a_2$

For this case, $\sqrt{a_1^2 - 4a_2} = 0$, the roots of the auxiliary equation are identical and are given by

$$s_1 = s_2 = -\frac{a_1}{2} \tag{2.29}$$

and only one solution

$$x_1 = C_1 e^{-(a_1/2)t} \tag{2.30}$$

is obtained. This is not the complete solution, since a second-order, homogeneous differential equation must have two independent solutions, each containing an arbitrary constant of integration. This dilemma will be resolved by recalling the following theorem from the mathematical theory of differential equations.

[2] Two complex numbers are said to be *conjugate* if the real parts of the two numbers are the same and if the imaginary parts differ only in sign. Thus, the two complex numbers $C_1 = 2 + 3i$ and $C_2 = 2 - 3i$ are complex conjugates. Further, it can be seen from this definition that the quantities $C_1 + C_2 = 4$ and $i(C_1 - C_2) = 6i^2 = -6$ are both real.

Theorem If we know one solution, say $x = x_1$, of the second-order differential equation $\ddot{x} + a_1\dot{x} + a_2 x = 0$, then the substitution $x = x_1 v$ will transform the given equation into an equation of the first order in $\dot{v}$ which may be solved for the function v. The complete solution to the original equation is then $x = x_1 v$.

To apply the theorem to case 3, let

$$x = x_1 v = C_1 e^{-(a_1/2)t} v \tag{2.31}$$

from which

$$\dot{x} = C_1\left(-\frac{a_1}{2}\right)e^{-(a_1/2)t}v + C_1 e^{-(a_1/2)t}\dot{v} \tag{2.32}$$

and

$$\ddot{x} = C_1\left(\frac{a_1}{2}\right)^2 e^{-(a_1/2)t}v + 2C_1\left(-\frac{a_1}{2}\right)e^{-(a_1/2)t}\dot{v} + C_1 e^{-(a_1/2)t}\ddot{v} \tag{2.33}$$

Substituting into the original equation

$$\ddot{x} + a_1\dot{x} + a_2 x = 0 \tag{2.34}$$

gives, after canceling and collecting terms,

$$\ddot{v} + \left(\frac{a_1^2}{4} - \frac{a_1^2}{2} + a_2\right)v = 0 \tag{2.35}$$

which is the same as

$$\ddot{v} + \left(\frac{4a_2 - a_1^2}{4}\right)\dot{v} = 0 \tag{2.36}$$

But for case 3, $4a_2 = a_1^2$, so the resulting equation for v is simply

$$\ddot{v} = 0 \tag{2.37}$$

which, after two integrations, gives

$$v = C_2 + C_3 t \tag{2.38}$$

Then, according to the theorem, the complete solution is

$$x(t) = C_1 e^{-(a_1/2)t}(C_2 + C_3 t) \tag{2.39}$$

or

$$x(t) = (A + Bt)e^{-(a_1/2)t} \tag{2.40}$$

which indeed contains two independent solutions as may be verified by substitution into the original equation.

Example 2.3 Determine the complete solution of

$$\ddot{x} + 2\dot{x} + x = 0$$

Solution Here, $a_1 = 2$, $a_2 = 1$, and the auxiliary equation is

$$s^2 + 2s + 1 = 0$$

which factors into

$$(s + 1)^2 = 0$$

giving the roots $s_1 = s_2 = -1$. Thus, we have a case of repeated roots, and the complete solution is

$$x(t) = (A + Bt)e^{-t}$$

2.2 INHOMOGENEOUS EQUATIONS, $f(t) \neq 0$

To obtain the complete solution to the differential equation

$$\ddot{x} + a_1\dot{x} + a_2 x = f(t) \tag{2.41}$$

where $f(t)$ is any nonzero function of time or a constant, we must determine both a *particular* solution, denoted by x_p, which satisfies Eq. (2.41), and the *homogeneous* solution, denoted by x_h, which satisfies Eq. (2.41) with the right-hand side set equal to zero. The particular solution is so called because it satisfies the differential equation for a particular form of the function $f(t)$. The particular solution is, by definition, such that

$$\frac{d^2x_p}{dt^2} + a_1\frac{dx_p}{dt} + a_2x_p = f(t) \tag{2.42}$$

while the homogeneous solution satisfies

$$\frac{d^2x_h}{dt^2} + a_1\frac{dx_h}{dt} + a_2x_h = 0 \tag{2.43}$$

Adding Eqs. (2.42) and (2.43) gives

$$\frac{d^2x_p}{dt^2} + a_1\frac{dx_p}{dt} + a_2x_p + \frac{d^2x_h}{dt^2} + a_1\frac{dx_h}{dt} + a_2x_h = f(t) + 0 \tag{2.44}$$

or

$$\frac{d^2}{dt^2}(x_p + x_h) + a_1\frac{d}{dt}(x_p + x_h) + a_2(x_p + x_h) = f(t) \tag{2.45}$$

The complete solution to the differential equation is then the sum of the particular solution and the homogeneous solution:

$$x(t) = x_p(t) + x_h(t) \tag{2.46}$$

The methods for obtaining the homogeneous solution were discussed in the previous section and will not be repeated here, except to note that the homogeneous solution will involve two arbitrary constants of integration. On the other hand, the particular solution will not contain any arbitrary constants, since it must satisfy the differential equation for a specific function $f(t)$.

2.3 METHODS FOR OBTAINING THE PARTICULAR SOLUTION

There are many methods whereby particular solutions can be obtained. One method of wide applicability, used extensively in engineering, is the method of *undetermined coefficients*. Since this method is applicable to all the differential equations to be encountered in this text, it is the only method which will be discussed here. In particular, the method of undetermined coefficients can be applied to differential equations of the form of Eq. (2.41) where $f(t)$ contains a polynomial, terms of the form $\sin pt$, $\cos pt$, and e^{pt} where p is constant, or combinations of sums and products of these terms. The method is illustrated by the examples which follow.

Example 2.4 Obtain the particular solution of

$$\ddot{x} + 12x = 3e^{2t}$$

SOLUTION The particular solution must be such that when multiplied by 12 and added to its own second derivative the result is $3e^{2t}$. Since the exponential functions do not change form in differentiation, it seems reasonable to assume the particular solution as

$$x_p(t) = ae^{2t}$$

where a is a constant whose value must be determined (thus the name for this method). Substituting the assumed solution into the differential equation gives

$$4ae^{2t} + 12ae^{2t} = 3e^{2t}$$

from which $4a + 12a = 3$ or $a = \frac{3}{16}$ yields the value of the undetermined coefficient. The particular solution is then $x_p = \frac{3}{16}e^{2t}$.

Example 2.5 Find the particular solution for

$$\ddot{x} + 2\dot{x} + x = 4 \sin 2t$$

SOLUTION Following the previous example, let us assume that $x_p(t) = a \sin 2t$ and substitute into the differential equation. This gives

$$-4a \sin 2t + 2(2a \cos 2t) + a \sin 2t = 4 \sin 2t$$

which clearly cannot be true for all values of time, due to the term containing $\cos 2t$ on the left-hand side (recall that a is *constant*). As a second possibility, assume $x_p(t) = a \sin 2t + b \cos 2t$, where a and b are both constant. Substituting this assumed particular solution into the original equation yields

$$-4a \sin 2t - 4b \cos 2t + 2(2a \cos 2t - 2b \sin 2t) + a \sin 2t + b \cos 2t = 4 \sin 2t$$

Collecting coefficients of similar terms, we have

$$(-4b + 4a + b) \cos 2t + (-4a - 4b + a) \sin 2t = 4 \sin 2t$$

Thus, the assumed form of the particular solution will satisfy the differential equation for all values of time if $4a - 3b = 0$ and $-3a - 4b = 4$. Simultaneous solution gives $a = -\frac{12}{25}$ and $b = -\frac{16}{25}$, so that the particular solution is

$$x_p(t) = -\tfrac{12}{25}\sin 2t - \tfrac{16}{25}\cos 2t$$

This example illustrates the fact that if the function $f(t)$ contains a sine or cosine term, the particular solution will, in general, contain both sine and cosine terms.

Example 2.6 Determine the particular solution of

$$\ddot{x} + 6\dot{x} + 3x = t^2e^t$$

SOLUTION In this instance, $f(t) = t^2e^t$ is a more complicated function than those encountered in the previous examples. In such cases, the proper form for the particular solution can be determined by differentiating $f(t)$ sequentially until no new functional forms appear as a result of differentiation. The assumed particular solution should then contain every functional form which appears in the various derivatives. For this problem, $f(t) = t^2e^t$ so that $\dot{f}(t) = 2te^t + t^2e^t$ and $\ddot{f}(t) = 2e^t + 4te^t + t^2e^t$, and the third derivative differs from $\ddot{f}$ only by constant values. Based on these observations, the particular solution should be assumed as

$$x_p(t) = (a + bt + ct^2)e^t$$

where a, b, and c are the undetermined constant coefficients. Differentiating, we obtain

$$\dot{x}_p(t) = (b + 2ct)e^t + (a + bt + ct^2)e^t$$

and $$\ddot{x}_p(t) = 2ce^t + 2(b + 2ct)e^t + (a + bt + ct^2)e^t$$

Substitution of these results into the original differential equation gives

$$2ce^t + 2(b + 2ct)e^t + (a + bt + ct^2)e^t$$
$$+6\left[(b + 2ct)e^t + (a + bt + ct^2)e^t\right] + 3(a + bt + ct^2)e^t = t^2e^t$$

from which a, b, and c must be determined. Grouping coefficients of similar time functions gives

$$(10a + 8b + 2c)e^t + (10b + 16c)te^t + 10ct^2e^t = t^2e^t$$

which will be satisfied for all values of t if

$$10a + 8b + 2c = 0$$
$$10b + 16c = 0$$
$$10c = 1$$

from which $a = \frac{27}{250}$, $b = -\frac{4}{25}$, and $c = \frac{1}{10}$. The particular solution is then

$$x_p(t) = \left(\tfrac{27}{250} - \tfrac{4}{25}t + \tfrac{1}{10}t^2\right)e^t$$

The method used to determine the particular solution in this example may be applied to many differential equations in which $f(t)$ is the product of a polynomial and an exponential or trigonometric function.

2.4 COMPLETE SOLUTION AND INITIAL CONDITIONS

The complete solution to a differential equation is the sum of the homogeneous solution and the particular solution. The homogeneous solution of a second-order differential equation contains two arbitrary constants. These two constants are arbitrary only in the sense that the solution will satisfy the differential equation regardless of the values of the constants. For the complete solution to represent the actual response of a physical problem such as a vibrating mechanical system, the two constants must be evaluated in terms of known quantities related to the system. In general, the values of the arbitrary constants are obtained by requiring the complete solution to satisfy two specific conditions. Since the specific conditions that must be satisfied are usually the values of $x(t)$ and $\dot{x}(t)$ at $t = 0$, the conditions are known as *initial conditions*. It should be noted, however, that the two conditions may be specified at any two values of time. The following examples will illustrate the application of initial conditions to the complete solution of second-order differential equations.

Example 2.7 Find the complete solution of

$$\ddot{x} + 2\dot{x} + x = 4 \sin 2t$$

subject to the initial conditions $x(0) = 0$, $\dot{x}(0) = 3$.

SOLUTION Since the complete solution is required, we must find both the homogeneous and particular solutions. The homogeneous solution is assumed to be $x_h(t) = Ce^{st}$, which gives the auxiliary equation

$$s^2 + 2s + 1 = 0$$

having the repeated roots $s_1 = s_2 = -1$. The homogeneous solution (case 3) is then given by

$$x_h(t) = (A + Bt)e^{-t}$$

Using the results of Example 2.5, the particular solution is found to be

$$x_p(t) = -\tfrac{12}{25}\sin 2t - \tfrac{16}{25}\cos 2t$$

Adding the homogeneous and particular solutions gives the complete solution as

$$x(t) = (A + Bt)e^{-t} - \tfrac{12}{25}\sin 2t - \tfrac{16}{25}\cos 2t$$

with the constants A and B still to be determined. To evaluate A and B, we invoke the initial conditions $x(0) = 0$ and $\dot{x}(0) = 3$ to obtain

$$x(0) = 0 = A - \tfrac{16}{25}$$

$$\dot{x}(0) = 3 = -A + B - 2 \times \tfrac{12}{25}$$

from which $A = \frac{16}{25}$ and $B = \frac{23}{5}$. Thus, the complete solution, which satisfies the differential equation and the given initial conditions, is

$$x(t) = \left(\tfrac{16}{25} + \tfrac{23}{5}t\right)e^{-t} - \tfrac{12}{25}\sin 2t - \tfrac{16}{25}\cos 2t$$

Example 2.8 Solve $\ddot{x} + x = t \cos 5t$ subject to $x(0) = 0$ and $\dot{x}(0) = 0$.

SOLUTION The homogeneous solution must satisfy $\ddot{x} + x = 0$, which has the auxiliary equation $s^2 + 1 = 0$. The roots of the auxiliary equation are $s_1 = i$ and $s_2 = -i$ so that the homogeneous solution (case 2) is

$$x_h(t) = A \sin t + B \cos t$$

Since the function $f(t) = t \cos 5t$ is the product of a polynomial and a trigonometric function, we shall attempt to determine the proper form of the particular solution by differentiating $f(t)$ twice and observing the form of the resulting functions. This gives $\ddot{f}(t) = -10 \sin 5t - 25t \cos 5t$, from which we are led to assume

$$x_p(t) = a \sin 5t + bt \cos 5t$$

Substitution of the assumed particular solution into the differential equation results in

$$(-24a - 10b) \sin 5t - 24bt \cos 5t = t \cos 5t$$

from which $a = \frac{5}{288}$ and $b = -\frac{1}{24}$.

Adding the homogeneous and particular solutions gives

$$x(t) = A \sin t + B \cos t + \tfrac{5}{288}\sin 5t - \tfrac{1}{24}t \cos 5t$$

Applying the initial conditions

$$x(0) = 0 = B$$

$$\dot{x}(0) = 0 = A + \tfrac{25}{288} - \tfrac{1}{24}$$

results in $A = -\frac{13}{288}$ and $B = 0$, so that the complete solution is

$$x(t) = -\tfrac{13}{288}\sin t + \tfrac{5}{288}\sin 5t - \tfrac{1}{24}t \cos 5t$$

2.5 CLOSING NOTE

This chapter presents a discussion of the solution of second-order, ordinary differential equations of the type most commonly encountered in elementary vibration analysis. No attempt is made to present theorems or rigorous mathematical proofs; rather, the basic methods of solution are discussed and their application illustrated in several example problems. This material is covered as a separate entity in the hope that understanding of the physical behavior of vibrating systems, as described in the following chapters, will not be overshadowed by the mathematics involved.

PROBLEMS

2.1 Find solutions for each of the following:

(*a*) $\ddot{x} + 8x = 0$

(*b*) $\ddot{x} + 8\dot{x} + 64x = 0$

(*c*) $\ddot{x} + 2\dot{x} + 3x = 0$

(*d*) $\ddot{x} + 5\dot{x} - 6x = 0$

(*e*) $3\ddot{x} - 7\dot{x} + 8x = 0$

2.2 Determine the particular solution for each of the following:

(*a*) $\ddot{x} + 4x = 4e^{2t}$

(*b*) $\ddot{x} + 8\dot{x} + 64x = 128 \cos 6t$

(*c*) $\ddot{x} + 6x = 2t^2 \sin 4t$

(*d*) $\ddot{x} - 4x = 8t^2$

(*e*) $\ddot{x} + 2\dot{x} + x = 2 \cos 2t + 3t + 2 + 3e^t$

2.3 For each of the following, determine the complete solution subject to the given conditions:

(*a*) $\ddot{x} + 20\dot{x} + 64x = 0$; $x(0) = \frac{1}{3}$, $\dot{x}(0) = 0$

(*b*) $\ddot{x} + x = te^t + 3 \cos 2t$; $x(0) = 1$, $\dot{x}(0) = 0$

(*c*) $\ddot{x} + 3\dot{x} + 4x = 6 \sin 4t$; $x(0) = 3$, $x(1) = 1$

(*d*) $\ddot{x} - 3\dot{x} + 2x = e^{-t}(1 + \sin 2t)$; $x(0) = 1$, $\dot{x}(0) = 2$

CHAPTER
THREE

FREE VIBRATIONS OF UNDAMPED SINGLE-DEGREE-OF-FREEDOM SYSTEMS

The simplest model of a vibrating mechanical system is composed of a single mass element which is connected to a rigid support through a linearly elastic spring as shown in Fig. 3.1*a*. If the mass element of such a system is displaced from its equilibrium position and released, the potential energy stored in the spring will be converted to kinetic energy of the mass as the system seeks to restore its equilibrium condition. However, owing to the kinetic energy acquired by the mass, it will pass through the equilibrium position, and the process of energy transfer will be reversed. That is, the kinetic energy of the mass will be transferred as potential energy to the spring, as the spring acts to stop the mass. In theory, this process will continue indefinitely, resulting in oscillation, or vibration, of the mass about its equilibrium position. Such an oscillation is called a *free vibration*, since the system is free of all external forces other than the weight of the mass element. Also, if the position of the mass element can be specified with a single coordinate, the system is a single-degree-of-freedom system. The term *undamped* is used to indicate the absence of any elements which remove, or dissipate, energy from the system.

3.1 SIMPLE HARMONIC OSCILLATION

To obtain a mathematical solution describing the oscillations of the system shown in Fig. 3.1*a*, it is necessary to apply Newton's second law to an appropriate free-body diagram of the mass. First let us consider the conditions

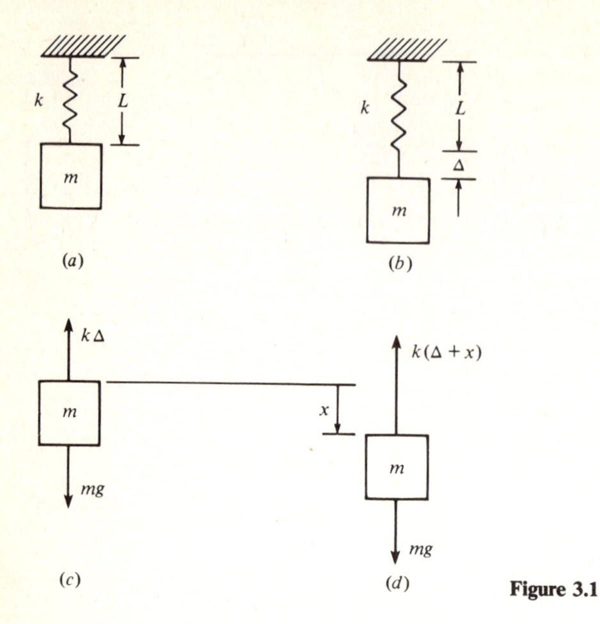

Figure 3.1

of static equilibrium of the system. Assuming that the mass is suspended by the spring in a vertical plane, the spring will be elongated a small amount Δ beyond its free length L as shown in Fig. 3.1*b*. With the linear spring constant denoted by k, the free-body diagram of the mass in the equilibrium position is as shown in Fig. 3.1*c*. Thus, for equilibrium, we have

$$\Sigma F_x = mg - k\Delta = 0 \tag{3.1}$$

where the x coordinate is taken as positive downward from the equilibrium position.

Let us now assume that the equilibrium condition is disturbed in some manner, and examine the free-body diagram of the mass at any position x as shown in Fig. 3.1*d*. Owing to an additional elongation (or compression) of the spring, the magnitude of the force exerted on the mass by the spring has changed by an amount kx, and the force system acting on the mass is no longer balanced. Writing Newton's second law for the mass at position x, we have

$$\Sigma F_x = -k(\Delta + x) + mg = m\ddot{x} \tag{3.2}$$

where $\ddot{x} = d^2x/dt^2$ is the acceleration of the mass. By Eq. (3.1) we have $mg = k\Delta$ from equilibrium considerations, so that Eq. (3.2) can be rewritten as

$$m\ddot{x} = -kx \tag{3.3}$$

which shows that the mass will have an acceleration that is directly proportional to the displacement of the mass but in the opposite direction to that displacement.

Rewriting Eq. (3.3) as

$$\ddot{x} + \frac{k}{m} x = 0 \tag{3.4}$$

we recognize it as a second-order, ordinary differential equation with constant coefficients. (Henceforth, a differential equation written such that the coefficient of the highest-order derivative is unity will be taken as a standard form and will be referred to as the *differential equation of motion* of the system to which it applies.) By the method of Chap. 2, the auxiliary equation corresponding to Eq. (3.4) is $s^2 + k/m = 0$, which has the roots $s_1 = i(k/m)^{1/2}$ and $s_2 = -i(k/m)^{1/2}$. The solution for the displacement of the mass as a function of time is then

$$x(t) = A \sin\left(\frac{k}{m}\right)^{1/2} t + B \cos\left(\frac{k}{m}\right)^{1/2} t \tag{3.5}$$

where A and B are constants which must be determined from the initial conditions.

Since the spring-mass system under consideration is free of external forces, motion will exist only if the mass is disturbed from its equilibrium position. Motion may be imparted to the system by displacing the mass through some distance x_o at which point the mass is released; by the mass acquiring an initial velocity $\dot{x}_o$ through an impact such as a hammer blow; or by a combination of these two. Consider the most general case, in which the mass is displaced an amount x_o and released from that position with a velocity $\dot{x}_o$. The initial conditions are then $x(0) = x_o$ and $\dot{x}(0) = \dot{x}_o$. Evaluating Eq. (3.5) at $t = 0$ and applying the first initial condition give

$$x(0) = x_o = A(0) + B(1) \tag{3.6}$$

from which we obtain the value of one constant as $B = x_o$. To determine the value of A we differentiate Eq. (3.5) with respect to time to obtain

$$\dot{x}(t) = A\left(\frac{k}{m}\right)^{1/2} \cos\left(\frac{k}{m}\right)^{1/2} t - B\left(\frac{k}{m}\right)^{1/2} \sin\left(\frac{k}{m}\right)^{1/2} t \tag{3.7}$$

which represents the velocity of the mass at any time. Evaluating Eq. (3.7) at $t = 0$ gives

$$\dot{x}(0) = \dot{x}_o = A\left(\frac{k}{m}\right)^{1/2} (1) - B\left(\frac{k}{m}\right)^{1/2} (0) \tag{3.8}$$

from which we find $A = \dot{x}_o/(k/m)^{1/2}$. The complete solution for the motion of the mass is

$$x(t) = \frac{\dot{x}_o}{(k/m)^{1/2}} \sin\left(\frac{k}{m}\right)^{1/2} t + x_o \cos\left(\frac{k}{m}\right)^{1/2} t \tag{3.9}$$

which is the sum of two functions which vary harmonically with time. Thus, the motion will be harmonic and will vary with time in a cyclic manner.

3.2 INTERPRETATION OF THE SOLUTION

The physical interpretation of Eq. (3.9) for the motion of a simple spring-mass system is more easily accomplished by rewriting the solution in an equivalent form containing only a single sine function or a single cosine function. With the notation $\omega = (k/m)^{1/2}$, Eq. (3.5) becomes

$$x(t) = A \sin \omega t + B \cos \omega t \tag{3.10}$$

Since A and B are constants, they can be replaced by any other set of two constants without affecting the validity of the solution. In particular, let us choose two new constants X and ϕ, such that $A = X \cos \phi$ and $B = X \sin \phi$, in which case Eq. (3.10) becomes

$$x(t) = X \cos \phi \sin \omega t + X \sin \phi \cos \omega t \tag{3.11}$$

or

$$x(t) = X(\cos \phi \sin \omega t + \sin \phi \cos \omega t) \tag{3.12}$$

Utilizing the trigonometric identity $\cos \phi \sin \omega t + \sin \phi \sin \omega t = \sin(\omega t + \phi)$, we obtain

$$x(t) = X \sin(\omega t + \phi) \tag{3.13}$$

as an equivalent (but simpler) form of the displacement equation for the mass. The constants X and ϕ may be determined by applying the initial conditions, but it is simpler to use the previously determined values of A and B as follows. To obtain X note that

$$X^2 \cos^2 \phi + X^2 \sin^2 \phi = A^2 + B^2 \tag{3.14}$$

and since $\cos^2 \phi + \sin^2 \phi = 1$, this gives

$$X = \sqrt{A^2 + B^2} = \sqrt{x_o^2 + \left(\frac{\dot{x}_o}{\omega}\right)^2} \tag{3.15}$$

Reference to Eq. (3.13) reveals that since the maximum value attained by the sine function is unity, X represents the maximum displacement of the mass from its equilibrium position, referred to as the *amplitude* of the oscillation. To determine ϕ, we form the ratio

$$\frac{B}{A} = \frac{X \sin \phi}{X \cos \phi} = \tan \phi \tag{3.16}$$

from which

$$\phi = \tan^{-1} \frac{B}{A} = \tan^{-1} \frac{x_o \omega}{\dot{x}_o} \tag{3.17}$$

The constant ϕ is known as the *phase angle*, whose physical significance will be discussed in the following paragraphs.

From Eq. (3.13), the motion of the mass is easily visualized as a cyclic oscillation about the equilibrium position between the values X and $-X$. Using a method from kinematics, the displacement-time curve shown in Fig. 3.2 may be obtained as follows: Lay off vertical and horizontal axes for displacement and time, respectively; to the left of the origin draw a circle having a radius

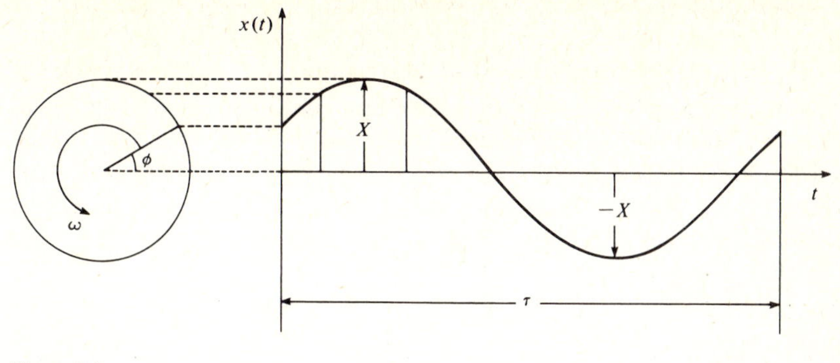

Figure 3.2

equal to the amplitude X; draw a radial line at an angle ϕ measured positive counterclockwise from the right-hand horizontal. The vertical distance from the horizontal to the tip of the radius drawn is $X \sin \phi$, which from Eq. (3.13) is the displacement of the mass at time zero. Thus, the position of the mass at time zero is found by projecting horizontally from the tip of the radius to the displacement axis at $t = 0$. This method of horizontal projection can be shown to produce the correct displacement at any time t by noting that any position of the mass will be repeated $2\pi/\omega$ s later. Now, if the radius is imagined to rotate in a counterclockwise direction with an angular velocity equal to ω, the radius will make one complete revolution in $2\pi/\omega$ s, so that the method of horizontally projecting the tip of the radius across to the corresponding vertical line will produce exactly the curve represented by Eq. (3.13).

Since the value of ω is the frequency with which a circular function (the sine function in this case) repeats, the quantity ω is known as the *circular frequency* of the system. Since $\omega = (k/m)^{1/2}$, it can be seen that a given system will vibrate at this frequency regardless of the initial displacement and/or initial velocity as long as no external forces act on the system.

Two other useful quantities may be ascertained from Fig. 3.2. The time for one complete cycle of the oscillation is called the *period* τ and is given by

$$\tau = \frac{2\pi}{\omega} \tag{3.18}$$

The period of oscillation is given in seconds since the circular frequency ω is generally specified in radians per second. The second quantity is known as the *natural frequency* f, which is the number of cycles of oscillation per unit time, given in hertz as

$$f = \frac{1}{\tau} = \frac{\omega}{2\pi}\ \text{Hz} \tag{3.19}$$

where one hertz is equal to one cycle per second. Note that ω, τ, and f depend only upon the elastic constant and mass and are fixed for any given physical system.

The velocity of the mass at any time is given by

$$\dot{x}(t) = X\omega \cos(\omega t + \phi) \tag{3.20}$$

from which it can be observed that the maximum velocity attained by the mass will be $\pm X\omega$, which occurs whenever $\cos(\omega t + \phi) = \pm 1$. At those times, $\sin(\omega t + \phi) = 0$ so that the maximum velocity occurs when the displacement is zero, that is, when the mass is passing through its equilibrium position. The velocity and displacement are said to be "90° out of phase." On the other hand, the acceleration of the mass is

$$\ddot{x}(t) = -X\omega^2 \sin(\omega t + \phi) \tag{3.21}$$

which shows that the acceleration attains maximum values of $\mp X\omega^2$ and is "180° out of phase" with the displacement. This means that maximum values of acceleration and displacement occur at the same time but have opposite directions.

Example 3.1 A block weighing 19.32 lb is suspended in a vertical plane from a spring having a constant of 25 lb/in. If the block is displaced downward from its equilibrium position through a distance of 1.5 in and released with an upward velocity of 2 in/s, determine (*a*) the circular frequency, (*b*) the period, (*c*) the maximum velocity, (*d*) the maximum acceleration, and (*e*) the phase angle.

SOLUTION The differential equation of motion is

$$\ddot{x} + \frac{k}{m}x = 0$$

with $x(0) = x_o = 1.5$ in, $\dot{x}_o = -2$ in/s, $k = 25$ lb/in, and $m = W/g = 19.32/386.4 = 0.05\ \text{lb}\cdot\text{s}^2/\text{in}$. Then (*a*) the circular frequency is

(*a*) $$\omega = \left(\frac{k}{m}\right)^{1/2} = \left(\frac{25}{0.05}\right)^{1/2} = 22.36\ \text{rad/s}$$

(*b*) $$\tau = \frac{2\pi}{\omega} = 0.281\ \text{s}$$

(*c*) $$\dot{x}_{\max} = X\omega = \left[(1.5)^2 + \left(\frac{2}{22.36}\right)^2\right]^{1/2}(22.36) = 33.6\ \text{in/s}$$

(*d*) $$\ddot{x}_{\max} = X\omega^2 = \dot{x}_{\max}\omega = 33.6 \times 22.36 = 751.3\ \text{in/s}^2$$

(*e*) $$\phi = \tan^{-1}\frac{x_o\omega}{\dot{x}_o} = \tan^{-1}\frac{1.5 \times 22.36}{-2} = 273.4°$$

3.3 ANGULAR OSCILLATIONS OF RIGID BODIES

Small rotational oscillations of a rigid body about a fixed point are described mathematically by equations similar to those for the simple spring-mass system previously discussed. If the magnitude of the oscillations becomes relatively large, however, the differential equation of motion becomes much more complex and requires numerical solution techniques. As an example of rigid-body oscillation, consider the simple pendulum shown in Fig. 3.3*a*. The simple pendulum is composed of a bob of mass m which is suspended in a vertical plane from a smooth pin attached to a rigid support at A. The connecting rod has length L and is assumed to be rigid and to have negligible mass in comparison to the mass of the bob. If the bob is rotated through an angle θ away from its equilibrium position as in Fig. 3.3*b* and released, rotational motion will occur as a result of unbalanced moment acting on the system from the weight of the bob. This is a case of plane rotation of a rigid body about a fixed point, for which the appropriate equation is

$$+\circlearrowleft \qquad \Sigma M_A = I_A \alpha \tag{3.22}$$

where $\alpha = d^2\theta/dt^2$ is the angular acceleration of the pendulum, taken as positive when counterclockwise. (Although the direction of positive rotation is arbitrary, once chosen the sign convention must be used consistently for all rotational quantities including the moments of the external forces.) For the simple pendulum, Eq. (3.22) becomes

$$-(L \sin \theta)mg = mL^2\ddot{\theta} \tag{3.23}$$

or

$$\ddot{\theta} + \frac{g}{L}\sin \theta = 0 \tag{3.24}$$

Equation (3.24) cannot be solved in closed form, owing to the presence of the $\sin \theta$ term, but an approximate solution can be obtained provided the maximum angular displacement from the vertical is small. The sine function has

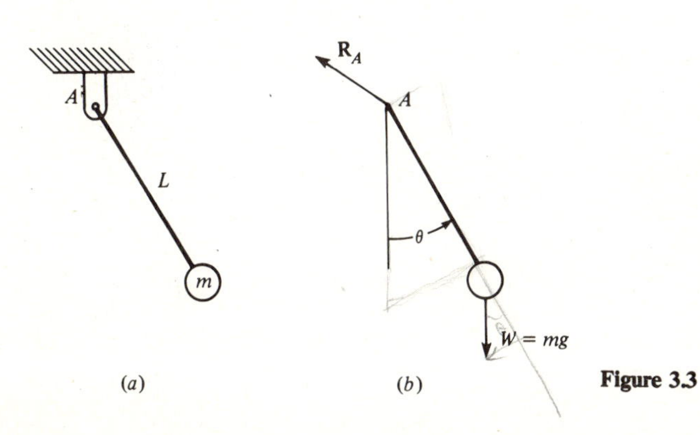

Figure 3.3

the power-series expansion

$$\sin\theta = \theta - \frac{\theta^3}{3!} + \frac{\theta^5}{5!} - \frac{\theta^7}{7!} \cdots \tag{3.25}$$

where θ is expressed in radians. Note that if θ is small, then θ^3, θ^5, and higher-order terms will be much smaller than θ, so that the approximation $\sin\theta = \theta$ could be utilized. It is easy to show that for values of θ less than about 10° (0.174 rad), the error introduced by this approximation is insignificant. Replacing $\sin\theta$ by θ in Eq. (3.24) results in

$$\ddot{\theta} + \frac{g}{L}\theta = 0 \tag{3.26}$$

which is much more amenable to solution, since it is simply a second-order, ordinary differential equation with constant coefficients. Having previously examined the solutions to equations having similar form, we can utilize those results to write the solution to Eq. (3.26) as

$$\theta(t) = A\sin\omega t + B\cos\omega t \tag{3.27}$$

where $\omega = (g/L)^{1/2}$ rad/s is the circular frequency of small oscillations of the pendulum. Rotational oscillations of other rigid-body systems may be handled in a similar fashion, as will be illustrated in the examples which follow.

Example 3.2 A uniform slender bar of weight W and length L is supported by a smooth pin connection at one end and a linear spring having a constant k at the other end, as shown in Fig. 3.4*a*. The free length of the spring is such that the bar is in equilibrium when it is in the horizontal position. Find the differential equation of motion and the circular frequency of small oscillations of the bar about its equilibrium position.

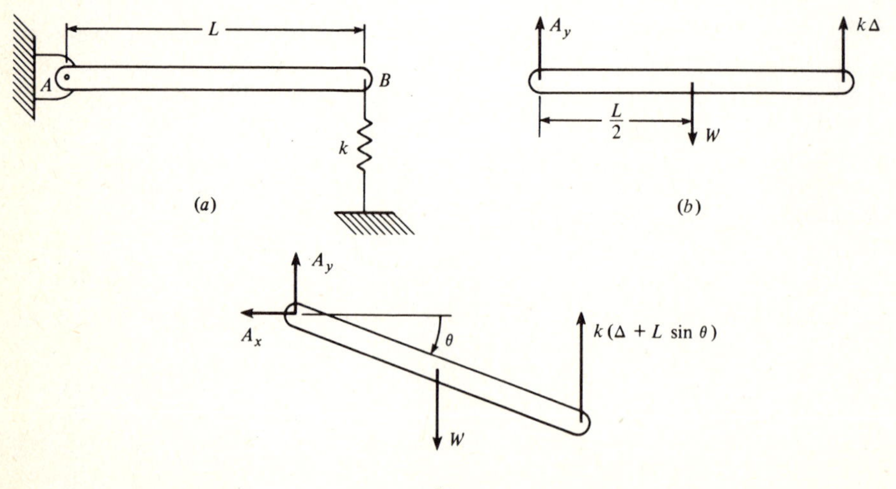

Figure 3.4

SOLUTION First consider the equilibrium free-body diagram as shown in Fig. 3.4*b*. At equilibrium, the spring will be compressed a small amount Δ, so that the moment equilibrium condition is

$$+\circlearrowright \quad \Sigma M_A = 0 = -Lk\Delta + \frac{L}{2}W$$

from which we obtain $\Delta = W/2k$.

Next imagine the bar to be displaced through a small rotation θ and in motion. The complete free-body diagram for this case is shown in Fig. 3.4*c*. The rotational motion is governed by

$$+\circlearrowright \quad \Sigma M_A = I_A\alpha$$

or

$$W\frac{L}{2}\cos\theta - (L\cos\theta)k(\Delta + L\sin\theta) = \frac{1}{3}mL^2\ddot{\theta}$$

where $m = W/g$.

The equation involves both $\sin\theta$ and $\cos\theta$ and is next to impossible to solve. However, for small oscillations, we may use the approximations[1] $\sin\theta = \theta$ and $\cos\theta = 1$ to obtain

$$W\frac{L}{2} - Lk\Delta - kL^2\theta = \frac{1}{3}mL^2\ddot{\theta}$$

and, since $WL/2 = Lk\Delta$ from equilibrium, the differential equation of motion becomes

$$\frac{1}{3}mL^2\ddot{\theta} + kL^2\theta = 0$$

which in standard form is

$$\ddot{\theta} + \frac{3k}{m}\theta = 0$$

The circular frequency is then $\omega = (3k/m)^{1/2}$ rad/s.

Example 3.3 The pendulum shown in Fig. 3.5*a* is such that the weight of the connecting rod is not negligible in comparison to the weight of the pendulum bob. If the weight of the rod and bob are each 3 lb and the length of the rod is 15 in, determine the natural frequency and the period for small oscillations of the pendulum.

SOLUTION The free-body diagram of the pendulum in a nonequilibrium position is as shown in Fig. 3.5*b*, where W_R and W_B denote the weights of the rod and bob, respectively. The governing equation is

$$+\circlearrowleft \quad \Sigma M_A = I_A\alpha$$

where I_A is the mass moment of inertia of the pendulum about an axis

[1] The approximation $\cos\theta = 1$ for small values of θ is drawn from the series expansion $\cos\theta = 1 - \theta^2/2! + \theta^4/4! - \theta^6/6! \cdots$.

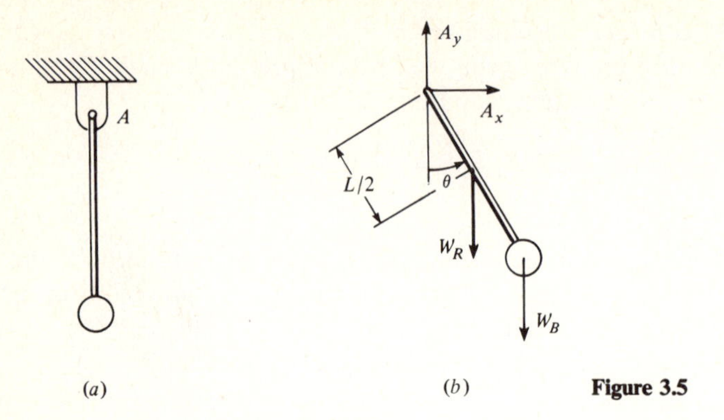

Figure 3.5

through A, including the effects of both the rod and the bob. By substituting, this becomes

$$-\left(\frac{L}{2}\sin\theta\right)W_R - (L\sin\theta)W_B = \left(\frac{1}{3}\frac{W_R}{g}L^2 + \frac{W_B}{g}L^2\right)\ddot{\theta}$$

which, for small oscillations, becomes

$$\ddot{\theta} + \frac{(L/2)W_R + LW_B}{(W_R/3g)L + (W_B/g)L}\theta = 0$$

or

$$\ddot{\theta} + \frac{3g(W_R + 2W_B)}{2L(W_R + 3W_B)}\theta = 0$$

The natural frequency is

$$f = \frac{\omega}{2\pi} = \frac{1}{2\pi}\left(\frac{3 \times 386.4 \times 9}{2 \times 15 \times 12}\right)^{1/2} = 0.857 \text{ Hz}$$

and the period is

$$\tau = \frac{2\pi}{\omega} = \frac{1}{f} = 1.167 \text{ s}$$

3.4 STABILITY CONDITIONS

In studying angular oscillations of the simple pendulum, we assumed the equilibrium position to be vertically downward. Let us now consider the inverted pendulum of Fig. 3.6a and examine the motion resulting from a disturbance away from the equilibrium position. Figure 3.6b shows the free-body diagram of the pendulum in an arbitrary position θ. Summing moments about point A gives the differential equation of motion as

$$mL^2\ddot{\theta} = mgL\sin\theta \tag{3.28}$$

Following the procedure used for the noninverted pendulum, we assume small

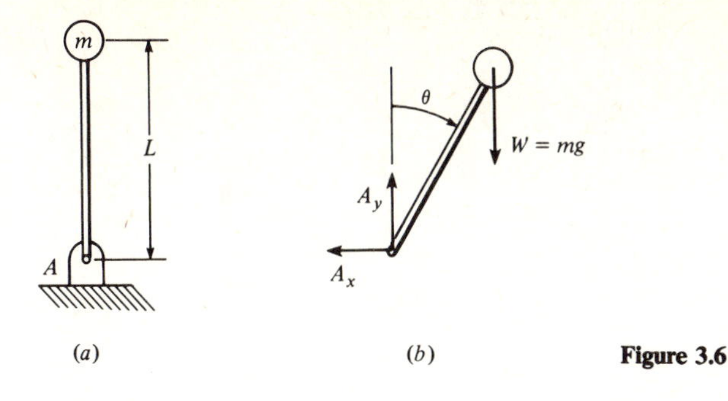

Figure 3.6

oscillations and use the approximation $\sin\theta = \theta$ to obtain

$$\ddot{\theta} - \frac{g}{L}\theta = 0 \tag{3.29}$$

By the method of Chap. 2, the corresponding auxiliary equation is $s^2 - g/L = 0$, which has the roots $s_1 = \sqrt{g/L} = \alpha$ and $s_2 = -\sqrt{g/L} = -\alpha$. Thus the solution is

$$\theta(t) = C_1 e^{\alpha t} + C_2 e^{-\alpha t} \tag{3.30}$$

If the initial conditions are $\theta(0) = \theta_o$ and $\dot{\theta}(0) = \dot{\theta}_o$, the solution becomes

$$\theta(t) = \frac{1}{2\alpha}\left[(\alpha\theta_o + \dot{\theta}_o)e^{\alpha t} + (\alpha\theta_o - \dot{\theta}_o)e^{-\alpha t}\right] \tag{3.31}$$

For any nonzero initial conditions, Eq. (3.31) shows that θ increases exponentially with time. Since θ was assumed to be small, this solution represents only the beginning of the motion. However, this solution for the early-time response is sufficient to show that the system is unstable. In this context a stable system is one which executes bounded oscillations about the equilibrium position. That this is not the case for the inverted pendulum is rather obvious from a second look at the free-body diagram of Fig. 3.6*b*. The moment of the gravity force about point A is always such as to accelerate the pendulum away from equilibrium. This is said to be a *nonrestoring* condition; a restoring condition is one which tends to restore equilibrium.

Rewriting Eqs. (3.26) and (3.29) in the common form

$$\ddot{\theta} + \omega^2\theta = 0 \tag{3.32}$$

we note that for the noninverted pendulum $\omega^2 = g/L$ is positive, and the system is stable. For the inverted pendulum $\omega^2 = -g/L$ is negative, and instability results. The third possibility, $\omega^2 = 0$, represents motion with constant angular velocity and is not physically meaningful for the pendulum. The following example will be used to illustrate stability conditions further.

Example 3.4 Determine the stability condition for the system shown in Fig. 3.7*a*. Consider the mass of the rod to be negligible.

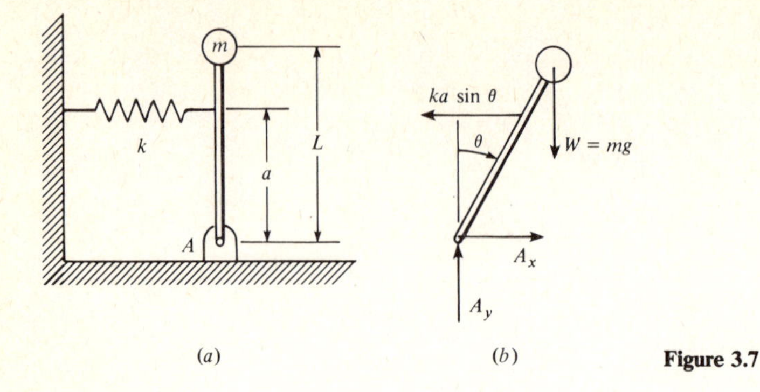

Figure 3.7

Solution The free-body diagram for an arbitrary displacement θ is as shown in Fig. 3.7*b*. Summing moments about A, we obtain

$$mL^2\ddot{\theta} = mgL \sin \theta - a(ka \sin \theta)$$

which for small oscillations may be written as

$$mL^2\ddot{\theta} = (mgL - ka^2)\theta$$

or

$$\ddot{\theta} + \frac{ka^2 - mgL}{mL^2}\theta = 0$$

So for this system

$$\omega^2 = \frac{ka^2 - mgL}{mL^2}$$

from which the stability condition $\omega^2 > 0$ becomes $ka^2 - mgL > 0$. Note that the stability condition simply indicates whether the moment of the spring force (restoring condition) is sufficient to overcome the moment of the gravity force (nonrestoring condition).

3.5 TORSIONAL OSCILLATIONS OF ELASTIC SHAFTING

Another type of rotary oscillation is that due to torsional deformations of shafting that is driving, or is driven by, flywheels, pulleys, sprockets, gears, and so forth. The simplest model of such a system is as shown in Fig. 3.8*a*, in which an elastic circular shaft is attached to a rigid support at one end and carries an inertial disk at the other end. The mass of the shaft is small in comparison to the mass of the disk and is neglected. If an external torque is applied to the disk, the shaft will twist, and rotation of the disk will occur. Owing to the torsional shear-stress distribution in the shaft, an equal and opposite internal reaction torque exists and will result in torsional oscillation of the system if the external torque is suddenly removed.

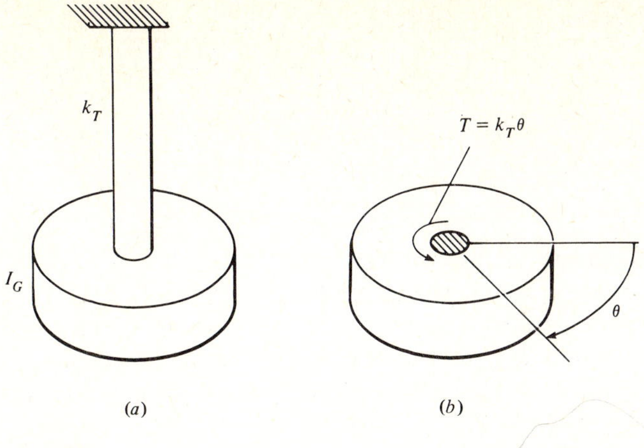

Figure 3.8

A torsional spring constant giving the torque required per unit angle of twist of a circular shaft is obtained from the strength of materials relation $\theta = TL/JG$, where θ is the total angle of twist of a circular shaft of length L, the cross section of which has a polar moment of inertia J, the material of the shaft has a shear modulus G, and the shaft is subjected to a torque T. The equivalent torsional spring constant k_T is then defined as

$$k_T = \frac{T}{\theta} = \frac{JG}{L} \tag{3.33}$$

Thus, if the disk is displaced through an angle θ, as shown in Fig. 3.8b, a restoring torque T is exerted on the disk by the shaft; the magnitude of T is

$$T = k_T\theta \tag{3.34}$$

and the sense of T is opposite to the angular displacement. If the disk is released, the restoring torque T results in angular acceleration of the disk, which causes rotation of the disk back toward the equilibrium position. In this process, the elastic energy stored in the shaft is transformed to rotational kinetic energy of the disk. As the disk reaches its equilibrium position, the kinetic energy acquired causes the disk to overshoot the equilibrium position, and the process of energy transformation reverses, creating oscillations of the disk.

Assuming the shaft centerline coincides with the mass center of the disk at the point of attachment (otherwise the shaft would also be in bending) and denoting the mass center by G, we have

$$+\circlearrowright \ \Sigma M_G = I_G\alpha \tag{3.35}$$

or

$$-k_T\theta = I_G\ddot{\theta} \tag{3.36}$$

where I_G is the mass moment of inertia of the disk about an axis through its mass center and perpendicular to the plane of motion. In standard form, the differential equation of motion is

$$\ddot{\theta} + \frac{k_T}{I_G}\theta = 0 \tag{3.37}$$

which has the solution

$$\theta(t) = A \sin \omega t + B \cos \omega t \tag{3.38}$$

where $\omega = (k_T/I_G)^{1/2}$ rad/s. Thus, the mathematics involved in describing torsional oscillations is similar to that for the types of oscillation discussed previously. One pertinent difference merits consideration, however. In arriving at Eq. (3.38), no small-angle approximation for the angle of twist was required. This would give the impression that there is no restriction on the magnitude of the angle of twist. This, however, is a false impression, since Eq. (3.33), which was used to obtain the torsional spring constant, is valid only for elastic deformations of the shaft. The true restriction is that the maximum angle of twist must be such that the shearing stress in the shaft does not exceed the elastic limit of the material in shear.

3.6 EQUIVALENT SPRING CONSTANTS

To this point, the examples discussed have involved only a single elastic element for energy storage. In actual mechanical systems this is seldom the case, since multiple supports generally exist. The purpose of this section is to develop a means of reducing multiple elastic supports to an equivalent single support to simplify the study of the oscillations of such a system.

Figure 3.9*a* depicts two linear springs having constants k_1 and k_2 which are used to suspend a mass from a rigid support. It is assumed that the spring placement is such that the mass is horizontal (see Prob. 3.14). We wish to find the elastic constant for a single spring such that the mechanical system of Fig. 3.9*b* is identical to that of Fig. 3.9*a*. This is accomplished most readily by considering the force equilibrium condition of each system. Owing to the weight of the mass element, a small elongation of the springs will exist at equilibrium. Denoting the equilibrium elongation by Δ, we note that Δ must be the same for the two systems if we are to have equivalence. Further, the arrangement of the original system is such that springs 1 and 2 are each elongated by the amount Δ. The equilibrium free-body diagrams are shown in Fig. 3.9*c* and *d*. These free-body diagrams show that the condition for equivalence is

$$k_e\Delta = (k_1 + k_2)\Delta \tag{3.39}$$

from which we obtain

$$k_e = k_1 + k_2 \tag{3.40}$$

In this case, the equivalent spring is one having an elastic spring constant equal to the sum of the constants of the two springs in the original system. If the configuration is such that any deformation of one elastic element is the same as

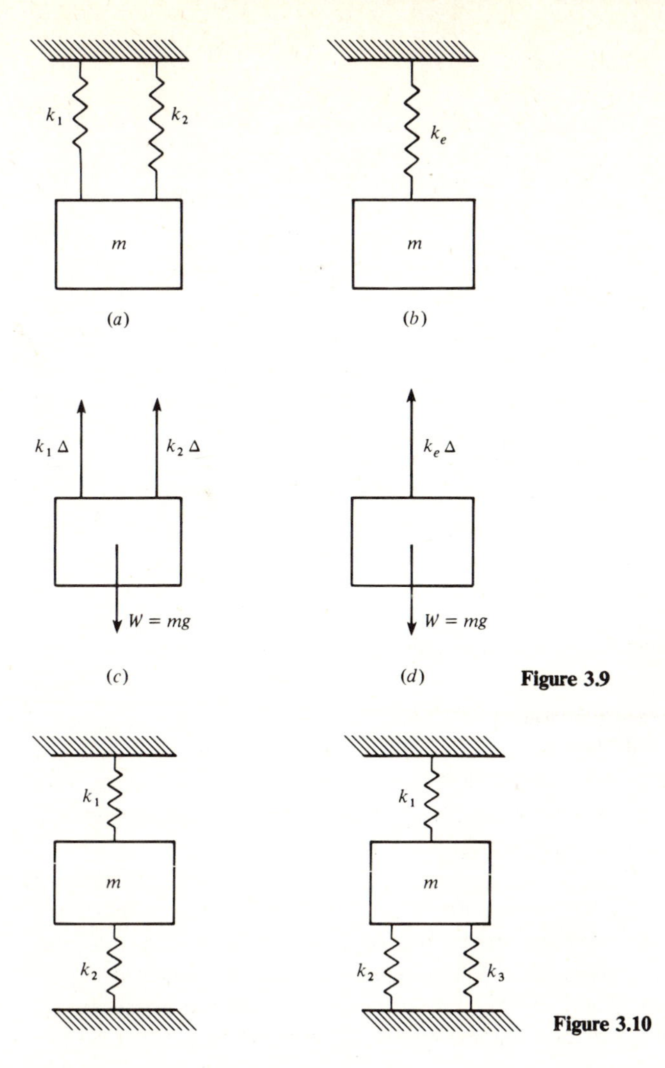

Figure 3.9

Figure 3.10

all other elastic elements in the system, then the elastic elements are said to be in *parallel*, and the equivalent elastic constant is the sum of the individual elastic constants. Other examples of parallel systems are shown in Fig. 3.10.

The springs in the system of Fig. 3.11 are said to be in *series*. A series combination of elastic elements is one in which the sum of the deformations of each element is equal to the total deformation. Again for equilibrium for the system, the displacement of the mass is

$$\Delta = \Delta_1 + \Delta_2 \tag{3.41}$$

where Δ_1 and Δ_2 are the elongations of springs 1 and 2, respectively. Since the weight ($W = mg$) is transferred through the springs to the rigid support, we note that the force in each spring is W. For a single equivalent spring with an elastic

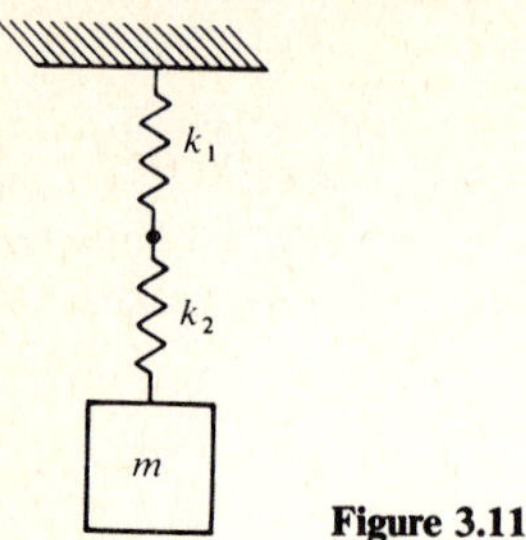

Figure 3.11

constant k_e we must have

$$\frac{W}{k_e} = \frac{W}{k_1} + \frac{W}{k_2} \tag{3.42}$$

from which

$$\frac{1}{k_e} = \frac{1}{k_1} + \frac{1}{k_2} \tag{3.43}$$

As a general rule, when elastic elements are in series, the reciprocal of the equivalent elastic constant is equal to the sum of the reciprocals of the elastic constants of the elements in the original system.

As another example of the equivalent spring constant for an elastic element, consider the simply supported beam loaded at midspan, as shown in Fig. 3.12. The deflection of the beam at the point of application of the load is

$$Y = \frac{FL^3}{48EI} \tag{3.44}$$

where E is the modulus of elasticity of the beam material, and I is the moment of inertia of the beam's cross-sectional area about the neutral axis. The equivalent spring constant for the beam is the ratio of the loading force to the deflection under the load, which, from Eq. (3.44), is found to be

$$k_e = \frac{F}{Y} = \frac{48EI}{L^3} \tag{3.45}$$

Equivalent spring constants for beams with other types of loading and/or other support conditions are obtained in a similar manner.

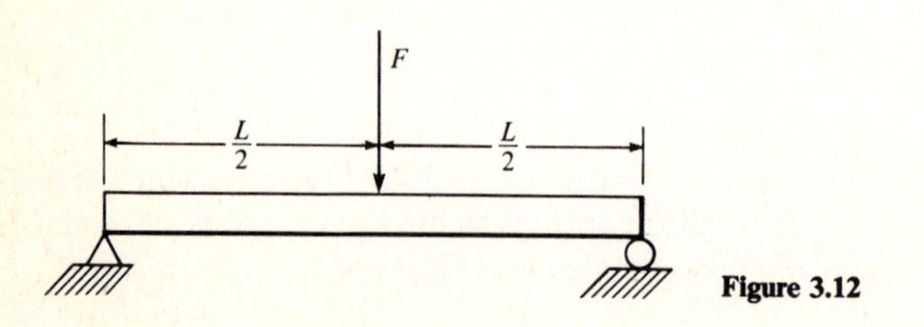

Figure 3.12

3.7 ENERGY METHODS

As discussed earlier, free vibration involves the cyclic interchange of kinetic and potential energy. For free vibrations without damping, no energy is removed from the system, and the principle of conservation of mechanical energy applies. If mechanical energy is conserved, the sum of the kinetic energy and potential energy is constant. This principle may be expressed as

$$T + V = \text{constant} \tag{3.46}$$

or

$$\frac{d}{dt}(T + V) = 0 \tag{3.47}$$

where T and V denote kinetic and potential energy, respectively. Equation (3.47) provides an alternative method of obtaining the differential equation of motion.

Consider the spring-mass system shown at a displaced position x in Fig. 3.13. The kinetic energy of the mass is

$$T = \tfrac{1}{2}m\dot{x}^2 \tag{3.48}$$

The potential energy is the sum of gravitational potential energy and elastic potential energy stored in the deformed spring. However, it is simpler to obtain an expression for V by recalling that change in potential energy is defined as the negative of the mechanical work done by all external forces as a system moves between two positions. By choosing the equilibrium position as the reference, the potential energy at position x is found to be

$$V = -\int_0^x \left[W - k(\Delta + x) \right] dx = \tfrac{1}{2}kx^2 \tag{3.49}$$

where the equilibrium condition $W = k\Delta$ has been utilized. Equation (3.47) then becomes

$$\frac{d}{dt}\left(\frac{1}{2}m\dot{x}^2 + \frac{1}{2}kx^2\right) = 0 \tag{3.50}$$

or

$$(m\ddot{x} + kx)\dot{x} = 0 \tag{3.51}$$

Equation (3.51) will be satisfied if

$$m\ddot{x} + kx = 0 \tag{3.52}$$

which we recognize as the differential equation of motion of the system. The

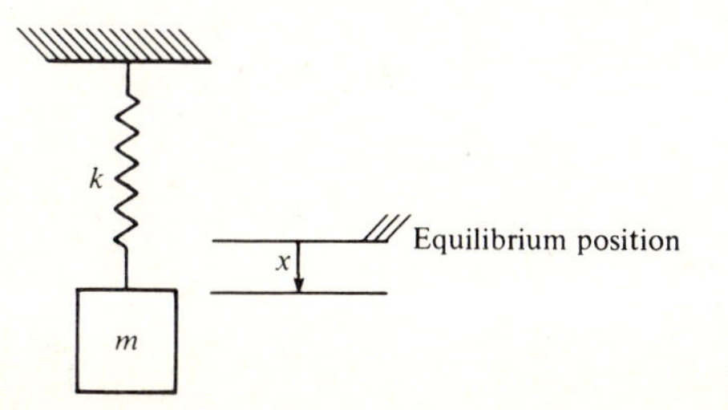

Figure 3.13

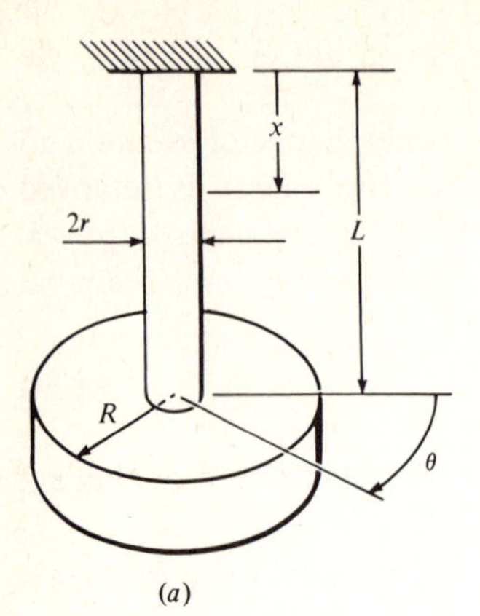

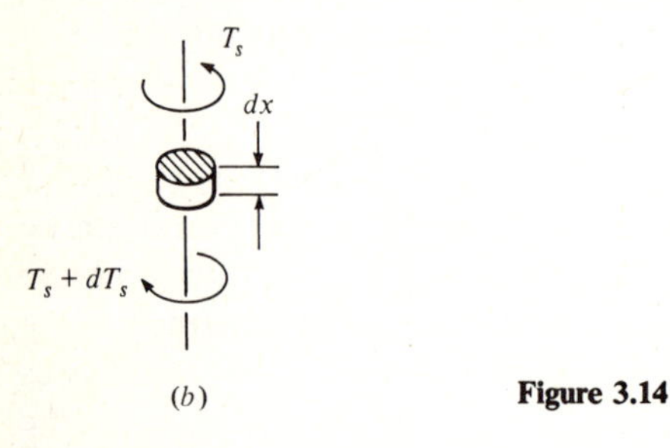

Figure 3.14

trivial solution $\dot{x} = 0$ is valid but not of interest, as it simply represents the static equilibrium condition.

The energy method is often used to obtain the approximate frequency of oscillation of systems in which the mass of the elastic element is not negligible. The torsional system of Fig. 3.8 will serve as an illustration of this approach. If the mass of the shaft is not negligible in comparison to that of the inertial disk, the kinetic energy of the shaft must be accounted for in the analysis.

Referring to Fig. 3.14*a*, let

x = position coordinate along shaft

L = length of shaft

θ_x = angle of twist of shaft at position x

r = shaft radius

γ = specific weight of shaft material

m = mass of shaft

M = mass of disk

R = radius of disk

θ = angular displacement of disk

G = shear modulus of shaft material

Assuming that twisting of the shaft is linearly distributed along its length, we can write

$$\theta_x = \frac{x}{L}\theta \tag{3.53}$$

and differentiate with respect to time to find the angular velocity at any section of the shaft as

$$\dot{\theta}_x = \frac{x}{L}\dot{\theta} \tag{3.54}$$

The kinetic energy of the shaft is obtained by considering a differential element of length dx as in Fig. 3.14*b*. For this element,

$$\begin{aligned} dT_s &= \frac{1}{2}I_s\dot{\theta}_x^2 = \frac{1}{2}\left(\frac{1}{2}r^2\,dm\right)\dot{\theta}_x^2 \\ &= \frac{1}{4}r^2\frac{\gamma}{g}(\pi r^2)\,dx\left(\frac{x}{L}\dot{\theta}\right)^2 \end{aligned} \tag{3.55}$$

The total kinetic energy of the shaft is obtained by integrating over the length of the shaft:

$$\begin{aligned} T_s &= \frac{1}{4}\frac{\pi r^4}{L^2}\frac{\gamma}{g}\dot{\theta}^2\int_0^L x^2\,dx \\ &= \frac{1}{12}\pi r^4 L\frac{\gamma}{g}\dot{\theta}^2 \\ &= \frac{1}{12}mr^2\dot{\theta}^2 \end{aligned} \tag{3.56}$$

The kinetic energy of the inertial disk is

$$T_d = \tfrac{1}{2}I_d\dot{\theta}^2 = \tfrac{1}{4}MR^2\dot{\theta}^2 \tag{3.57}$$

so that total kinetic energy is

$$T = T_d + T_s = \left(\tfrac{1}{4}MR^2 + \tfrac{1}{12}mr^2\right)\dot{\theta}^2 \tag{3.58}$$

The potential energy is the torsional strain energy of the shaft, which can be expressed as

$$V = \frac{1}{2}K_T\theta^2 = \frac{1}{2}\frac{JG}{L}\theta^2 = \frac{\pi r^4 G}{4L}\theta^2 \tag{3.59}$$

Taking the time derivative of $T + V$ then yields the differential equation of motion:

$$\left(\frac{1}{2}MR^2 + \frac{1}{6}mr^2\right)\ddot{\theta} + \frac{\pi r^4 G}{2L}\theta = 0 \tag{3.60}$$

or

$$\ddot{\theta} + \frac{3\pi r^4 G}{(3MR^2 + mr^2)L}\theta = 0 \tag{3.61}$$

The approximate circular frequency as given by this method is then

$$\omega = \left[\frac{3\pi r^4 G}{(3MR^2 + mr^2)L} \right]^{1/2} \tag{3.62}$$

For purposes of comparison, Eq. (3.62) can be rewritten as

$$\omega = \left[\frac{JG}{(I_d + I_s/3)L} \right]^{1/2} \tag{3.63}$$

where I_d and I_s are the mass moments of inertia of the disk and shaft about the axis of rotation, respectively. Denoting by ω_o the frequency corresponding to $m = I_s = 0$, we obtain

$$\omega_o = \left(\frac{JG}{I_d L} \right)^{1/2} \tag{3.64}$$

which is identical to the result of Sec. 3.5. If the mass of the shaft is such that $I_s = 0.5I_d$, say, Eq. (3.62) gives $\omega = 0.926\omega_o$, which shows the effect of the inertia of the shaft in reducing the frequency of oscillation. A more rigorous analysis[2] shows that, for the same inertia ratio, the frequency given by the approximate technique is in error by less than 1 percent. In many cases, this order of accuracy may be sufficient when consideration is given to the accuracy of the physical parameters involved.

PROBLEMS

3.1 A spring-mass system is composed of a 5-lb weight suspended by a linear spring having an elastic constant of 20 lb/in. The weight is given an initial displacement of 3.4 in and is released with zero initial velocity. Determine (*a*) the differential equation of motion, (*b*) the natural frequency, (*c*) the period, and (*d*) the maximum velocity.

3.2 A 6.8-kg mass is suspended by a linear spring. The mass receives an impact such that its motion begins with an initial velocity but no initial displacement. In the ensuing motion, the period is measured to be 0.25 s, and the amplitude is 50 mm. Find (*a*) the spring constant and (*b*) the initial velocity. (Use $g = 9.81\ \text{m/s}^2$.)

3.3 The vibration of a simple spring-mass system is such that the mass attains a maximum acceleration of 40 in/s^2 and has a natural frequency of 60 Hz. Determine (*a*) the amplitude and (*b*) the maximum velocity.

3.4 A spring-mass system is such that the static equilibrium deflection of the spring is 10 mm. Determine the natural frequency of free vibration of this system.

3.5 An unknown mass m is attached to a linear spring for which the elastic constant is also unknown. The period of free oscillation is observed to be 1.5 s. When a 0.5-kg mass is added, the period changes by 15 percent. Calculate the mass and elastic constant of the original system.

3.6 The maximum velocity attained by the mass of a simple harmonic oscillator is observed to be 5 in/s, and the period of oscillation is 2 s. If the mass was released with an initial displacement of 2 in, find the (*a*) amplitude, (*b*) maximum acceleration, (*c*) initial velocity, and (*d*) phase angle.

3.7 A spring-mass system is arranged on an inclined plane as shown in Fig. 3.15. Assuming that the plane is frictionless, determine the differential equation of motion of the mass.

[2] See Sec. 11.3.

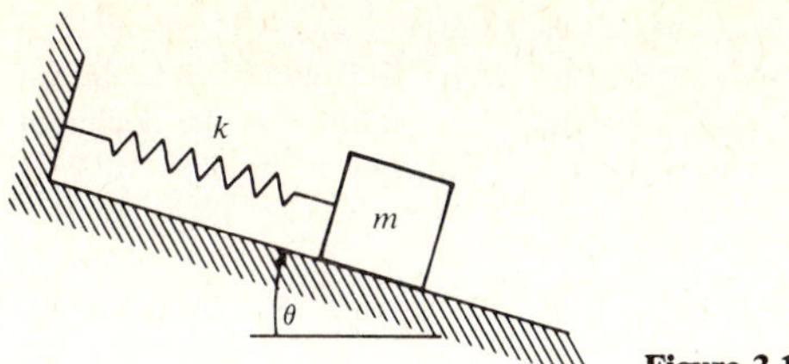

Figure 3.15

3.8 Determine the required length of a simple pendulum if the period of small oscillations is to be 1.5 s.

3.9 A uniform circular plate having a mass of 5.5 kg and a diameter of 0.75 m is supported in the vertical plane by a smooth pin at A (Fig. 3.16). Obtain the differential equation of motion, and calculate the natural frequency of small oscillations about equilibrium.

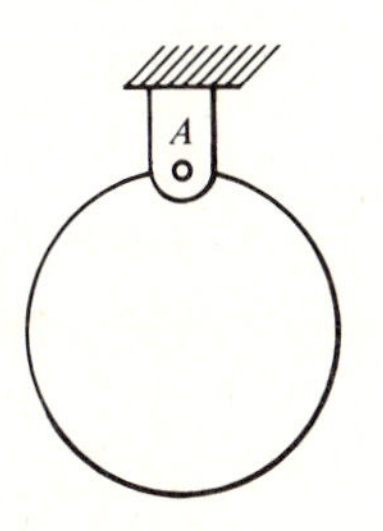

Figure 3.16

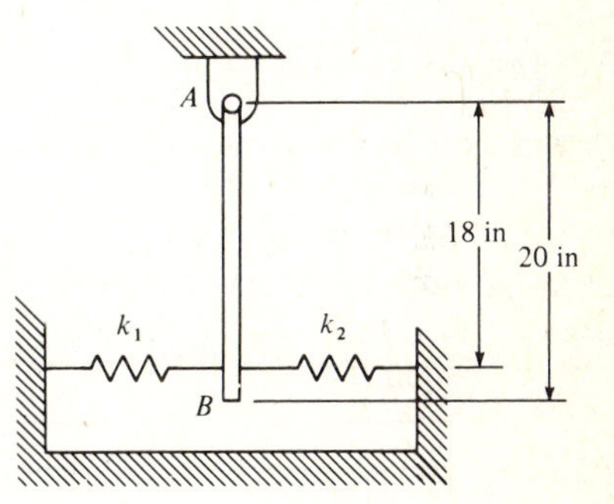

Figure 3.17

3.10 In the system shown in Fig. 3.17, the uniform rod AB weighs 10 lb, and $k_1 = k_2 = 20$ lb/in. If the rod is given a clockwise rotation of 5° and released, find (*a*) the differential equation of motion, (*b*) the circular frequency, (*c*) the period, and (*d*) the maximum angular velocity.

3.11 Figure 3.18 shows a uniform square plate of weight W which is supported by a linear spring k and a smooth pin at B. Side AB is horizontal at equilibrium. If corner A is given a small displacement and released, determine the period of the resulting motion.

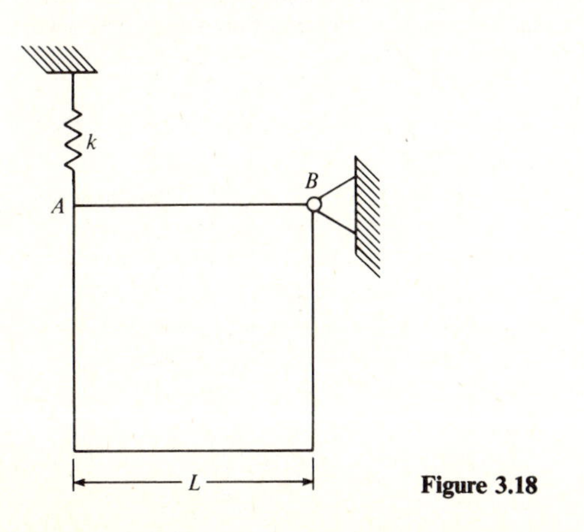

Figure 3.18

3.12 The uniform bar shown in Fig. 3.19 weighs 25 lb, has a length of 25 in, and is pinned at A. At equilibrium the bar is horizontal. If the bar is given a clockwise rotation of 5° and released with no initial angular velocity, find the (*a*) period, (*b*) natural frequency, (*c*) amplitude of the oscillation, and (*d*) angular velocity after 5 s. Use the following data: $a = 10$ in, $b = 15$ in, $k_1 = 10$ lb/in, $k_2 = 65$ lb/in.

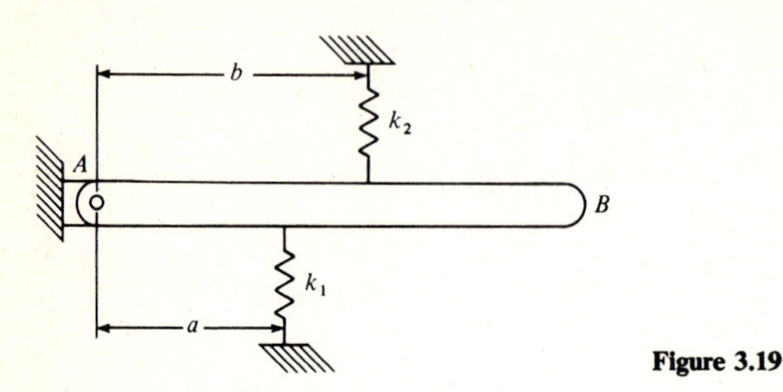

Figure 3.19

3.13 The uniform slender bar shown in Fig. 3.20 weighs 25 lb and is pinned such that it may pivot about point A. If the left end is displaced downward 2 in and released, determine (*a*) the differential equation of motion, (*b*) the period of oscillation, and (*c*) the maximum velocity of point B.

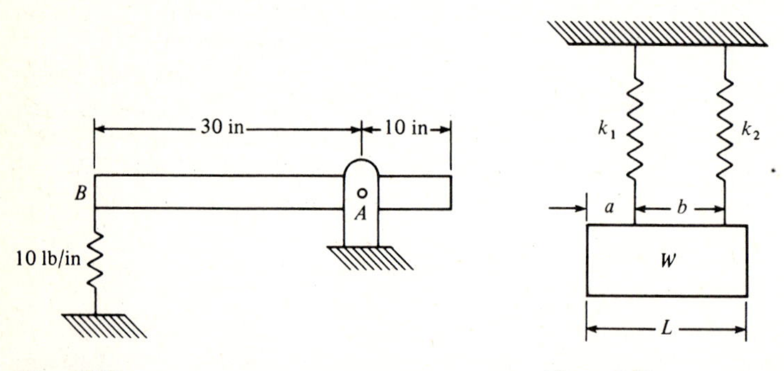

Figure 3.20

Figure 3.21

3.14 Determine the relation between L, a, b, k_1, k_2, and W if the system in Fig. 3.21 is to exhibit only vertical motion if displaced vertically and released.

3.15 For the torsional system shown in Fig. 3.22, the inertial disk has a weight of 193.2 lb, $L = 31.4$ in, $D = 2$ in, $R = 4$ in, and it is known that the natural frequency of torsional oscillation is 87.2 Hz. What is the shear modulus of the shaft material?

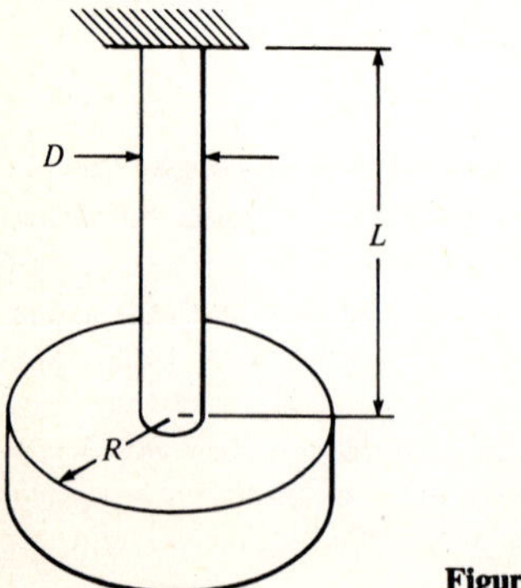

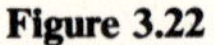

Figure 3.22

3.16 In the system shown in Fig. 3.23, $k_1 = 20$ lb/in, $k_2 = 30$ lb/in, $k_3 = 12$ lb/in, $k_4 = k_5 = 6$ lb/in, $W = 3.864$ lb, and the weight is constrained to move only in the vertical direction. If the system receives an impact such that it acquires an initial downward velocity of 1800 in/s, determine the displacement after 4 s.

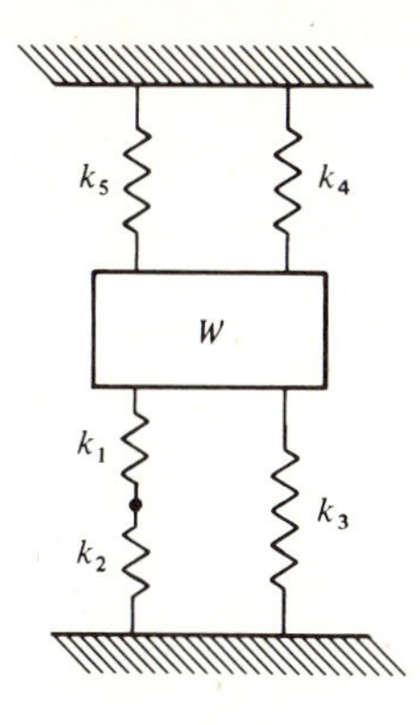

Figure 3.23

3.17 The pendulum of Example 3.3 is released from rest with $\theta = 7°$. At $t = 5$ s the bob falls off, after which the rod continues to oscillate. What is the angular velocity of the rod after 8 s?

3.18 The torsional pendulum of Fig. 3.24 has a linear spring attached tangentially to the inertial disk. If the disk is rotated through an angle of 4° and released, find the maximum velocity of point A on the rim of the disk. Given: $k = 500$ lb/in, $D = 2$ in, $L = 100$ in, $R = 6$ in, $G = 10 \times 10^6$ lb/in^2, and $I = 160$ lb · s^2 · in.

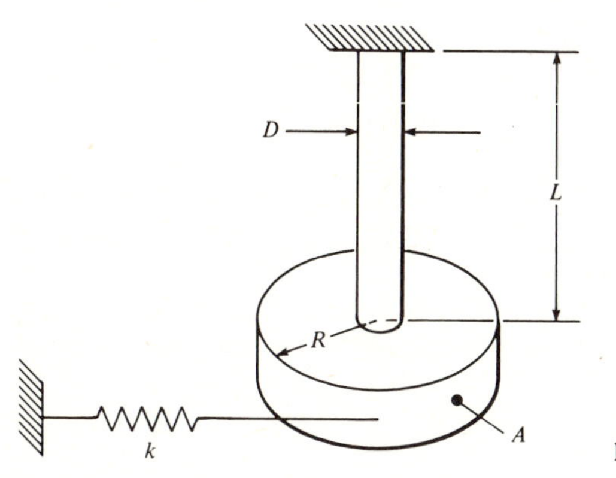

Figure 3.24

3.19 Member AB of the system shown in Fig. 3.25 is a cantilever beam of negligible weight. The free end is supported by series springs k_1 and k_2 and supports a mass m. Determine (a) the differential equation for small vertical oscillations of m and (b) the period of small oscillations.

3.20 A 45-kg mass is attached to rod AB and to three springs as in Fig. 3.26. Determine the natural frequency of small oscillations of the system if the mass of rod AB is negligible. Under what conditions will stable oscillations occur?

3.21 As in Fig. 3.27, a massless bar of length L is pinned at its center. It carries a mass m at its top end and a mass $3m$ at the bottom. At the quarter point is attached a linear spring having a constant k. Find the differential equation of motion and the natural frequency of small oscillations using the energy method.

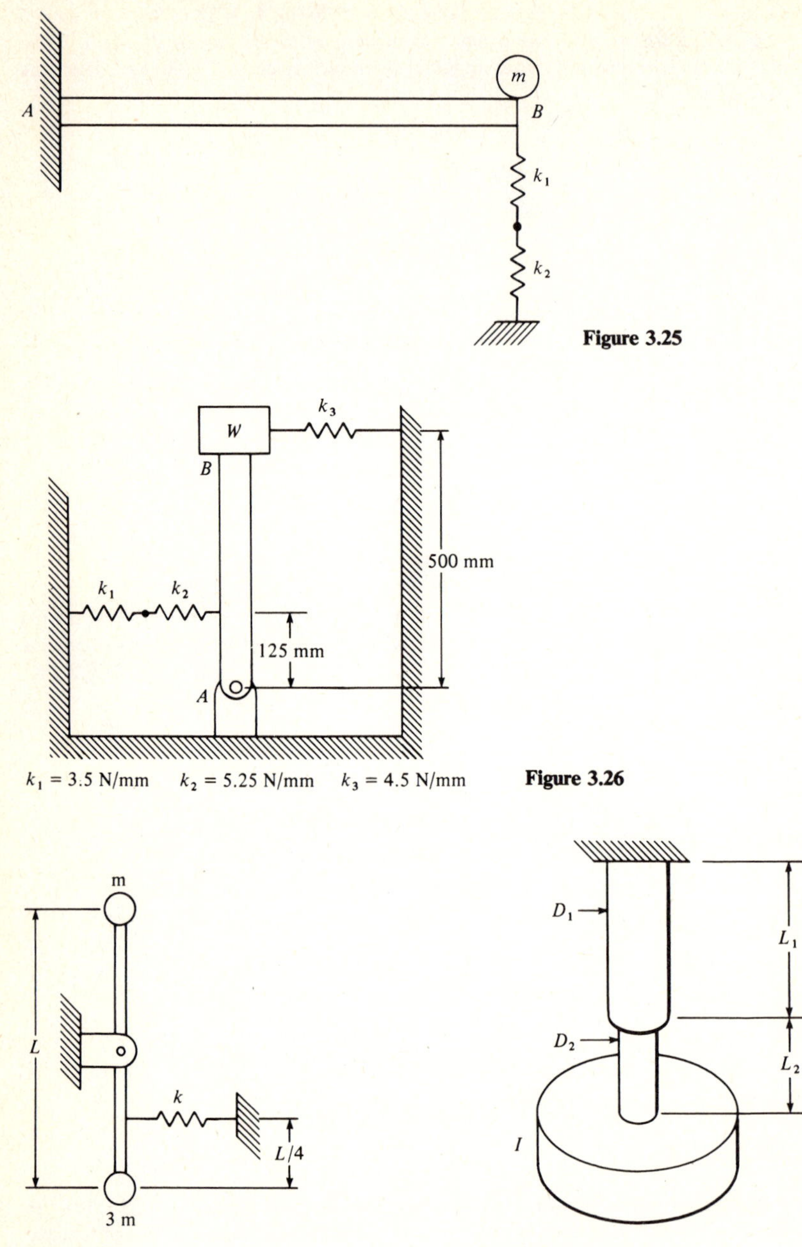

Figure 3.25

Figure 3.26

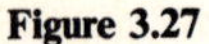

Figure 3.27

Figure 3.28

3.22 The stepped shaft shown in Fig. 3.28 supports a disk having a mass moment of inertia I. Use the energy method to obtain the differential equation of motion for the disk.

3.23 A uniform angled rod is pinned at one end as shown in Fig. 3.29. If the total weight of the rod is $2W$, use the energy method to find the natural frequency of small oscillations about the equilibrium position.

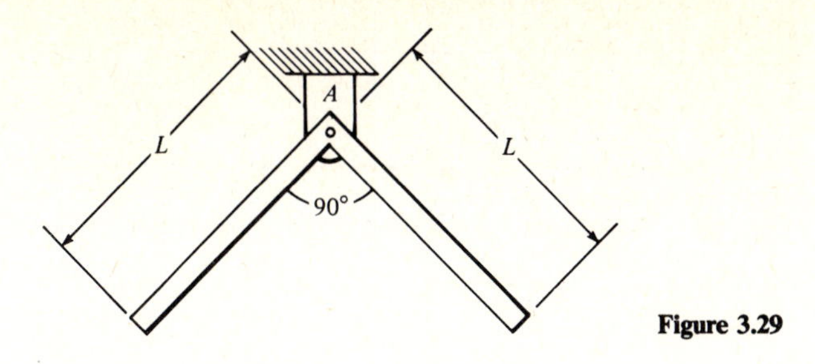

Figure 3.29

3.24 A cylinder of mass m and radius r rolls without slip inside a circular trough of radius R, as shown in Fig. 3.30. Determine the differential equation of motion for small oscillations about the vertical centerline of the trough by the energy method.

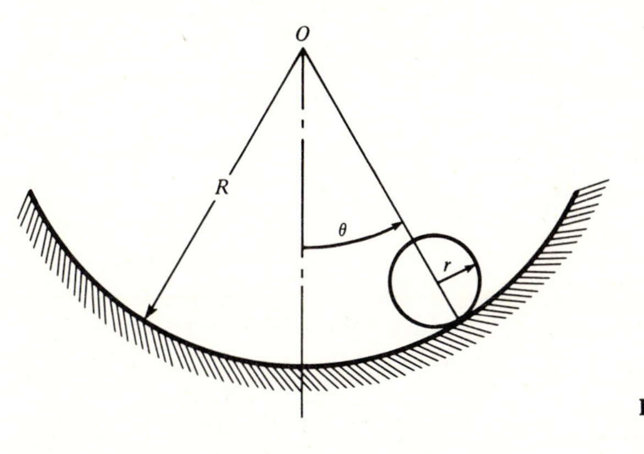

Figure 3.30

3.25 Solve Prob. 3.9 by the energy method.

CHAPTER
FOUR

FORCED VIBRATIONS OF UNDAMPED SINGLE-DEGREE-OF-FREEDOM SYSTEMS

The vibrations which occur in mechanical equipment most often result from forces which arise from the functional operation of the equipment. While there exist many cases in which forces in machinery arise in a random, unpredictable manner, most mechanical equipment is designed to operate in a definite, repeated pattern so that the forces generated will follow a predictable pattern. In the mathematical study of forced vibrations, the exciting forces are often modeled as sinusoidally varying functions of time, and a similar approach will be used here.

4.1 RESPONSE OF A SPRING-MASS SYSTEM TO A HARMONIC FORCING FUNCTION

Figure 4.1*a* depicts a simple spring-mass system which is subjected to a time-varying external force given by $F = F_o \sin \omega_f t$, where F_o is the force amplitude, and ω_f is the circular frequency of the forcing function. To obtain the differential equation of motion, Newton's second law is written for the free-body diagram of the mass in some displaced position, as shown in Fig. 4.1*b*, to obtain

$$\Sigma F_x = m\ddot{x} = -k(\Delta + x) + mg + F \tag{4.1}$$

Consideration of the equilibrium condition of the system will show that $k\Delta = mg$ so that Eq. (4.1) may be written in standard form as

$$\ddot{x} + \frac{k}{m}x = \frac{F}{m} \tag{4.2}$$

or
$$\ddot{x} + \frac{k}{m}x = \frac{F_o}{m}\sin\omega_f t \tag{4.3}$$

Equation (4.3) is an inhomogeneous, second-order differential equation with constant coefficients. From Chap. 2, the complete solution is the sum of a homogeneous solution $x_h(t)$ and a particular solution $x_p(t)$ such that

$$x(t) = x_h(t) + x_p(t) \tag{4.4}$$

where
$$\ddot{x}_h + \frac{k}{m}x_h = 0 \tag{4.5}$$

and
$$\ddot{x}_p + \frac{k}{m}x_p = \frac{F_0}{m}\sin\omega_f t \tag{4.6}$$

Equation (4.5) is the differential equation of motion for free vibration of a simple harmonic oscillator, for which the solution has previously been determined as

$$x_h(t) = A\sin\omega t + B\cos\omega t \tag{4.7}$$

where $\omega = (k/m)^{1/2}$ as before.

To obtain the particular solution corresponding to Eq. (4.6), note that the differential equation involves $x_p(t)$, its second derivative, and a sine function. Since the second derivative of a sine function is itself a sine function, it is logical to assume

$$x_p(t) = X'\sin\omega_f t \tag{4.8}$$

where X' is an unknown constant which must be determined such that the assumed solution does in fact satisfy the differential equation. Substituting $x_p(t)$ into Eq. (4.6) yields

$$-X'\omega_f^2\sin\omega_f t + X'\frac{k}{m}\sin\omega_f t = \frac{F_o}{m}\sin\omega_f t \tag{4.9}$$

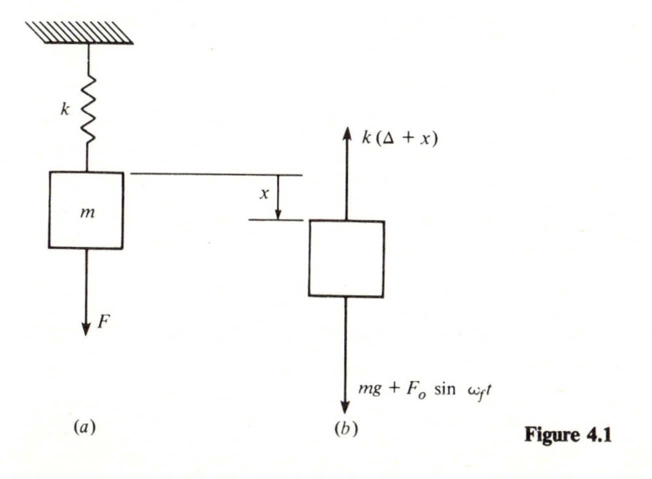

Figure 4.1

from which the value of X' must be

$$X' = \frac{F_o/m}{k/m - \omega_f^2} \tag{4.10}$$

to satisfy the differential equation. Multiplying both numerator and denominator of Eq. (4.10) by m/k gives

$$X' = \frac{F_o/k}{1 - (m/k)\omega_f^2} = \frac{F_o/k}{1 - (\omega_f/\omega)^2} \tag{4.11}$$

since $m/k = 1/\omega^2$. The particular solution is then

$$x_p(t) = \frac{F_o/k}{1 - (\omega_f/\omega)^2} \sin \omega_f t \tag{4.12}$$

and the complete solution for the motion of the mass is

$$x(t) = A \sin \omega t + B \cos \omega t + \frac{F_o/k}{1 - (\omega_f/\omega)^2} \sin \omega_f t \tag{4.13}$$

The constants A and B which arise from the homogeneous solution are evaluated by applying the initial conditions $x(0) = x_o$ and $\dot{x}(0) = \dot{x}_o$ to Eq. (4.13) and its first derivative, respectively.

4.2 INTERPRETATION OF THE SOLUTION

To discuss the physical meaning of the solution obtained in the previous section, it is convenient to replace Eq. (4.13) with the equivalent form

$$x(t) = X \sin(\omega t + \phi) + \frac{F_o/k}{1 - (\omega_f/\omega)^2} \sin \omega_f t \tag{4.14}$$

where X and ϕ are the amplitude and phase angle of the free-vibration response, respectively. To interpret the physical significance of the forced response, note that F_o/k is the deflection which would be produced by a statically applied force having a magnitude equal to the amplitude F_o of the forcing function. The quantity ω_f/ω is the ratio of the circular frequency of the forcing function to the inherent circular frequency of the spring-mass system and is referred to as the *frequency ratio*. By introducing the notation $X_o = F_o/k$ and $r = \omega_f/\omega$, the amplitude of the forced response is obtained as

$$X' = \frac{X_0}{1 - r^2} \tag{4.15}$$

In this form, the amplitude of the forced response is observed to be the equivalent static deflection X_o multiplied by the *magnification factor* $1/(1 - r^2)$. The value of X' may be less than, equal to, or greater than the static deflection, depending on the value of the frequency ratio; each possibility will be considered.

As a final form of the solution for the motion of the mass, we introduce X_o and r into Eq. (4.14) to obtain

$$x(t) = X \sin(\omega t + \phi) + \frac{X_o}{1 - r^2} \sin \omega_f t \tag{4.16}$$

which shows that the motion of the mass is the sum of a free vibration and a forced response, each of which varies sinusoidally with time. Three distinct types of motion are possible, depending upon the value of the frequency ratio r. Two of the cases will be discussed here, and the special case $r = 1$ will be considered in the next section.

Frequency Ratio $r < 1$

If the frequency ratio is less than unity, then $\omega_f < \omega$, which means that the natural frequency of the free response is larger than the natural frequency of the forced response. In other words, the free-vibration portion of the motion will complete several cycles in the time required for one cycle of the forced response. The motion is as depicted by the solid line of Fig. 4.2*a* and is seen to be a sinusoidal variation about a slowly varying base curve represented by the dashed line. Also, since $r < 1$, the magnification factor $1/(1 - r^2)$ is greater than

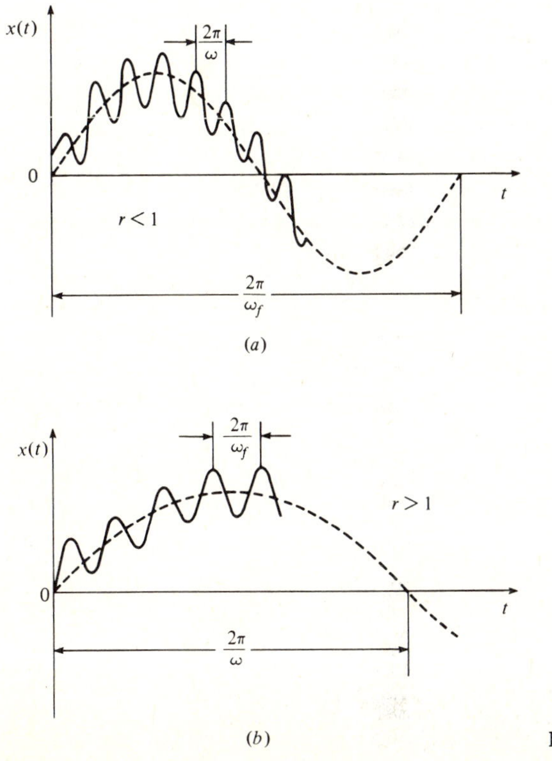

Figure 4.2

unity, so that the amplitude of the forced response is greater than the static deflection X_o.

Frequency Ratio $r > 1$

If the frequency ratio is greater than unity, then $\omega_f > \omega$ and the motion is that of the forced response oscillating about the free response as in Fig. 4.2*b*. Further, if $r > \sqrt{2}$, the amplitude of the forced response is less than X_o.

Example 4.1 For the system of Fig. 4.1, $k = 40$ lb/in, $m = 0.1$ lb $\cdot$ s^2/in, and $F_o = 12$ lb. The forcing function has a circular frequency of 25 rad/s. The initial conditions are $x(0) = 2$ in and $\dot{x}(0) = 1$ in/s. Determine (*a*) the frequency ratio, (*b*) the amplitude of the forced response, and (*c*) the displacement at $t = 6$ s.

Solution The circular frequency of the system is

$$\omega = \left(\frac{k}{m}\right)^{1/2} = \left(\frac{40}{0.1}\right)^{1/2} = 20 \text{ rad/s}$$

(*a*) The frequency ratio is then

$$r = \left(\frac{\omega_f}{\omega}\right) = \frac{25}{20} = 1.25$$

(*b*) The frequency ratio is used to determine the amplitude of the forced response as

$$|X'| = \left|\frac{F_o/k}{1 - r^2}\right| = \left|\frac{12/40}{1 - 1.25^2}\right| = 0.533 \text{ in}$$

where the amplitude is taken as the absolute value.

(*c*) To find the displacement at $t = 6$ s we must evaluate the constants in the complete solution by applying the initial conditions. The complete solution is

$$x(t) = X\sin(\omega t + \phi) + \frac{F_o/k}{1 - r^2}\sin\omega_f t$$

or

$$x(t) = X\sin(20t + \phi) - 0.533\sin 25t$$

Applying the initial conditions $x(0) = 2$ and $\dot{x}(0) = 1$, we have

$$x(0) = 2 = X\sin\phi$$

and

$$\dot{x}(0) = 1 = 20X\cos\phi - 0.533 \times 25$$

which by simultaneous solution give $X = 2.12$ in and $\phi = 70.3°$ (1.23 rad). The complete solution is then

$$x(t) = 2.12\sin(20t + 1.23) - 0.533\sin 25t$$

The displacement at 6 s is

$$x(6) = 2.12\sin 121.23 - 0.533\sin 150 = 2.42 \text{ in}$$

where the arguments of the sine functions have been expressed in radians.

4.3 RESONANCE

The case $r = 1$, for which the circular frequency of the forcing function is identical to the circular frequency of the spring-mass system, is referred to as *resonance* and requires special consideration. Reference to the solution given by Eq. (4.16) shows that for $r = 1$ the amplitude of the forced response is infinite for all values of time. This result is neither physically acceptable nor mathematically correct, as the system could not possibly attain an infinite displacement instantaneously.

To identify the source of the problem, let us examine the method by which the particular solution was obtained, and specifically the effect on the procedure of setting $\omega_f = \omega$. In this case, the differential equation becomes

$$\ddot{x}_p + \omega^2 x_p = \frac{F_o}{m} \sin \omega t \tag{4.17}$$

where the substitution $\omega^2 = k/m$ has been utilized. Assuming as before that

$$x_p(t) = X' \sin \omega t \tag{4.18}$$

and substituting into the differential equation give

$$-X'\omega^2 \sin \omega t + X'\omega^2 \sin \omega t = \frac{F_o}{m} \sin \omega t \tag{4.19}$$

or

$$0 = \frac{F_o}{m} \sin \omega t \tag{4.20}$$

which is impossible except for the trivial case $F_o = 0$! We must conclude that the particular solution previously obtained is not valid for $r = 1$, and a new particular solution must be found.

Although it may not be obvious, this situation is somewhat analogous to the case of repeated roots, and the dilemma will be resolved if we assume the particular solution as

$$x_p(t) = X't \cos \omega t \tag{4.21}$$

Substitution of this assumed $x_p(t)$ into Eq. (4.17) gives

$$-X'(2\omega \sin \omega t + t\omega^2 \cos \omega t) + X't\omega^2 \cos \omega t = \frac{F_o}{m} \sin \omega t \tag{4.22}$$

from which

$$X' = -\frac{F_o}{2m\omega} = -\frac{X_o \omega}{2} \tag{4.23}$$

Therefore the solution for the resonant case is

$$x(t) = X \sin(\omega t + \phi) - \frac{X_o \omega}{2} t \cos \omega t \tag{4.24}$$

which shows that the forced response is a sinusoidally varying function having an amplitude which increases linearly with time. The amplitude of the forced response grows with time as in Fig. 4.3 and, in theory, will eventually become infinite. In actuality, the amplitude will grow to a point at which the spring fails and the vibration problem will have been eliminated in a less than desirable manner.

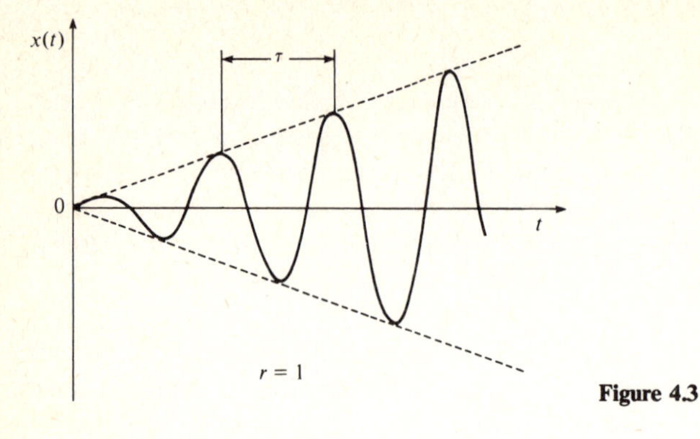

Figure 4.3

Example 4.2 A spring-mass system consists of a 10-lb weight and a spring having a modulus of 20 lb/in. The system is driven by a force having a maximum value of 5 lb and a circular frequency which corresponds to resonance of the system. Determine the amplitude of the forced motion after (*a*) one-half cycle, (*b*) $5\frac{1}{2}$ cycles, and (*c*) $10\frac{1}{2}$ cycles.

SOLUTION For $r = 1$, the forced response is

$$x_f(t) = -\frac{X_o \omega}{2} t \cos \omega t$$

where $X_o = F_o/k = 5/20 = 0.25$ in, and $\omega = (k/m)^{1/2} = (20 \times 386.4/10)^{1/2} = 27.8$ rad/s. Substitution for X_o and ω gives

$$x_f(t) = -3.475t \cos 27.8t$$

(*a*) At one-half cycle, $t = \tau/2 = \pi/\omega$, which gives

$$x_f\left(\frac{\tau}{2}\right) = -3.475\frac{\pi}{27.8}\cos \pi = 0.393 \text{ in}$$

(*b*) At $5\frac{1}{2}$ cycles, $t = 11\tau/2 = 11\pi/\omega$ and

$$x_f\left(\frac{11\tau}{2}\right) = -3.475\frac{11\pi}{27.8}\cos 11\pi = 4.32 \text{ in}$$

(*c*) Similarly, at $10\frac{1}{2}$ cycles, the displacement is

$$x_f\left(\frac{21\tau}{2}\right) = -3.475\frac{21\pi}{27.8}\cos 21\pi = 8.25 \text{ in}$$

4.4 BEATING

An interesting phenomenon, known as *beating*, occurs for an undamped forced system if the frequency ratio is near, but not equal to, unity. That is, the forcing frequency ω_f and the system frequency ω are nearly the same. Recalling the

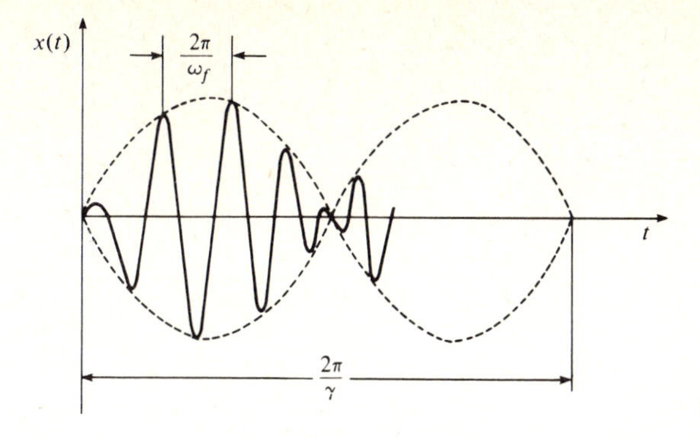

Figure 4.4

general solution in the form

$$x(t) = A \sin \omega t + B \cos \omega t + \frac{X_o}{1 - r^2} \sin \omega_f t \tag{4.25}$$

and applying the general initial conditions $x(0) = x_o$ and $\dot{x}(0) = \dot{x}_o$ result in (see Prob. 4.1)

$$x(t) = \frac{\dot{x}_o}{\omega} \sin \omega t + x_o \cos \omega t + \frac{X_o}{1 - r^2} (\sin \omega_f t - r \sin \omega t) \tag{4.26}$$

To illustrate beating, we consider the case in which the initial conditions are $x_o = \dot{x}_o = 0$ so that the complete solution is

$$x(t) = \frac{X_o}{1 - r^2} (\sin \omega_f t - r \sin \omega t) \tag{4.27}$$

which is the same as

$$x(t) = \frac{X_o \omega^2}{\omega^2 - \omega_f^2} \left(\sin \omega_f t - \frac{\omega_f}{\omega} \sin \omega t \right) \tag{4.28}$$

If ω_f and ω are nearly equal, this can be written as

$$x(t) = -\left(\frac{X_o \omega}{2\gamma} \cos \omega_f t \right) \sin \gamma t \tag{4.29}$$

where $\gamma = (\omega_f - \omega)/2$ is a small quantity. [The details of the derivation of Eq. (4.29) may be found in App. A.3.] Since ω_f is much larger than γ, the term in parentheses varies much more rapidly than $\sin \gamma t$ such that the cosine term varies within an envelope defined by $\sin \gamma t$, as depicted in Fig. 4.4. The displacement of the mass varies cyclically between zero and some maximum value. In mechanical equipment the beating phenomenon is often detectable as an emitted sound which has a similar cyclically varying magnitude.

4.5 FORCED ANGULAR OSCILLATIONS OF RIGID BODIES

The analysis of forced vibrations of rigid bodies is similar to that of the spring-mass system, as will be illustrated for the system shown in Fig. 4.5*a*. The

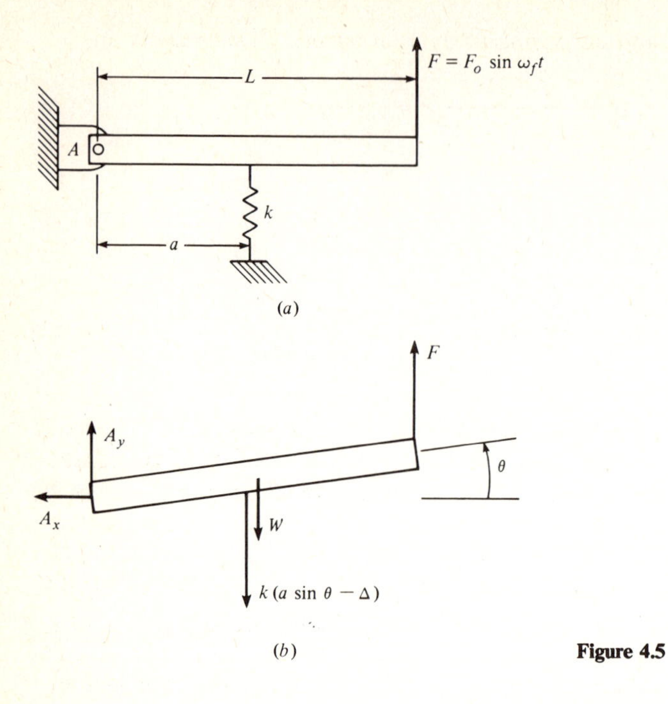

Figure 4.5

uniform rigid bar of weight W is pinned at A, supported at an intermediate point by a linear spring having a constant k, and subjected to a harmonically varying force as shown. If the bar is horizontal at equilibrium, the differential equation of motion for small oscillations of the bar is

$$+\circlearrowleft \quad I_A \ddot{\theta} = -ka^2\theta + F_o L \sin \omega_f t \tag{4.30}$$

or, since $I_A = mL^2/3$, Eq. (4.30) may be written in standard form as

$$\ddot{\theta} + \frac{3ka^2}{mL^2}\theta = \frac{3F_o}{mL} \sin \omega_f t \tag{4.31}$$

The homogeneous solution is

$$\theta_h(t) = A \sin \omega t + B \cos \omega t \tag{4.32}$$

where $\omega = (3ka^2/mL^2)^{1/2}$.

For the particular solution, assume

$$\theta_p = Q \sin \omega_f t \tag{4.33}$$

and substitute to obtain

$$-Q\omega_f^2 \sin \omega_f t + Q\omega^2 \sin \omega_f t = \frac{3F_o}{mL} \sin \omega_f t \tag{4.34}$$

from which

$$Q = \frac{3F_o/mL}{\omega^2 - \omega_f^2} \tag{4.35}$$

Multiplying numerator and denominator by $1/\omega^2 = mL^2/3ka^2$ results in

$$Q = \frac{F_o L/ka^2}{1 - (\omega_f/\omega)^2} = \frac{\Theta_o}{1 - r^2} \tag{4.36}$$

where $r = \omega_f/\omega$ is the frequency ratio, and $\Theta_o = F_o L/ka^2$ is the angle of rotation which would result from the static application of a force of magnitude F_o. The complete solution is then

$$\theta(t) = A \sin \omega t + B \cos \omega t + \frac{\Theta_o}{1 - r^2} \sin \omega_f t \tag{4.37}$$

where the analogy with the solution for the forced response of a simple spring-mass system is readily apparent. It should be noted that the solution obtained is not valid for $r = 1$, but the solution for the resonant case would be obtained using the method of Sec. 4.4.

Example 4.3 For the system shown in Fig. 4.6*a*, $k_1 = k_2 = 10$ lb/in, $F = 15 \sin 3t$, $L = 18$ in, and AB is a uniform slender rod weighing 20 lb. Determine (*a*) the differential equation of motion for small oscillations of the rod and (*b*) the forced response. The rod is horizontal at equilibrium.

SOLUTION By referring to the dynamic free-body diagram shown in Fig. 4.6*b*, the differential equation of motion is obtained from

$$\circlearrowleft + \Sigma M_A = I_A \alpha$$

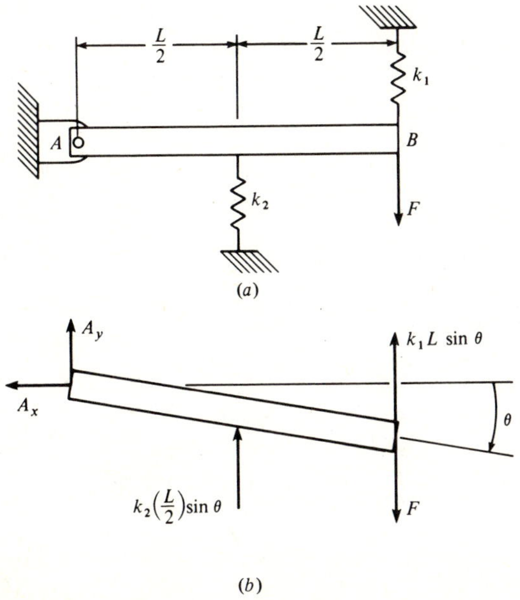

(*a*)

(*b*)

Figure 4.6

as $FL \cos\theta - (k_1 L \sin\theta)(L\cos\theta) - \left(k_2 \frac{L}{2}\sin\theta\right)\left(\frac{L}{2}\cos\theta\right) = \frac{1}{3} mL^2\ddot{\theta}$

Utilizing the small-angle approximations and rearranging, we obtain

$$\frac{1}{3} mL^2\ddot{\theta} + \left(k_1 L^2 + \frac{k_2 L^2}{4}\right)\theta = FL$$

or

$$\ddot{\theta} + \frac{12k_1 + 3k_2}{4m}\theta = \frac{3F}{mL}$$

Substituting the given values gives the differential equation of motion as

$$\ddot{\theta} + 724.5\theta = 48.3 \sin 3t$$

To obtain the forced response, we seek a particular solution of the form

$$\theta_p = a \sin 3t$$

which, when substituted into the differential equation, gives

$$-9a \sin 3t + 724.5a \sin 3t = 48.3 \sin 3t$$

from which $a = 0.067$. The forced response is then

$$\theta_f = 0.067 \sin 3t$$

PROBLEMS

4.1 For the system shown in Fig. 4.1, the initial conditions are $x(0) = x_o$ and $\dot{x}(0) = \dot{x}_o$. Show that the response of the system is given by Eq. (4.26).

4.2 Determine the initial conditions required so that $A = B = 0$ in Eq. (4.25).

4.3 A spring-mass system is subjected to a harmonic force having an amplitude of 75 N and a frequency of 6 Hz. The mass is 4.5 kg, and the spring constant is 2 N/mm. Find the amplitude of the forced vibration.

4.4 The forced-vibration amplitude of a spring-mass system is observed to be 0.35 in. If the spring constant is 15 lb/in and the applied force is 30 sin $2t$ lb, find the weight of the mass.

4.5 A system consists of a 20-lb/in spring which supports a 15-lb weight. The system is subjected to a harmonic force having an amplitude of 25 lb and frequency of 125 Hz.

(*a*) What is the amplitude of the forced response?

(*b*) To what value should the weight be changed if the forced amplitude is to be reduced by 50 percent?

4.6 A harmonic force having an amplitude of 35 N and a frequency of 8 Hz is applied to a spring-mass system and produces a forced amplitude of 10 mm. If the spring constant is 25 N/mm, determine the mass.

4.7 A spring-mass system having a mass of 4.5 kg and a spring constant of 3.5 N/mm is excited by a harmonic force having an amplitude of 100 N and a circular frequency of 18 rad/s. If the initial conditions are $x(0) = 15$ mm and $\dot{x}(0) = 150$ mm/s, find (*a*) the displacement of the mass at $t = 2$ s, and (*b*) the velocity at $t = 4$ s.

4.8 Determine the equation describing the motion of a spring-mass system acted upon by the force $F = F_o \cos \omega_f t$ for the initial conditions $x(0) = x_o$ and $\dot{x}(0) = x_o$ for (*a*) $r \neq 1$ and (*b*) $r = 1$.

4.9 The system shown in Fig. 4.7 has $W = 96.6$ lb and $k = 225$ lb/in. The external forces are $F_1 = 400 \sin 5t$ lb and $F_2 = 300 \sin 28t$ lb. If the initial displacement and velocity are both zero, determine (*a*) $x(t)$ and (*b*) the velocity of the mass at $t = \pi/8$ s.

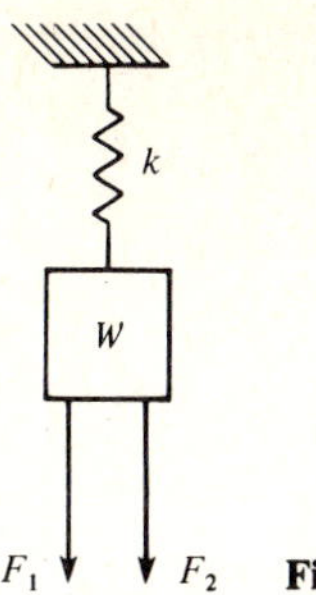

Figure 4.7

4.10 Repeat Prob. 4.9 for $F_2 = 300 \sin 30t$.

4.11 A 5-kg mass is suspended from a spring having an elastic constant of 5 N/mm. The mass is driven at resonant frequency by a harmonic force having an amplitude of 15 N. Calculate the forced amplitude at the end of (*a*) 1.5 cycles, (*b*) 4.5 cycles, and (*c*) 7.5 cycles.

4.12 A uniform slender rod of length L and mass m is acted upon by an external force $F_o \sin \omega_f t$, as shown in Fig. 4.8.

(*a*) Determine the differential equation of motion of the rod.

(*b*) For small oscillations, find the angular position of the rod as a function of time subject to the initial conditions $\theta(0) = \theta_o$ and $\dot{\theta}(0) = 0$.

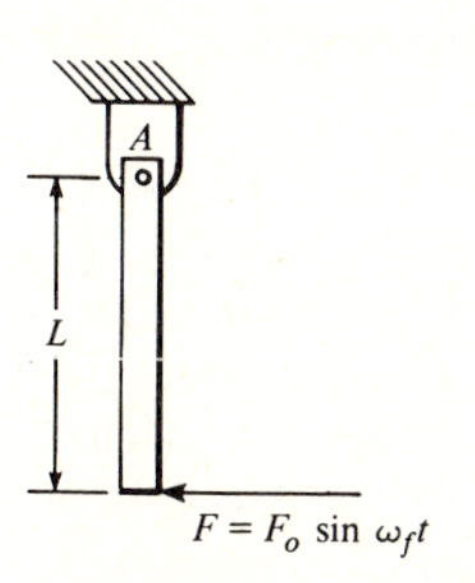

Figure 4.8

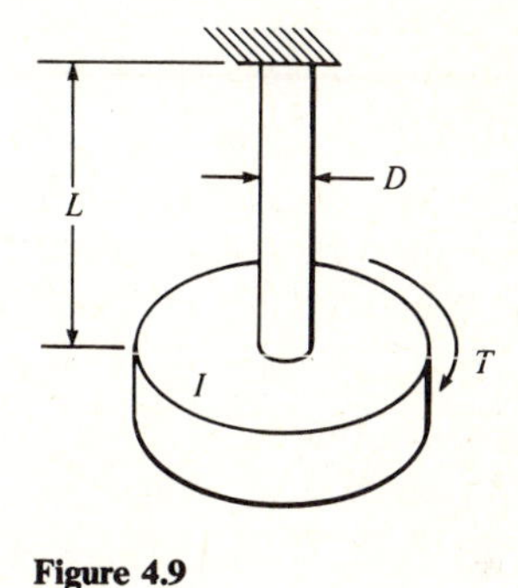

Figure 4.9

4.13 The torsional system of Fig. 4.9 is excited by a harmonic torque $T = T_o \sin \omega_f t$. Derive the differential equation of motion, and obtain the solution for zero initial conditions.

4.14 Repeat Prob. 4.13 for the system shown in Fig. 4.10.

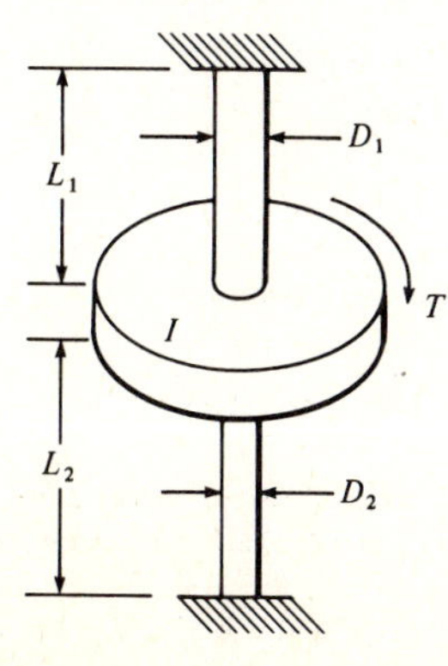

Figure 4.10

4.15 A uniform circular cylinder rolls without slip on a horizontal plane and is restrained by a linear spring as shown in Fig. 4.11. Determine the differential equation of motion if the cylinder is driven by the harmonic force $F = F_o \sin \omega_f t$.

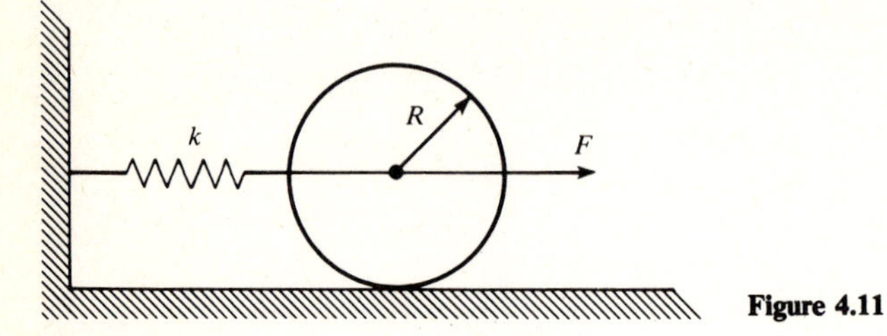

Figure 4.11

4.16 A uniform rod of mass m and length L rests on the two pulleys which rotate in opposite directions, as shown in Fig. 4.12. The rod is given a small displacement x_o to the left and released. Denoting the coefficient of friction between the rod and the pulleys as f, determine the differential equation of motion and the frequency of the resulting vibration.

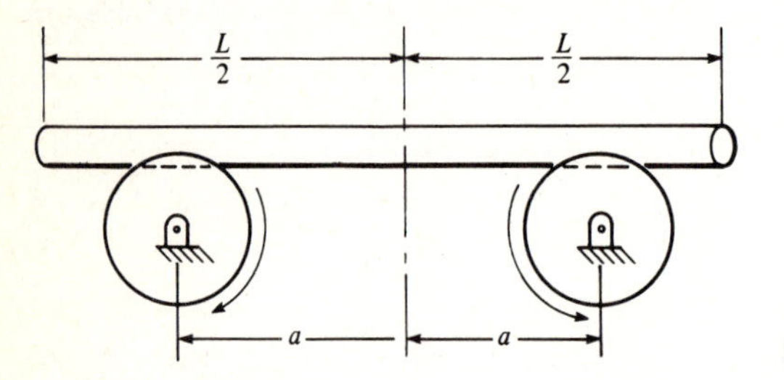

Figure 4.12

CHAPTER

FIVE

VIBRATIONS OF DAMPED SINGLE-DEGREE-OF-FREEDOM SYSTEMS

To this point, we have considered the vibrations of mechanical systems composed only of mass (inertia) elements and elastic elements for the cyclic exchange of kinetic and potential energy. In such systems, any vibration would, in theory, continue forever, since no component that dissipates energy is included. In actuality, even the simple spring-mass system will eventually dissipate its mechanical energy through air resistance on the mass and heat generation in the spring. Similarly, in real mechanical equipment, there is always energy dissipation in one form or another.

The process of energy dissipation is generally referred to in the study of vibrations as *damping*. The three main forms of damping are *viscous damping*, *Coulomb* or *dry-friction damping*, and *hysteresis damping*. We shall consider viscous damping in detail first; the other types will be discussed in turn.

5.1 ENERGY-DISSIPATING ELEMENTS

The most common type of energy-dissipating element found in the study of vibrations is the *viscous damper*, which is also often referred to as a *dashpot*. The viscous damper is characterized by the resistive force exerted on a body moving in a viscous fluid, and hence the name. The common automobile shock absorber is a viscous damper.

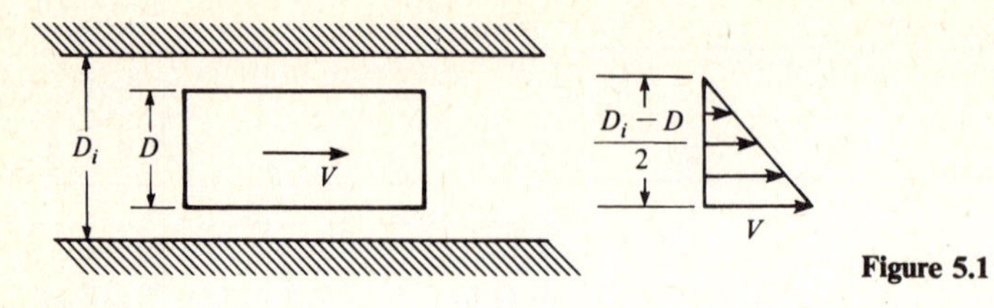

Figure 5.1

To study the phenomenon of viscous damping, we consider Fig. 5.1, which depicts a piston of diameter D moving with a velocity V through a fixed fluid-filled cylinder having an inside diameter D_i. With the viscosity of the fluid denoted by μ, Newton's equation of viscosity gives the shearing stress in the fluid as

$$\tau = \mu \frac{dv}{dy} \tag{5.1}$$

where dv/dy is the velocity gradient in the fluid. Since the velocity of the fluid at the surface of the piston is V, while the velocity of the fluid at the inside surface of the cylinder is zero, we may approximate the velocity distribution across the clearance $(D_i - D)/2$ as a linear distribution to obtain the velocity gradient as

$$\frac{dv}{dy} = \frac{2V}{D_i - D} \tag{5.2}$$

so that the shearing stress is

$$\tau = \mu \frac{2V}{D_i - D} \tag{5.3}$$

As a result of the shearing stress at the surface of the piston, a resistive force will be exerted on the piston by the fluid. The resistive force is given by

$$F_R = \tau A_P \tag{5.4}$$

where $A_P = \pi DL$ is the surface area of the piston. Combining Eqs. (5.3) and (5.4), we obtain

$$F_R = \frac{2\pi DL\mu}{D_i - D} V \tag{5.5}$$

as the resistive force exerted on the piston. Thus, an equal and opposite force would be required to maintain the velocity of the piston. In the absence of such a sustaining force, the resistive, or viscous, force acts to decrease the velocity of the piston, and in time the motion will cease.

While this is a greatly oversimplified model of viscous damping, it should nevertheless serve to illustrate the basic idea of viscous resistance to motion. In actual viscous dampers, the fluid flow is through holes in the piston rather than around the piston through the clearance space. The damping characteristics can be changed by varying the number and size of the holes, the diameter and length of the piston, and/or the fluid viscosity.

The fraction in Eq. (5.5) is a constant for a given dashpot, since it contains only physical dimensions and the viscosity of the damping fluid.[1] Thus, the resistive force is equal to a constant multiple of the velocity of the piston. The constant is known as the *damping constant* and will be denoted henceforth as c. For dimensional uniformity the damping constant is expressed in pound seconds per inch or newton seconds per millimeter.

To include the effects of viscous damping in vibrating mechanical systems, it is assumed that the mass element is directly connected to the piston of a dashpot similar to that just described. Then, the velocity of the piston at any time is the same as that of the vibrating mass, and the resistive force exerted on the piston is also exerted on the mass. Thus, the mass will be subjected to a damping force having a magnitude given by

$$F_R = c\dot{x} \tag{5.6}$$

where $\dot{x}$ is the velocity of the mass, and acting in the direction *opposite to the velocity of the mass*. The effects of viscous damping on the vibrations of mechanical systems will be examined in detail in the remainder of this chapter.

5.2 FREE VIBRATIONS WITH VISCOUS DAMPING

Consider the system shown in Fig. 5.2*a*, in which a viscous damper has been added to the simple spring-mass system. The differential equation of motion for the system will be obtained with the aid of the free-body diagram shown in Fig. 5.2*b*. The displacement coordinate x is taken as positive downward from the equilibrium position, and it is assumed that the velocity and acceleration $\dot{x}$ and $\ddot{x}$ are positive. The damping force $c\dot{x}$ is thus included as an upward, or negative, force so that a reversal of direction of the velocity will automatically account for reversal of the damping force. Applying Newton's second law gives

$$m\ddot{x} = -k(\Delta + x) + mg - c\dot{x} \tag{5.7}$$

Since the damper exerts no force at equilibrium, the equilibrium condition $mg = k\Delta$ is still valid so that Eq. (5.7) may be written as

$$m\ddot{x} = -kx - c\dot{x} \tag{5.8}$$

or

$$\ddot{x} + \frac{c}{m}\dot{x} + \frac{k}{m}x = 0 \tag{5.9}$$

which is the differential equation of motion for free vibration of a damped spring-mass system.

Applying the method of Chap. 2 to Eq. (5.9), we assume a solution in the form $x(t) = Ce^{st}$ to obtain the auxiliary equation

$$s^2 + \frac{c}{m}s + \frac{k}{m} = 0 \tag{5.10}$$

[1] The viscosity of the fluid may be considered constant in the absence of widely fluctuating temperatures.

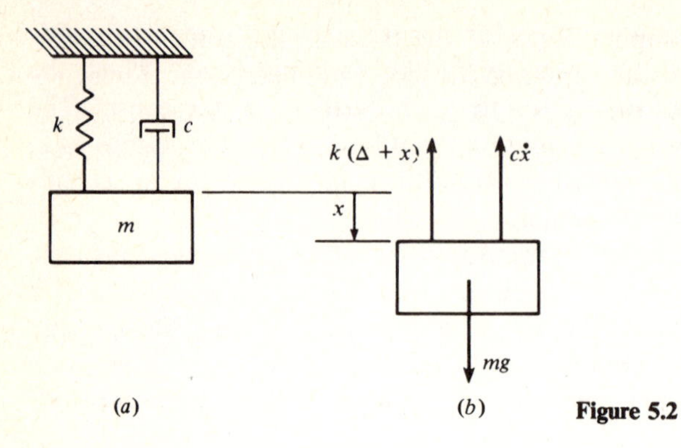

Figure 5.2

which has the roots

$$s_{1,2} = \frac{1}{2}\left[-\frac{c}{m} \pm \sqrt{\left(\frac{c}{m}\right)^2 - 4\frac{k}{m}}\right] \tag{5.11}$$

or

$$s_{1,2} = -\frac{c}{2m} \pm \sqrt{\left(\frac{c}{2m}\right)^2 - \frac{k}{m}} \tag{5.12}$$

From previous experience, we know that the solution takes one of three forms, depending on whether the quantity $(c/2m)^2 - k/m$ is zero, positive, or negative. Note that if this quantity is zero, we have

$$\frac{c}{2m} = \sqrt{\frac{k}{m}} = \omega \tag{5.13}$$

or

$$c = 2m\omega \tag{5.14}$$

in which case we have the repeated roots $s_1 = s_2 = -c/2m$, and the solution is

$$x(t) = (A + Bt)e^{-(c/2m)t} \tag{5.15}$$

As the case in which repeated roots occur has special significance, we shall refer to the corresponding value of the damping constant as the *critical damping constant*, denoted by $C_c = 2m\omega$. Utilizing the notation for the critical damping constant, we may rewrite the roots given by Eq. (5.12) as

$$s_{1,2} = -\frac{c}{C_c}\omega \pm \omega\sqrt{\left(\frac{c}{C_c}\right)^2 - 1} \tag{5.16}$$

or

$$s_{1,2} = \left(-\zeta \pm \sqrt{\zeta^2 - 1}\right)\omega \tag{5.17}$$

where $\omega = (k/m)^{1/2}$ is the circular frequency of the corresponding undamped system, and

$$\zeta = \frac{c}{C_c} = \frac{c}{2m\omega} \tag{5.18}$$

is the *damping factor*. The damping factor is the ratio of the actual damping present in a system to the critical damping constant for the system. Equation (5.17) shows that the nature of the roots of the auxiliary equation depends on whether the value of the damping factor ζ is less than, equal to, or greater than unity.

Case 1: $\zeta < 1$ or $c < 2m\omega$

If $\zeta < 1$, the roots of the auxiliary equation are both imaginary and are given by

$$s_{1,2} = \left(-\zeta \pm i\sqrt{1-\zeta^2}\right)\omega \tag{5.19}$$

and the solution for the motion of the mass becomes

$$x(t) = e^{-\zeta\omega t}\left(A \cos\sqrt{1-\zeta^2}\,\omega t + B \sin\sqrt{1-\zeta^2}\,\omega t\right) \tag{5.20}$$

The solution represented by Eq. (5.20) is more easily written and interpreted in the form

$$x(t) = Xe^{-\zeta\omega t}\sin(\omega_d t + \phi) \tag{5.21}$$

where

$$\omega_d = \sqrt{1-\zeta^2}\,\omega \tag{5.22}$$

is the damped circular frequency, and ϕ is the phase angle of the damped oscillations. The displacement is a harmonic function having an amplitude which decays exponentially with time. The general form of the motion is shown in Fig. 5.3. For motion of this type, the system is said to be *underdamped*, and the damping is below critical.

Case 2: $\zeta = 1$ or $c = C_c = 2m\omega$

If $\zeta = 1$, the damping constant is equal to the critical damping constant, and the system is said to be *critically damped*. The displacement is given by Eq. (5.15), which can also be written as

$$x(t) = (A + Bt)e^{-\omega t} \tag{5.23}$$

The solution is the product of a linear function of time and a decaying exponential. Depending on the values of A and B, many forms of motion are

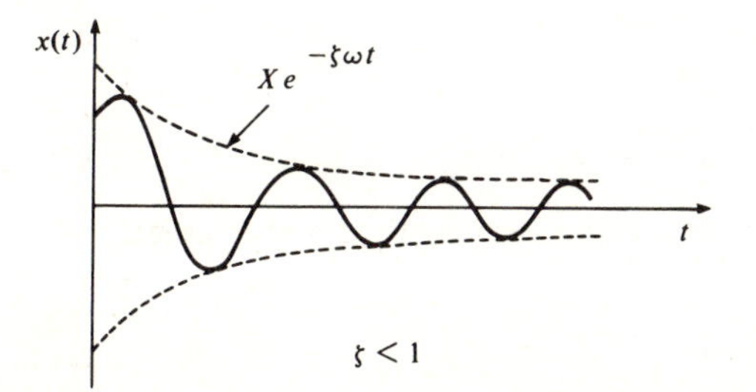

Figure 5.3

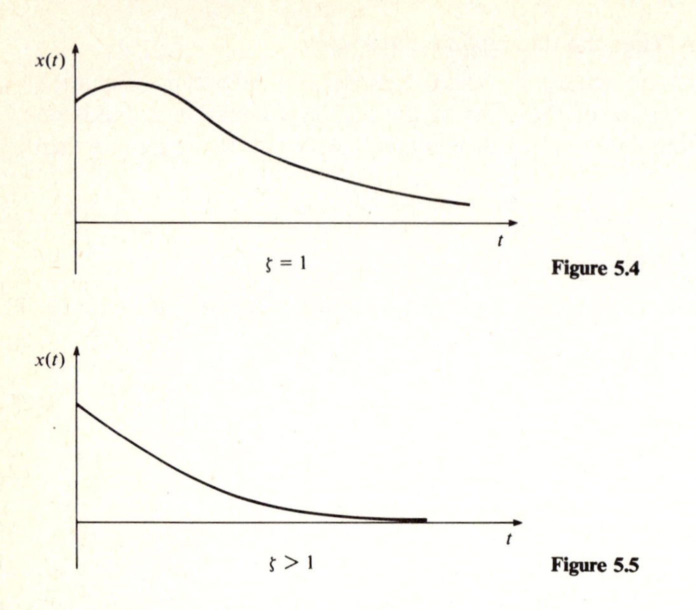

Figure 5.4

Figure 5.5

possible, but each form is characterized by an amplitude which decays without oscillations, such as is shown in Fig. 5.4.

Case 3: $\zeta > 1$ or $c > 2m\omega$

Finally, consider the case $\zeta > 1$, for which the system is said to be *overdamped*. The roots of the auxiliary equation are both real and are given by

$$s_{1,2} = \left(-\zeta \pm \sqrt{\zeta^2 - 1}\right)\omega \tag{5.24}$$

Since $\sqrt{\zeta^2 - 1} < \zeta$, it can be seen that both s_1 and s_2 are negative so that the displacement is the sum of two decaying exponentials given by

$$x(t) = C_1 e^{\left(-\zeta + \sqrt{\zeta^2 - 1}\right)\omega t} + C_2 e^{\left(-\zeta - \sqrt{\zeta^2 - 1}\right)\omega t} \tag{5.25}$$

The motion will be nonoscillatory and will be similar to that shown in Fig. 5.5.

Example 5.1 The damped spring-mass system of Fig. 5.2 has $m = 8$ kg, $k = 5$ N/mm, and $c = 0.2$ N · s/mm. Determine the equation for the displacement of the mass.

SOLUTION The circular frequency of the undamped system is

$$\omega = \sqrt{\frac{k}{m}} = \sqrt{\frac{5 \times 1000}{8}} = 25 \text{ rad/s}$$

The critical damping constant is

$$C_c = 2m\omega = 2 \times 8 \times 25 = 400 \text{ N} \cdot \text{s/m}$$

or 0.4 N · s/mm. Thus the damping factor is

$$\zeta = \frac{c}{C_c} = \frac{0.2}{0.4} = 0.5$$

so that the system is underdamped. The damped circular frequency is

$$\omega_d = \sqrt{1 - (0.5)^2} \times 25 = 21.65 \text{ rad/s}$$

and
$$\zeta\omega = 0.5 \times 25 = 12.5$$

Substituting known values into Eq. (5.21) for underdamped motion gives

$$x(t) = Xe^{-12.5t} \sin(21.65t + \phi)$$

5.3 LOGARITHMIC DECREMENT

As shown by Eq. (5.21) and Fig. 5.3, the displacement of an underdamped system is a sinusoidal oscillation with decaying amplitude. A quite useful property of an underdamped system can be obtained by comparing the amplitudes of any two successive cycles of the displacement. If x_i is the amplitude of the ith cycle, which occurs at t_i, then by Eq. (5.21)

$$x_i = Xe^{-\zeta\omega t_i} \tag{5.26}$$

The amplitude x_{i+1} of the next cycle occurs one period later, at $t_i + \tau$, and is given by

$$x_{i+1} = Xe^{-\zeta\omega(t_i+\tau)} \tag{5.27}$$

The ratio of successive amplitudes is then

$$\frac{x_i}{x_{i+1}} = \frac{Xe^{-\zeta\omega t_i}}{Xe^{-\zeta\omega(t_i+\tau)}} = e^{\zeta\omega\tau} = \text{constant} \tag{5.28}$$

Since the amplitudes used to obtain this result were chosen arbitrarily, Eq. (5.28) shows that the ratio of *any* two successive amplitudes is constant. Most commonly, the rate of decay of amplitude is expressed as the natural logarithm of the amplitude ratio, known as the *logarithmic decrement* and denoted by δ. Therefore,

$$\delta = \ln \frac{x_i}{x_{i+1}} = \ln e^{\zeta\omega\tau} = \zeta\omega\tau \tag{5.29}$$

Substituting $\tau = 2\pi/\omega_d = 2\pi/\omega\sqrt{1-\zeta^2}$ gives

$$\delta = \frac{2\pi\zeta}{\sqrt{1-\zeta^2}} \tag{5.30}$$

Note that if ζ is small, then $\delta \simeq 2\pi\zeta$.

The logarithmic decrement can also be obtained from the amplitudes of nonsuccessive cycles. If we form the ratio of amplitudes x_i and x_{i+n}, where n is any integer, we have

$$\ln\frac{x_i}{x_{i+n}} = \ln\frac{Xe^{-\zeta\omega t_i}}{Xe^{-\zeta\omega(t_i+n\tau)}} = n\zeta\omega\tau = n\delta \tag{5.31}$$

This expression is useful for measurement purposes when it may not be feasible to measure successive amplitudes or when an average δ is to be obtained from several amplitude ratios.

The primary usefulness of the concept of the logarithmic decrement is in the experimental determination of system damping. If any two amplitudes of a damped oscillation are obtained by measurement, the logarithmic decrement can be obtained with Eq. (5.29) or Eq. (5.31) as appropriate. With δ known, Eq. (5.30) is used to obtain the damping factor as

$$\zeta = \frac{\delta}{\sqrt{(2\pi)^2 + \delta^2}} \tag{5.32}$$

This approach is particularly useful where the actual damping mechanism in a system is not precisely known and is to be modeled by an equivalent viscous damping factor.

Example 5.2 A damped spring-mass system has $W = 2$ lb, $k = 4$ lb/in, and an unknown damping constant. If amplitude measurements show a decay rate of 10 percent per cycle, determine the value of the damping constant.

SOLUTION Using the amplitude data, we have

$$\delta = \ln\frac{x_i}{x_{i+1}} = \ln\frac{1}{0.9} = 0.105$$

The damping factor is then

$$\zeta = \frac{0.105}{\sqrt{(2\pi)^2 + (0.105)^2}} = 0.017$$

Using the definition of the damping factor gives

$$c = \zeta C_c = \zeta(2m\omega) = 2\zeta\sqrt{mk} = 2 \times 0.105 \times \sqrt{\frac{2 \times 4}{386.4}}$$
$$= 0.03 \text{ lb} \cdot \text{s/in}$$

5.4 STRUCTURAL DAMPING

Unlike the simple models studied thus far, real mechanical systems are not often composed of discrete elastic, inertial, and damping elements. In fact, structural components act simultaneously as springs, masses, and dampers. Solid materials, particularly metals, exhibit what is known as *structural* or *hysteretic damping*. This type of damping is due to internal friction in the material as relative

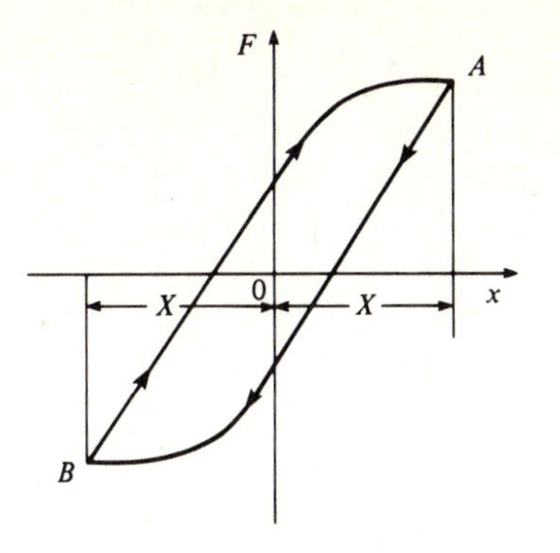

Figure 5.6

slipping or sliding of internal planes occurs during deformation. The result is a phase lag between force and deformation, as depicted in Fig. 5.6. This curve is known as a *hysteresis loop*, and the enclosed area is the energy loss per loading cycle. Mathematically,

$$\Delta U = \int F\,dx \tag{5.33}$$

Experiment has shown that in structural damping the energy loss per cycle is proportional to the stiffness of the material and the square of the displacement amplitude but is independent of frequency. Thus we can write

$$\Delta U = \pi \beta k X^2 \tag{5.34}$$

where β is a dimensionless structural damping constant, k is the equivalent spring constant, and X is the displacement amplitude. The factor π is included for convenience, as will be shown.

Although structural damping is the most common type, it is difficult to treat mathematically since it is defined in terms of energy loss and is a nonlinear function of displacement. For mathematical analysis, it is more convenient to define an *equivalent viscous damping constant* such that the energy loss per cycle is the same as that given by Eq. (5.34). For harmonic motion, the force exerted by a viscous damper is

$$F_v = c\dot{x} = c\omega X \cos(\omega t + \phi) \tag{5.35}$$

and the energy loss per cycle is

$$\begin{aligned}\Delta U &= \int c\dot{x}\,dx = \int_0^{2\pi/\omega} c\dot{x}^2\,dt \\ &= \int_0^{2\pi/\omega} c\omega^2 X^2 \cos^2(\omega t + \phi)\,dt \\ &= \pi c \omega X^2\end{aligned} \tag{5.36}$$

Equating the expressions for ΔU and denoting by C_e the equivalent viscous damping constant, we obtain

$$C_e = \frac{\beta k}{\omega} \tag{5.37}$$

The value of the structural damping constant β can be determined experimentally by determining the logarithmic decrement in a manner similar to that of Sec. 5.3. If the motion is essentially harmonic as in Fig. 5.7, the energy equation for the half cycle between t_1 and t_2 is

$$\frac{kX_1^2}{2} - \frac{\pi\beta kX_1^2}{4} - \frac{\pi\beta kX_{1.5}^2}{4} = \frac{kX_{1.5}^2}{2} \tag{5.38}$$

where we have further assumed that energy loss in a part cycle is proportional to that in a full cycle. Equation (5.38) can be rearranged to obtain

$$\frac{X_1^2}{X_{1.5}^2} = \frac{1 + \pi\beta/2}{1 - \pi\beta/2} \tag{5.39}$$

Similar consideration of the next half-cycle yields

$$\frac{X_{1.5}^2}{X_2^2} = \frac{1 + \pi\beta/2}{1 - \pi\beta/2} \tag{5.40}$$

Combining Eqs. (5.39) and (5.40) gives the ratio of successive amplitudes as

$$\frac{X_1}{X_2} = \frac{1 + \pi\beta/2}{1 - \pi\beta/2} \tag{5.41}$$

For many materials, β is very small so that the right-hand side of Eq. (5.41) can be replaced by the first two terms of its partial-fraction expansion to yield

$$\frac{X_1}{X_2} \simeq 1 + \pi\beta \tag{5.42}$$

The logarithmic decrement is then

$$\delta = \ln\frac{X_1}{X_2} = \ln(1 + \pi\beta) \simeq \pi\beta \tag{5.43}$$

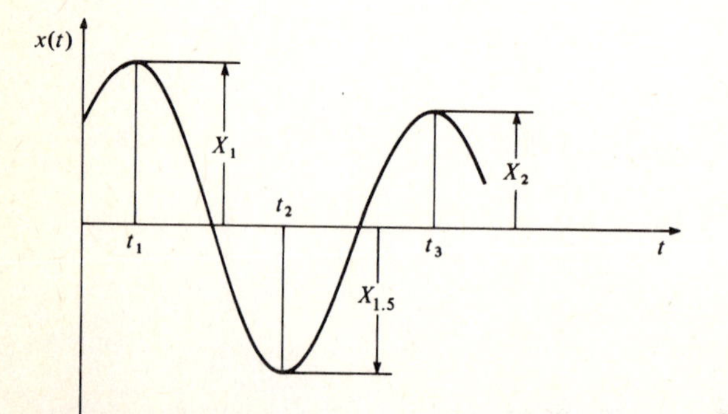

Figure 5.7

By measuring successive amplitudes, β is determined from Eq. (5.42) or Eq. (5.43), and the equivalent viscous damping constant can then be calculated using Eq. (5.37).

Example 5.3 A system which exhibits structural damping has a mass of 10 kg and an equivalent spring constant of 6 N/mm. Measurement shows the ratio of successive amplitudes to be 1.05. Determine the structural damping constant and the equivalent viscous damping constant.

SOLUTION From Eq. (5.43), the structural damping constant is

$$\beta = \frac{\delta}{\pi} = \frac{1}{\pi}\ln\left(\frac{X_1}{X_2}\right) = \frac{1}{\pi}\ln 1.05 = 0.0155$$

The equivalent viscous damping constant is then given by Eq. (5.37) as

$$C_e = \frac{\beta k}{\omega} = \frac{0.0155 \times 6}{\sqrt{6 \times 1000/10}} = 0.0038 \text{ N} \cdot \text{s/mm}$$

5.5 COULOMB DAMPING

Coulomb or *dry-friction* damping occurs when sliding contact exists between parts of a system. Consider a spring-mass system in which the mass slides on a horizontal surface, as in Fig. 5.8*a*. In such a system, energy is dissipated as heat because of sliding friction. The friction force acts always in a direction such as to oppose the motion, and its magnitude is proportional to the normal force existing between the contact surfaces. With the coefficient of sliding friction denoted by f, the friction force is

$$F_f = fN \tag{5.44}$$

where N is the normal contact force.

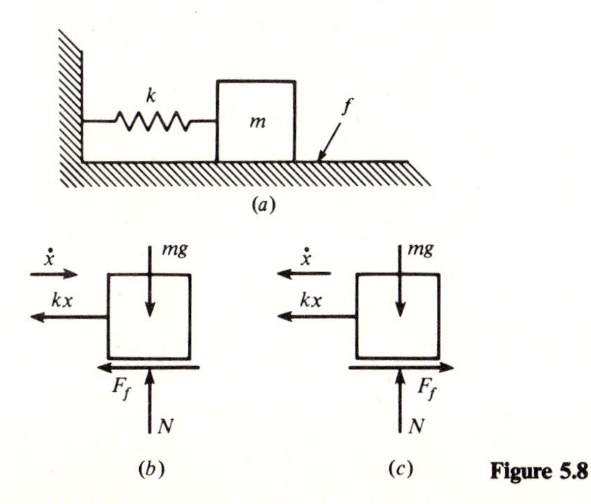

Figure 5.8

Owing to the directional nature of the friction force, two separate free-body diagrams are required, depending on whether motion is the right or left at a given instant. Figure 5.8*b* and *c* show the free-body diagrams for motion to the right and left, respectively. The corresponding differential equations of motion are

$$m\ddot{x} = -kx - F_f \qquad \text{if } \dot{x} > 0 \tag{5.45}$$

$$m\ddot{x} = -kx + F_f \qquad \text{if } \dot{x} < 0 \tag{5.46}$$

Equations (5.45) and (5.46) can be written in combined form as

$$\ddot{x} + \frac{k}{m}x = \mp\frac{F_f}{m} \tag{5.47}$$

where the negative sign applies to motion to the right, and conversely. Equations (5.47) are inhomogeneous, owing to the presence of the term F_f/m, which does not contain the displacement x or any of its derivatives. Thus its solutions are composed of both homogeneous and particular solutions. As may be shown by direct substitution, the solutions are

$$x(t) = A_1 \sin \omega t + B_1 \cos \omega t - \frac{F_f}{k} \qquad \text{if } \dot{x} > 0 \tag{5.48}$$

$$x(t) = A_2 \sin \omega t + B_2 \cos \omega t + \frac{F_f}{k} \qquad \text{if } \dot{x} < 0 \tag{5.49}$$

where $\omega = (k/m)^{1/2}$. The constants A_1 and B_1 depend on the initial conditions of motion to the right, while A_2 and B_2 depend on the initial conditions of leftward motion.

Assume that the mass is displaced to the right a distance x_o and released with zero initial velocity. In this case, motion is initially to the left with $x(0) = x_o$ and $\dot{x}(0) = 0$. Applying the initial conditions to Eq. (5.49) gives $A_2 = 0$ and $B_2 = x_o - F_f/k$ so that

$$x(t) = \left(x_o - \frac{F_f}{k}\right) \cos \omega t + \frac{F_f}{k} \tag{5.50}$$

The displacement given by Eq. (5.50) is valid only until such time as the direction of motion reverses. This will occur when the velocity is zero or

$$\dot{x}(t) = -\omega\left(x_o - \frac{F_f}{k}\right) \sin \omega t = 0 \tag{5.51}$$

which corresponds to $t = \pi/\omega = \tau/2$. At this time the displacement is

$$x\left(\frac{\pi}{\omega}\right) = \left(x_o - \frac{F_f}{k}\right)(-1) + \frac{F_f}{k} = -\left(x_o - 2\frac{F_f}{k}\right) \tag{5.52}$$

Equation (5.52) shows that the maximum displacement to the left is $2F_f/k$ units less than the original amplitude x_o. This reduction is a direct result of dissipation of energy by friction.

The next half cycle (motion to the right) is described by Eq. (5.48) with the initial conditions $x(\pi/\omega) = -(x - F_f/k)$ and $\dot{x}(\pi/\omega) = 0$. Applying these ini-

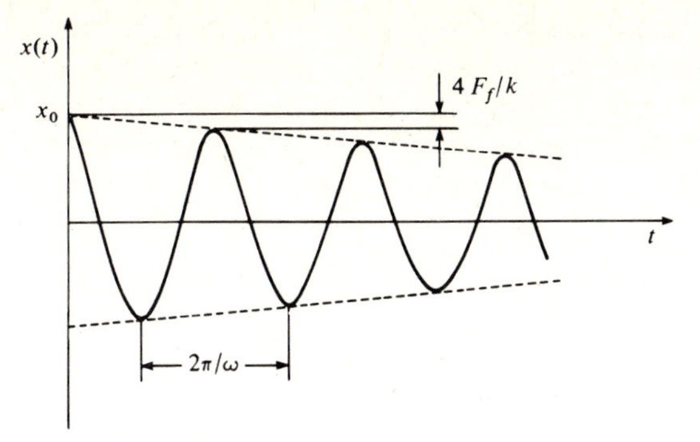

Figure 5.9

tial conditions results in

$$x(t) = \left(x_o - 3\frac{F_f}{k}\right)\cos\omega t - \frac{F_f}{k} \tag{5.53}$$

This expression is valid until the velocity again becomes zero at the extreme right position. To find the corresponding time, we set the velocity equal to zero

$$\dot{x}(t) = -\omega\left(x_o - 3\frac{F_f}{k}\right)\sin\omega t = 0 \tag{5.54}$$

to obtain $t = 2\pi/\omega = \tau$ as we might expect. The corresponding displacement is

$$x\left(\frac{2\pi}{\omega}\right) = \left(x_o - 3\frac{F_f}{k}\right)(1) - \frac{F_f}{k} = x_o - 4\frac{F_f}{k} \tag{5.55}$$

At this time, one complete cycle of motion has occurred ($t = \tau$), but the system has not returned to its starting position; instead it has returned to

$$x = x_o - 4\frac{F_f}{k} \tag{5.56}$$

By continuing in this manner we find a constant amplitude loss of $2F_f/k$ units per half cycle of motion.

The motion is a linearly decaying harmonic function of time as shown in Fig. 5.9. The frequency of oscillation is unaffected by Coulomb damping, as contrasted to the reduction in frequency observed for a viscously damped system. Note also that the system will not necessarily come to rest at the undeformed spring position, since another equilibrium position is possible. This would occur at some amplitude X_i when the restoring force kX_i is insufficient to overcome the friction force F_f.

Example 5.4 The following data apply to the Coulomb-damped system shown in Fig. 5.8*a*: $k = 10$ lb/in, $W = 19.32$ lb, $f = 0.1$, $x(0) = 5$ in, and $\dot{x}(0) = 0$. Determine the number of cycles of motion completed before the mass comes to rest.

SOLUTION The magnitude of the friction force is

$$F_f = fN = fW = 0.1 \times 19.32 = 1.932 \text{ lb}$$

so that motion will cease when the amplitude of the ith cycle is such that $kX_i \leq 1.932$ or $X_i \leq 0.1932$ in. Since the amplitude loss per half cycle is $2F_f/k = 2 \times 1.932/10 = 0.3864$ in, the number of half cycles n is given by

$$X_o - n\left(2\frac{F_f}{k}\right) \leq 0.1932$$

or

$$5 - n(0.3864) \leq 0.1932$$

Solving for n as the smallest integer satisfying this inequality gives $n = 13$ half cycles or 6.5 full cycles completed.

5.6 FORCED VIBRATIONS WITH VISCOUS DAMPING

A damped system subjected to a harmonically varying force is shown in Fig. 5.10*a*. Writing Newton's second law for the free-body diagram shown in Fig. 5.10*b* and utilizing the equilibrium condition $mg = k\Delta$, we obtain

$$m\ddot{x} = -kx - c\dot{x} + F_o \sin \omega_f t \tag{5.57}$$

or

$$\ddot{x} + \frac{c}{m}\dot{x} + \frac{k}{m}x = \frac{F_o}{m}\sin \omega_f t \tag{5.58}$$

as the differential equation of motion. Since the equation is inhomogeneous, the complete solution will be the sum of a homogeneous and a particular solution. Since the homogeneous solution must satisfy Eq. (5.58) with the right-hand side equal to zero, it will be one of the free-vibration solutions discussed in Sec. 5.2. Thus the homogeneous solution will not be discussed here except to note that it decays with time and will eventually become insignificant.

Let us assume a particular solution to Eq. (5.58) in the form

$$x_p(t) = Y \sin \omega_f t + Z \cos \omega_f t \tag{5.59}$$

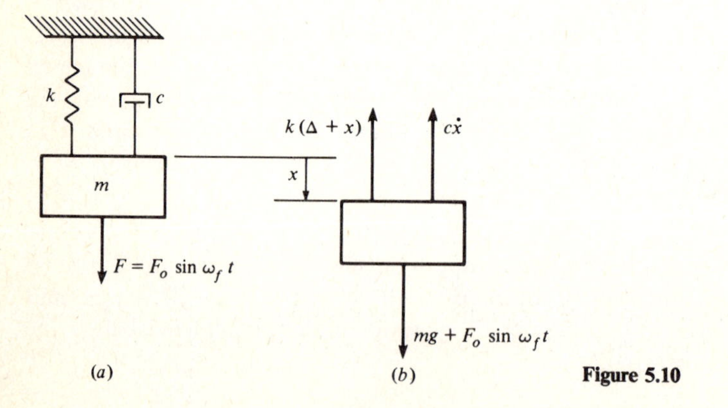

Figure 5.10

where Y and Z are constants. Substitution of the assumed solution into the differential equation of motion yields

$$-Y\omega_f^2 \sin \omega_f t - Z\omega_f^2 \cos \omega_f t + \frac{c}{m}\omega_f Y \cos \omega_f t$$

$$-\frac{c}{m}\omega_f Z \sin \omega_f t + \frac{k}{m} Y \sin \omega_f t + \frac{k}{m} Z \cos \omega_f t = \frac{F_o}{m} \sin \omega_f t \tag{5.60}$$

Equating coefficients of similar functions of time results in the algebraic equations

$$\left(\frac{k}{m} - \omega_f^2\right) Y - \frac{c}{m}\omega_f Z = \frac{F_o}{m} \tag{5.61}$$

and

$$\frac{c}{m}\omega_f Y + \left(\frac{k}{m} - \omega_f^2\right) Z = 0 \tag{5.62}$$

from which Y and Z may be determined. Simultaneous solution gives

$$Y = \frac{(k/m - \omega_f^2)(F_o/m)}{(k/m - \omega_f^2)^2 + [(c/m)\omega_f]^2} \tag{5.63}$$

$$Z = \frac{-(c/m)\omega_f(F_o/m)}{(k/m - \omega_f^2)^2 + [(c/m)\omega_f]^2} \tag{5.64}$$

These expressions are simplified by noting that

$$\left(\frac{k}{m} - \omega_f^2\right)^2 = (\omega^2 - \omega_f^2)^2 = \omega^4(1 - r^2)^2 \tag{5.65}$$

$$\left(\frac{c}{m}\omega_f\right)^2 = \left(\frac{2c\omega^2}{2m\omega}\frac{\omega_f}{\omega}\right)^2 = \omega^4(2\zeta r)^2 \tag{5.66}$$

and

$$\frac{F_o}{m} = \frac{F_o}{m}\frac{m}{k}\omega^2 = \frac{F_o}{k}\omega^2 = X_o\omega^2 \tag{5.67}$$

Substitution of Eqs. (5.65) through (5.67) into Eqs. (5.63) and (5.64) results in

$$Y = \frac{(1 - r^2)X_o}{(1 - r^2)^2 + (2\zeta r)^2} \tag{5.68}$$

and

$$Z = \frac{-2\zeta r X_o}{(1 - r^2)^2 + (2\zeta r)^2} \tag{5.69}$$

Thus, the particular solution is found to be

$$x_p(t) = \frac{X_o}{(1 - r^2)^2 + (2\zeta r)^2}\left[(1 - r^2) \sin \omega_f t - 2\zeta r \cos \omega_f t\right] \tag{5.70}$$

where X_o is the equivalent static deflection as previously defined. The motion described by Eq. (5.70) represents the response of the damped system to the harmonic force and will persist as long as the forcing function continues to act. Specifically, this motion will exist after the free vibrations represented by the homogeneous solutions have died out due to exponential decay. For this reason, the forced response is known as the *steady-state solution*.

To discuss the behavior of the forced response, it will be convenient to rewrite Eq. (5.70) in the form

$$x(t) = X' \sin(\omega_f t - \psi) \tag{5.71}$$

where X' is the forced amplitude, and ψ is the phase angle of the forced response. From previous results, we have

$$X' = \sqrt{Y^2 + Z^2} = \frac{X_o}{\sqrt{(1 - r^2)^2 + (2\zeta r)^2}} \tag{5.72}$$

and

$$\psi = \tan^{-1}\frac{-Z}{Y} = \tan^{-1}\frac{2\zeta r}{1 - r^2} \tag{5.73}$$

The steady-state solution is then

$$x(t) = \frac{X_o}{\sqrt{(1 - r^2)^2 + (2\zeta r)^2}} \sin(\omega_f t - \psi) \tag{5.74}$$

which shows that the amplitude of the forced vibration is equal to the equivalent static deflection X_o multiplied by the *magnification factor*

$$\frac{1}{\sqrt{(1 - r^2)^2 + (2\zeta r)^2}}$$

The effects of various combinations of r and ζ will be discussed in detail in the next section.

Example 5.5 A damped spring-mass system having $W = 19.32$ lb, $k = 45$ lb/in, and $c = 3$ lb · s/in is subjected to a harmonic force $F = 90 \sin 15t$ lb. The initial conditions are $x(0) = 1.5$ in and $\dot{x}(0) = 0$. Determine the equation that describes the displacement of the mass as a function of time.

SOLUTION The differential equation of motion is

$$\ddot{x} + \frac{c}{m}\dot{x} + \frac{k}{m}x = \frac{F_o}{m}\sin \omega_f t$$

which is inhomogeneous, so we must find both the homogeneous and particular solutions. The homogeneous solution corresponds to free damped vibrations, and its form depends upon the value of the damping factor.

From the above data we have

$$m = \frac{W}{g} = 0.05 \text{ lb} \cdot \text{s}^2/\text{in}$$

$$\omega_n = \sqrt{\frac{k}{m}} = 30 \text{ rad/s}$$

$$\zeta = \frac{c}{C_c} = \frac{c}{2m\omega} = \frac{3}{2 \times 0.05 \times 30} = 1$$

indicating that the system is critically damped. The homogeneous solution is

$$x_h(t) = (A + Bt)e^{-30t}$$

from Eq. (5.23).

The particular solution which represents the steady-state or forced response is given by Eq. (5.74). The equivalent static deflection is

$$X_o = \frac{F_o}{k} = \frac{90}{45} = 2 \text{ in}$$

and the frequency ratio is

$$r = \frac{\omega_f}{\omega_n} = \frac{15}{30} = 0.5$$

The steady-state amplitude is calculated, using Eq. (5.72), as

$$X' = \frac{2}{\sqrt{[1 - (0.5)^2]^2 + (2 \times 1 \times 0.5)^2}} = 1.6 \text{ in}$$

and the phase angle of the forced response is

$$\psi = \tan^{-1} \frac{2 \times 1 \times 0.5}{1 - (0.5)^2} = 0.93 \text{ rad}$$

Thus the forced response (particular solution) is

$$x_p(t) = 1.6 \sin(15t - 0.93) \text{ in}$$

The complete solution for the displacement is

$$x(t) = (A + Bt)e^{-30t} + 1.6 \sin(15t - 0.93)$$

Applying the initial conditions gives

$$x(0) = 1.5 = A - 1.28$$
$$\dot{x}(0) = 0 = -30A + B + 14.42$$

from which $A = 2.78$ and $B = 68.98$. The displacement is then

$$x(t) = (2.78 + 68.98t)e^{-30t} + 1.6 \sin(15\text{t} - 0.93)$$

5.7 INFLUENCE OF FREQUENCY RATIO AND DAMPING FACTOR ON STEADY-STATE RESPONSE

The effects of the frequency ratio and the damping factor on the steady-state amplitude X' may be observed by plotting the magnification factor

$$\mathrm{MF} = \frac{X'}{X_o} = \frac{1}{\sqrt{(1 - r^2)^2 + (2\zeta r)^2}} \tag{5.75}$$

as a function of r for several values of ζ as in Fig. 5.11. As expected from the results of Chap. 4, the maximum value of the magnification factor occurs at $r = 1$ for $\zeta = 0$ and represents the undamped resonant condition. For nonzero values of ζ, the maximum magnification is seen to occur for values of r less than unity, and the magnification decreases with increased damping as would be expected on physical grounds.

Before discussing further the curves of Fig. 5.11, let us formally consider the extreme points of the MF curve by assuming ζ to be held fixed at some value and setting the derivative of the magnification factor with respect to r equal to zero. The result is

$$\frac{d(\mathrm{MF})}{dr} = \frac{r(1 - r^2 - 2\zeta^2)}{\left[(1 - r^2)^2 + (2\zeta r)^2\right]^{3/2}} = 0 \tag{5.76}$$

Equation (5.76) will be satisfied if $r = 0$, or $r = \infty$, or $1 - r^2 - 2\zeta^2 = 0$. Again referring to Fig. 5.11, note that the case in which $r = 0$ simply represents the

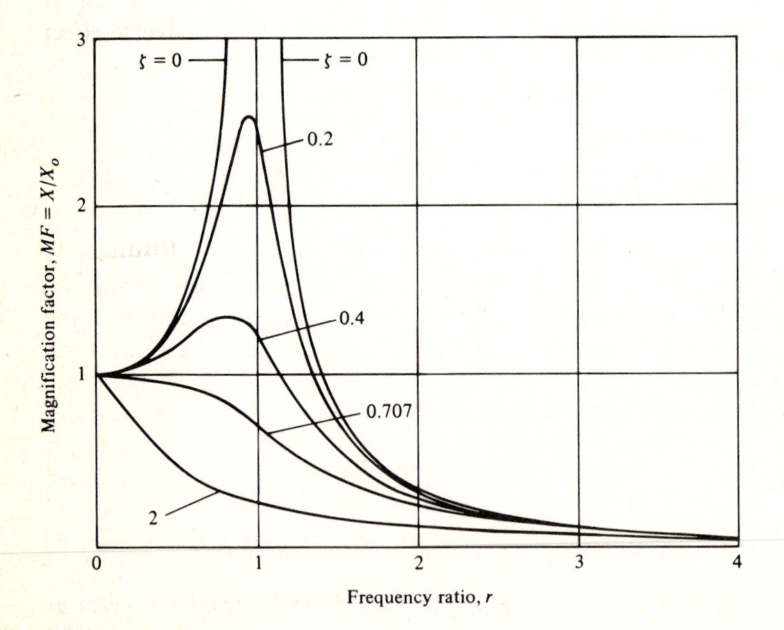

Figure 5.11

displacement of the system due to a statically applied load and is the initial point on each curve. The case in which $r = \infty$ results in zero magnification factor, and hence no displacement, and represents the case in which the forcing frequency is so large in comparison to the frequency of the system that there is insufficient time for the system to respond. This case gives the absolute minimum point for each curve. Finally, consider the case $1 - r^2 - 2\zeta^2 = 0$, which gives a maximum magnification factor at

$$r = \sqrt{1 - 2\zeta^2} \tag{5.77}$$

provided $\zeta \leq \sqrt{2}/2$. The curves show that an unbounded amplitude cannot occur in a damped system but that relatively large amplitudes can result for values of r which satisfy Eq. (5.77). The maximum steady-state amplitude is found by setting $r^2 = 1 - 2\zeta^2$ in Eq. (5.72) to obtain

$$X'_{\text{max}} = \frac{X_o}{2\zeta\sqrt{1 - \zeta^2}} \tag{5.78}$$

As a practical note, we observe that regardless of the amount of damping present, the magnification factor becomes increasingly small as the frequency ratio increases beyond a value of about 2. In mechanical equipment, the operating speed often corresponds to ω_f and cannot be changed. In such cases, an increase in the frequency ratio can be accomplished only by decreasing the circular frequency $\omega = (k/m)^{1/2}$. This could be accomplished by (1) decreasing the equivalent spring constant for the equipment or (2) increasing its mass. The former is neither recommended nor easily accomplished, as changing the equivalent stiffness may involve considerable design effort. However, in many cases it is a simple matter to increase the mass of a machine without any adverse effect on its functional characteristics. Thus, a severe vibration in a machine could be reduced by simply adding some "dead weight."

5.8 FORCE TRANSMISSION AND VIBRATION ISOLATION

The vibratory forces generated by mechanical equipment cannot be eliminated in many cases. In fact, in equipment such as screening devices and materials-testing machines, vibration of the equipment is an essential feature of the functioning of the device. Whatever the case, the forces associated with the vibrations of a machine will be transmitted to its support structure, and these transmitted forces may produce undesirable effects. For instance, vibrations in heating and air-conditioning equipment in an office building may result in vibration of the associated ductwork or piping and result in an intolerable noise problem. To reduce the transmitted forces, mechanical equipment may be mounted on properly designed flexible supports called *vibration isolators* or simply *isolators*.

In Fig. 5.12 is shown the free-body diagram of a damped spring-mass system subjected to a harmonic force; in addition, the forces acting on the

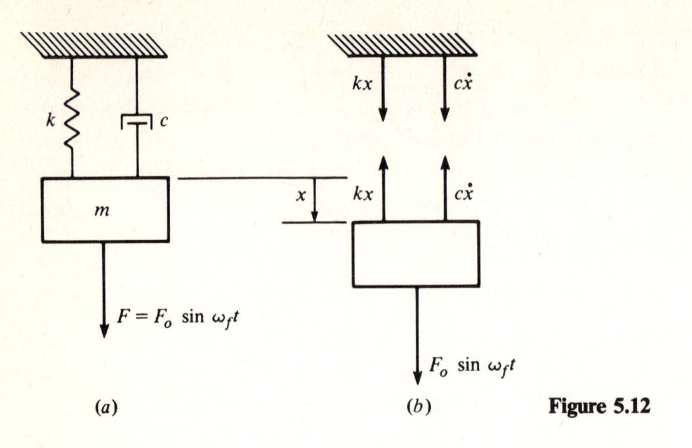

Figure 5.12

support are shown. The force transmitted to the support is

$$F_T = kx + c\dot{x} \tag{5.79}$$

which may be written as

$$F_T = kX' \sin(\omega_f t - \psi) + c\omega_f X' \cos(\omega_f t - \psi) \tag{5.80}$$

The two terms of Eq. (5.80) may be combined to obtain

$$F_T = X'\sqrt{k^2 + (c\omega_f)^2} \sin(\omega_f t - \psi') \tag{5.81}$$

where ψ' is the phase angle for the transmitted force and is given by

$$\psi' = \psi + \tan^{-1}\left(-\frac{c\omega_f}{k}\right) \tag{5.82}$$

Thus the amplitude, or maximum value, of the transmitted force is

$$(F_T)_{\max} = X'\sqrt{k^2 + (c\omega_f)^2} \tag{5.83}$$

or

$$(F_T)_{\max} = \frac{X_o\sqrt{k^2 + (c\omega_f)^2}}{\sqrt{(1 - r^2)^2 + (2\zeta r)^2}} \tag{5.84}$$

The *transmissibility* T is defined as the ratio of the maximum transmitted force to the amplitude of the applied force and is obtained from Eq. (5.84) as

$$T = \frac{(F_T)_{\max}}{F_o} = \frac{\frac{1}{k}\sqrt{k^2 + (c\omega_f^2)}}{\sqrt{(1 - r^2)^2 + (2\zeta r)^2}} \tag{5.85}$$

by using the substitution $X_o = F_o/k$. In terms of the frequency ratio and the damping factor, the transmissibility is

$$T = \frac{\sqrt{1 + (2\zeta r)^2}}{\sqrt{(1 - r^2)^2 + (2\zeta r)^2}} \tag{5.86}$$

in which form the effects of r and ζ on the transmitted force may be studied.

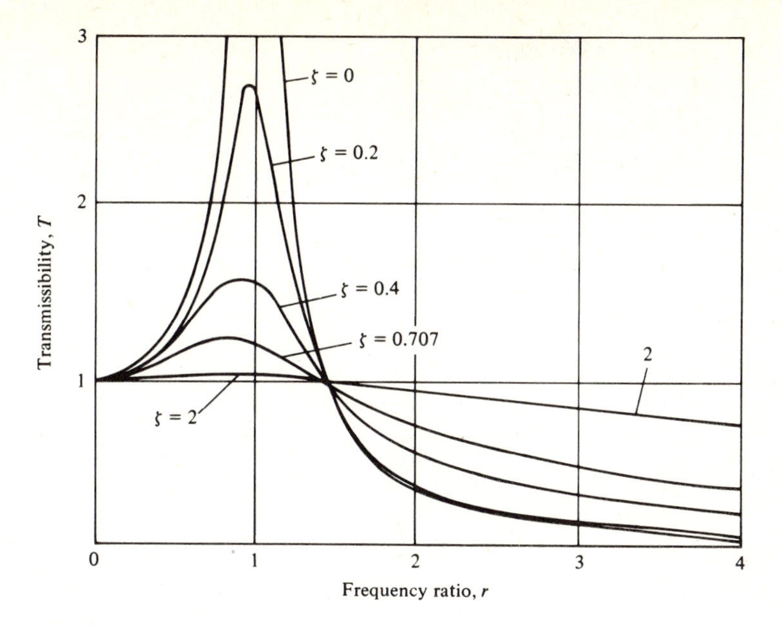

Figure 5.13

Considering the spring and dashpot as a vibration isolator, we desire to select the spring constant and the damping constant, hence r and ζ, so as to minimize the transmissibility in a given situation. Owing to the complexity of Eq. (5.86), the behavior of the transmissibility is most easily studied by obtaining curves for various combinations of r and ζ as in Fig. 5.13. For values of the frequency ratio in the resonant range (near $r = 1$), the transmissibility is greater than unity, meaning that the force transmitted to the support is greater than the applied force. The region should be avoided if at all possible, since it results in force multiplication rather than isolation. A more pertinent observation is that at $r = \sqrt{2}$ every curve passes through the point $T = 1$ and becomes asymptotic to zero as the frequency ratio is increased. Two conclusions may be drawn: (1) vibration isolation ($T < 1$) is possible only if $r > \sqrt{2}$ regardless of the value of the damping factor, and (2) for a given $r > \sqrt{2}$ the transmissibility decreases as the damping factor is decreased. Thus, vibration isolation is best accomplished by an isolator composed only of spring elements for which $r > \sqrt{2}$ with no damping element included.

Vibration isolators are commercially available from many firms and may vary from simple corrugated rubber pads to complex actively controlled air springs. For design and application details, consult the appropriate manufacturer's literature.

Example 5.6 Given the data of Example 5.5, determine the (*a*) transmissibility and (*b*) magnitude of the maximum force transmitted to the support.

SOLUTION From Example 5.5 we have $r = 0.5$, $\zeta = 1$ and $F_o = 90$ lb. The transmissibility is found with Eq. (5.86) as

$$T = \frac{\sqrt{1 + (2 \times 1 \times 0.5)^2}}{\sqrt{(1 - 0.5^2)^2 + (2 \times 1 \times 0.5)^2}} = 1.131$$

The maximum transmitted force is

$$(F_T)_{\max} = TF_o = 1.131 \times 90 = 101.79 \text{ lb}$$

5.9 ROTATING UNBALANCE

One of the most common sources of vibration in mechanical equipment is unbalance of a rotating component. Unbalance frequently occurs in fans, blower and pump impellers, flywheels, rotors, and so forth. Most of us have experienced the shaky steering and roughness of ride which accompany unbalance in automobile wheels. Unbalance in rotating mechanical equipment results in excess vibration of the equipment and in the generation of forces which may lead to premature failure of components such as bearings and couplings, or even failure of the unbalanced component itself.

The mechanical model which will be used to illustrate the vibration resulting from rotating unbalance is shown in Fig. 5.14. The rotating component is assumed to be mounted in bearings at O and to rotate with a counterclockwise angular velocity ω_f rad/s. The mass distribution of the rotor is such that there results an equivalent eccentric mass m located a distance R from the center of the rotation. The total mass of the system is M, and the system is constrained so as to allow vertical motion only.

To determine the force resulting from rotation of the eccentric mass, we recall from dynamics that a mass moving in a circular path with constant speed will be subjected to an acceleration having the direction of the inward normal to

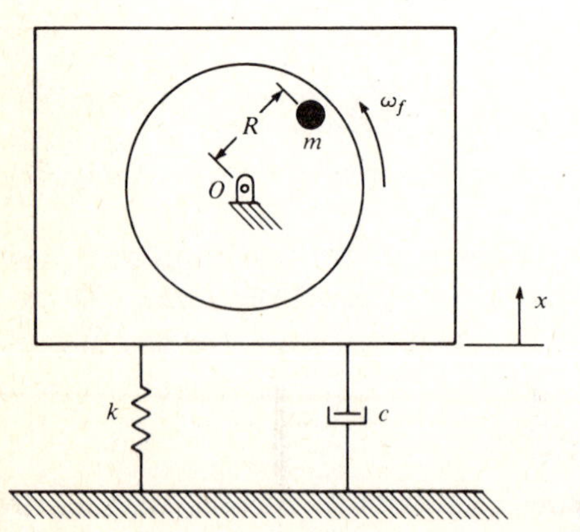

Figure 5.14

the path. In other words, the mass experiences a normal acceleration which is given in this case by

$$A_n = \frac{V^2}{R} = R\omega_f^2 \tag{5.87}$$

Figure 5.15*a* shows the equal and opposite forces exerted on the eccentric mass by the rotor and on the rotor by the mass when the rotor is in an arbitrary position $\theta = \omega_f t$ from the horizontal. By applying Newton's second law to the eccentric mass in the normal direction, the magnitude of the force is found to be

$$F = mA_n = mR\omega_f^2 \tag{5.88}$$

which is also the magnitude of the force exerted on the machine as a whole through the bearings at O. Thus, as the rotation occurs, the machine is subjected to a constant force of varying direction. The dynamic free-body diagram of the machine in some displaced position x is as shown in Fig. 5.15*b*, from which we obtain

$$\Sigma F_x = M\ddot{x} = -kx - c\dot{x} + F\sin\theta \tag{5.89}$$

or

$$M\ddot{x} = -kx - c\dot{x} + mR\omega_f^2 \sin\omega_f t \tag{5.90}$$

The effect of the unbalanced mass is a harmonically varying force having a magnitude $mR\omega_f^2$.

In standard form, the differential equation of motion is

$$\ddot{x} + \frac{c}{M}\dot{x} + \frac{k}{M}x = \frac{mR\omega_f^2}{M}\sin\omega_f t \tag{5.91}$$

which is completely analogous to Eq. (5.58). The solution to Eq. (5.91) is the sum of a homogeneous solution and a particular solution, with the homogeneous solution corresponding to one of the cases discussed in Sec. 5.2. Here we shall concentrate only on the steady-state response, since it is known that the free

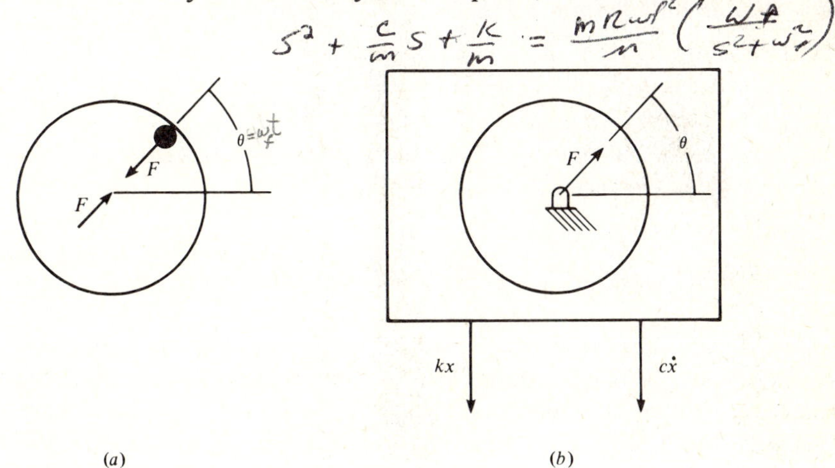

Figure 5.15

vibrations decay exponentially and in time disappear. The steady-state response is the same as that given by Eq. (5.74) if we make the substitution $X_o = F_o/k = mR\omega_f^2/k$ to obtain

$$x(t) = \frac{mR\omega_f^2/k}{\sqrt{(1 - r^2)^2 + (2\zeta r)^2}} \sin(\omega_f t - \psi) \tag{5.92}$$

which can also be written as

$$x(t) = \frac{(mR/M)r^2}{\sqrt{(1 - r^2)^2 + (2\zeta r)^2}} \sin(\omega_f t - \psi) \tag{5.93}$$

Equation (5.93) shows that the steady-state vibration resulting from a rotating unbalance varies directly with the amount of unbalance m and its distance R from the center of rotation, and generally increases as the square of the rotating speed. We also note that the circular frequency ω_f of the forced vibration is the same as the rotating speed, and that the maximum displacement of the system lags the maximum value of the forcing function by the phase angle ψ. Since the maximum upward force occurs when the eccentric mass is in the upward vertical position, this means that the maximum displacement will not occur until the mass has rotated through an angle ψ past the vertical. This observation is pertinent to the balancing procedures to be discussed in Chap. 6.

Example 5.7 A machine having a total mass of 400 kg contains an unbalanced rotating component. When operating at resonance, the steady-state vibration amplitude is 0.3 mm. If the system is critically damped, determine the unbalance mR.

SOLUTION At resonance $r = 1$ and for critical damping $\zeta = 1$, so that the steady-state amplitude from Eq (5.93) is

$$0.3 = \frac{(mR/400)(1)^2}{\sqrt{(1 - 1^2)^2 + (2 \times 1 \times 1)^2}}$$

which gives $mR = 240$ kg · mm. This is equivalent to an unbalance mass of 1.6 kg at a radial distance of 150 mm, for example.

5.10 BASE EXCITATION

The systems studied thus far were all assumed to be mounted via springs and damping elements to a fixed support. Another significant class of vibrations corresponds to the case of equipment mounted on a moving support or base. The motion of the base results in forces being transmitted to the mounted equipment, and the resulting vibrations can be severe unless care is exercised in the design of the mounting devices. This type of induced vibration would be of concern, for example, in the mounting of instruments to a support which is subjected to vibration from other sources, as in aircraft.

The model to be used for the study of movable-base-excited vibrations is shown in Fig. 5.16*a*. The support executes oscillatory vertical motion given by $y = Y_o \sin \omega_f t$, and the displacement of the mass is measured from its equilibrium position as usual. In this case, since each end of both spring and dashpot may move, the spring force and damping force will depend upon the *difference* in displacements and velocities, respectively. The free-body diagram of the mass is as shown in Fig. 5.16*b*, where x, $\dot{x}$, y, and $\dot{y}$ are assumed positive with $x > y$ and $\dot{x} > \dot{y}$. In this manner the proper direction for each force is automatically obtained.

Applying Newton's second law to the free-body diagram of the mass gives

$$m\ddot{x} = -k(x - y) - c(\dot{x} - \dot{y}) \tag{5.94}$$

or

$$m\ddot{x} + c\dot{x} + kx = ky + c\dot{y} \tag{5.95}$$

Substitution for y and $\dot{y}$ results in

$$m\ddot{x} + c\dot{x} + kx = kY_o \sin \omega_f t + c\omega_f Y_o \cos \omega_f t \tag{5.96}$$

which shows that the motion of the support results in the mass being subjected to two harmonic forcing functions. The forcing functions may be combined into an equivalent single force to give

$$m\ddot{x} + c\dot{x} + kx = Y_o\sqrt{k^2 + (c\omega_f)^2}\,\sin(\omega_f t + \beta) \tag{5.97}$$

where the phase angle β is found from

$$\beta = \tan^{-1}\frac{c\omega_f}{k} = \tan^{-1} 2\zeta r \tag{5.98}$$

In standard form, the differential equation of motion of the mass is then

$$\ddot{x} + \frac{c}{m}\dot{x} + \frac{k}{m}x = \frac{Y_o}{m}\sqrt{k^2 + (c\omega_f)^2}\,\sin(\omega_f t + \beta) \tag{5.99}$$

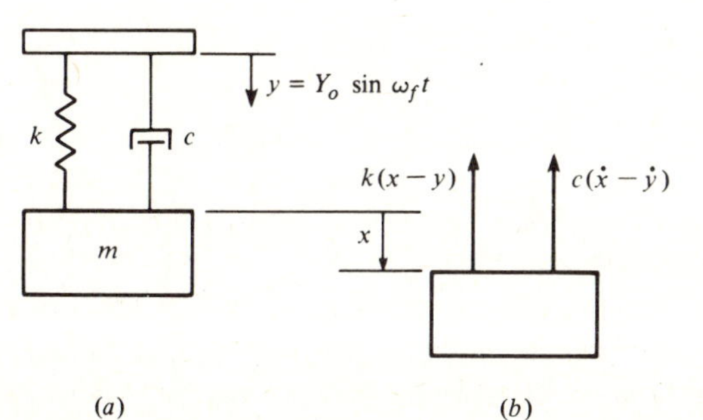

Figure 5.16

By analogy, the steady-state solution to Eq. (5.99) is given by Eq (5.74) with

$$X_o = F_o/k = (k/Y_o)\sqrt{k^2 + (c\omega_f)^2} = Y_o\sqrt{1 + (2\zeta r)^2}$$

The solution then is

$$x(t) = \frac{Y_o\sqrt{1 + (2\zeta r)^2}}{\sqrt{(1 - r^2)^2 + (2\zeta r)^2}}\sin(\omega_f t + \beta - \psi) \tag{5.100}$$

where the phase angle ψ is again given by

$$\psi = \tan^{-1}\frac{2\zeta r}{1 - r^2} \tag{5.101}$$

An examination of the displacement given by Eq. (5.100) shows that the amplitude of the steady-state response is the product of the amplitude of the base motion and the transmissibility. Thus the curves of Fig. 5.13 and the associated discussion are applicable to the problem of base excitation. Note that the transmissibility refers to the transmission of motion rather than force.

Example 5.8 A tape player is mounted to the dashboard of an automobile with a flexible mount which has an equivalent spring constant of 40 lb/in and negligible damping. At a certain engine speed, the dashboard vibrates with a steady-state amplitude of 0.2 in at a frequency of 115 Hz. Determine the steady-state amplitude of the vibrations of the tape player if it weighs 8 lb.

SOLUTION To determine the frequency ratio we first calculate

$$\omega = \sqrt{\frac{k}{m}} = \sqrt{\frac{40 \times 386.4}{8}} = 43.95 \text{ rad/s}$$

and

$$f = \frac{\omega}{2\pi} = \frac{43.95}{2\pi} = 6.99 \text{ Hz}$$

The frequency ratio is then

$$r = \frac{115}{6.99} = 16.45$$

Since $\zeta = 0$ and $Y_o = 0.2$, the steady-state amplitude from Eq. (5.100) is

$$X' = \frac{0.2}{(16.45)^2 - 1} = 0.0007 \text{ in}$$

showing that the flexible mount used is a quite effective isolator.

5.11 SELF-EXCITED VIBRATIONS

If a mechanical system is such that the exciting force is proportional to displacement or one of its time derivatives, vibrations of such a system are said to be *self-excited*. Examples of self-excited vibrations include airplane wing flutter, galloping of power transmission lines, and tool chatter in machine tools. Under certain conditions, self-excited vibrations are unstable and can grow to the point of structural failure.

Consider a damped spring-mass system driven by a force which is proportional to velocity and has the same direction as the velocity. The differential equation of motion of this system is

$$m\ddot{x} + c\dot{x} + kx = F_o\dot{x} \tag{5.102}$$

which can be rearranged as

$$\ddot{x} + \frac{c - F_o}{m}\dot{x} + \frac{k}{m}x = 0 \tag{5.103}$$

From this form it is apparent that a velocity-proportional force is equivalent to negative viscous damping, since energy is *added* in proportion to velocity.

Following the method of Sec. 2.1, we assume the solution of Eq. (5.103) as $x(t) = Ce^{st}$ to obtain the auxiliary equation

$$s^2 + \frac{c - F_o}{m}s + \frac{k}{m} = 0 \tag{5.104}$$

which has the roots

$$s_{1,2} = -\frac{c - F_o}{2m} \pm \sqrt{\frac{c - F_o}{2m} - 4\frac{k}{m}} \tag{5.105}$$

We note that if $c - F_o = 0$, the system behaves as an undamped simple harmonic oscillator and the resulting vibration is stable. If $c - F_o > 0$, we have a damped spring-mass system with an effective damping constant $C_e = c - F_o$, and the resultant motion is also stable as described in Sec. 5.2.

The final possibility, $c - F_o < 0$, is of particular interest here. In this event, at least one of the roots given by Eq. (5.105) is *real* and *positive*. Thus the displacement exhibits exponential growth and is said to be dynamically unstable. This behavior can occur in structures subjected to wind loading unless care is exercised in evaluating the aerodynamic properties of the structure. For details of several specific examples of this phenomenon, refer to the text by Den Hartog.[2]

PROBLEMS

5.1 A 3-kg mass is suspended from a fixed support by a spring having an elastic modulus of 2 N/mm. A viscous-damping element is included in the system, but the damping constant is unknown. Damped free oscillations of the system are observed to occur at a frequency of 3.8 Hz. Find (*a*) the damping factor and (*b*) the damping constant.

5.2 A damped spring-mass system has $m = 0.013$ lb · s²/in, $k = 10$ lb/in, and $c = 0.9$ lb · s/in. The mass is set in motion with initial conditions $x(0) = 2$ in and $\dot{x}(0) = 4$ in/s. Determine the displacement after 0.2 s.

[2] J. P. Den Hartog, *Mechanical Vibrations*, 4th ed., McGraw-Hill, New York, 1956, chap. 7.

5.3 A damped spring-mass system is subjected to intial conditions $x(0) = x_o$ and $\dot{x}(0) = \dot{x}_o$. Write the equation of motion for (*a*) $\zeta = 3$, (*b*) $\zeta = 1$, and (*c*) $\zeta = 0.5$.

5.4 A viscously damped system has a damping constant of 1.5 lb · s/in, spring constant of 5 lb/in, and mass of 0.15 lb · s^2/in. Find (*a*) the damping factor, (*b*) the frequency of damped oscillations, and (*c*) the spring constant needed to obtain critical damping.

5.5 A critically damped system has an undamped natural frequency of 0.5 Hz. Plot the displacement versus time for the following initial conditions: (*a*) $x(0) = 4$, $\dot{x}(0) = 0$; (*b*) $x(0) = 0$, $\dot{x}(0) = 4$; (*c*) $x(0) = 0$, $\dot{x}(0) = -4$.

5.6 Repeat Prob. 5.5 if the system is overdamped with $\zeta = 1.2$.

5.7 The torsional system shown in Fig. 5.17 contains a torsional viscous damper having a damping constant C_T in · lb · s/rad. Write the differential equation of motion for the system, and obtain an expression for the critical damping constant.

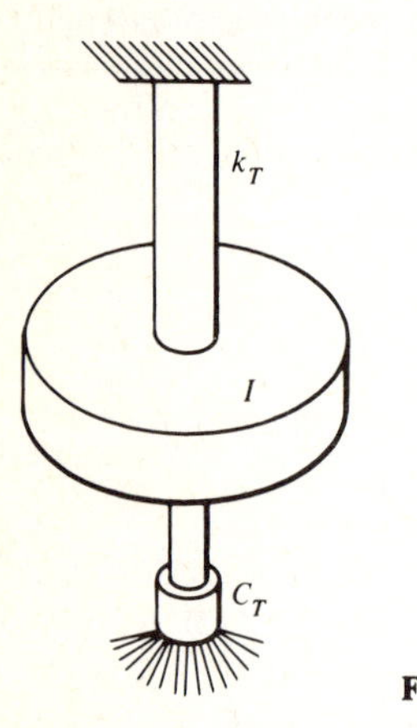

Figure 5.17

5.8 Bar AB is of negligible mass and pivots about a smooth pin at A as shown in Fig. 5.18. The spring constant is 1.5 N/mm, and the mass is 2 kg. If $a = b = L/3 = 300$ mm, calculate the damping constant required if the system is to be critically damped.

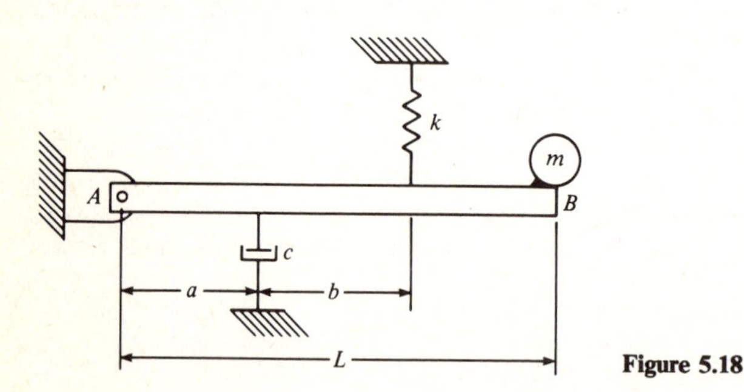

Figure 5.18

5.9 For the system shown in Fig. 5.18, $m = 0.05$ lb · s^2/in, $k = 20$ lb/in, $c = 1.25$ lb · s/in, $a = 6$ in, $b = 12$ in, and $L = 18$ in. Determine the equation describing small angular oscillations of the system.

5.10 Determine the equivalent viscous damping constant for two dashpots c_1 and c_2 connected (*a*) in series and (*b*) in parallel.

5.11 A damped spring-mass system has a spring constant of 3 lb/in. Free oscillations of the system exhibit a period of 1.8 s, and the ratio of successive amplitudes is 4.2:1. Determine the mass and damping constant of the system.

5.12 The damped free oscillations of a system are such that the amplitude of the fourteenth cycle is 55 percent that of the fifth cycle. If the spring constant is 1.4 N/mm and the mass is 9 kg, find the damping constant and the frequency of the damped oscillations.

5.13 A spring-mass system is damped such that an amplitude loss of 1 percent occurs in each full cycle of oscillation. Calculate the damping factor for the system.

5.14 The physical constants in a damped system are $m = 4.5$ kg, $k = 4$ N/mm, and $c = 0.2$ N · s/mm. If the mass is released from the rest with an initial displacement of 100 mm, calculate the amplitude after (*a*) 2 cycles, (*b*) 10 cycles, and (*c*) 100 cycles.

5.15 Using the data of Prob. 5.14, determine the time required for an amplitude reduction of 50 percent.

5.16 A simply supported beam carries a concentrated mass at its center, as shown in Fig. 5.19. The beam properties are $E = 30 \times 10^6$ lb/in^2, $I = 2.5$ in^4, and $L = 48$ in, and the concentrated mass is 0.15 lb · s^2/in. Free oscillations of the system exhibit an amplitude decay of 0.5 percent per cycle. If the mass of the beam is negligible, find (*a*) the structural damping constant and (*b*) the equivalent viscous damping constant.

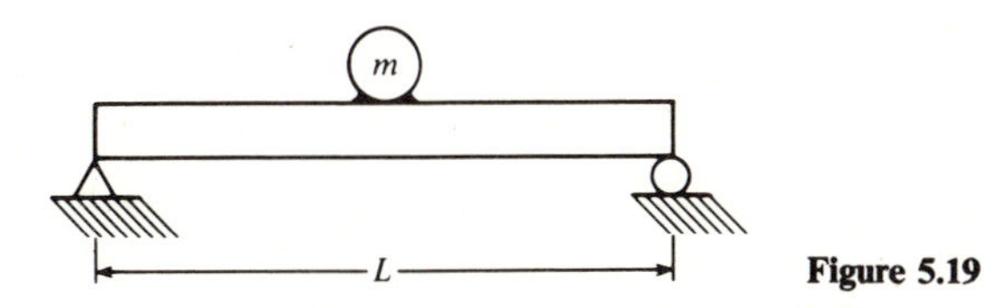

Figure 5.19

5.17 Show that for small structural damping, the equivalent viscous damping constant given by Eq. (5.37) can also be obtained by equating the logarithmic decrements.

5.18 A torsional pendulum similar to Fig. 3.8 has as its elastic element an aluminum shaft having $D = 30$ mm, $L = 500$ mm, and $G = 28 \times 10^3$ N/mm^2. Free oscillations of the system exhibit structurally damped decay with each amplitude measured as 99 percent of the preceding amplitude. Determine (*a*) the equivalent viscous damping constant and (*b*) the equation describing small oscillations of the equivalent system.

5.19 Show that for Coulomb damping the amplitude loss F_f/k per half cycle is also obtained by equating the loss of potential energy to the work done by friction during a half cycle.

5.20 In the system shown in Fig. 5.20, $m = 5$ kg, $k = 2$ N/mm, $c = 0.15$ N · s/mm, and the coefficient of friction between the mass and horizontal surface is $f = 0.1$. The mass is displaced 50 mm to the right and released from the rest. Determine (*a*) the differential equation of motion, (*b*) the displacement $x(t)$, and (*c*) the number of cycles of motion completed before the mass reaches equilibrium.

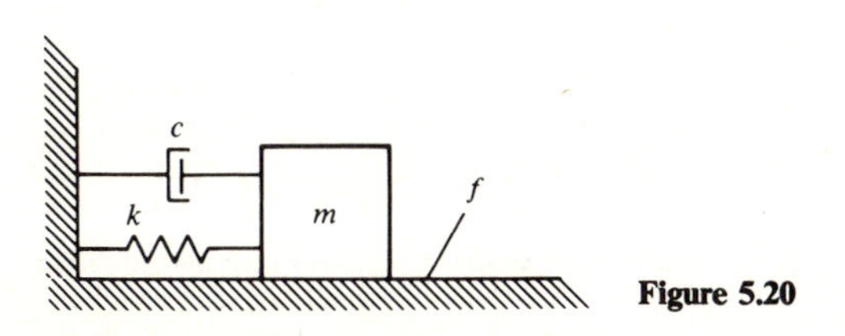

Figure 5.20

5.21 A damped system is composed of a 10-lb weight suspended by a spring having an elastic constant of 12 lb/in and a dashpot having a damping constant of 0.95 lb · s/in. The mass is subjected to a harmonic force $F = 25 \sin 18t$ lb. Initial conditions are $x(0) = 1.5$ in and $\dot{x}(0) = 0$. Determine (*a*) the displacement $x(t)$, including both transient and steady-state response, (*b*) the time required for the transient to decay to 1 percent of its initial value, and (*c*) the time lag between maximum amplitude and maximum exciting force considering only the steady-state response.

5.22 Repeat Prob. 5.21 if the frequency of the harmonic force corresponds to resonance.

5.23 A damped spring-mass system is excited at resonance by a harmonic force which has an amplitude of 100 N. The steady-state amplitude of the resulting vibration is observed to be 2 mm. When the frequency of the exciting force is doubled, the steady-state amplitude becomes 1.4 mm. Determine (*a*) the damping factor and (*b*) the spring constant.

5.24 The torsional pendulum of Prob. 5.18 is subjected to a harmonic torque $T = 4500 \sin 400t$ N · mm. Determine the amplitude of the resulting steady-state oscillations.

5.25 In Fig. 5.21, bar AB has negligible mass and pivots freely about point A. For small oscillations about the horizontal equilibrium, determine (*a*) the differential equation of motion, (*b*) the expression for the circular frequency of the system, (*c*) the expression for the critical damping constant, and (*d*) the expression for the steady-state amplitude.

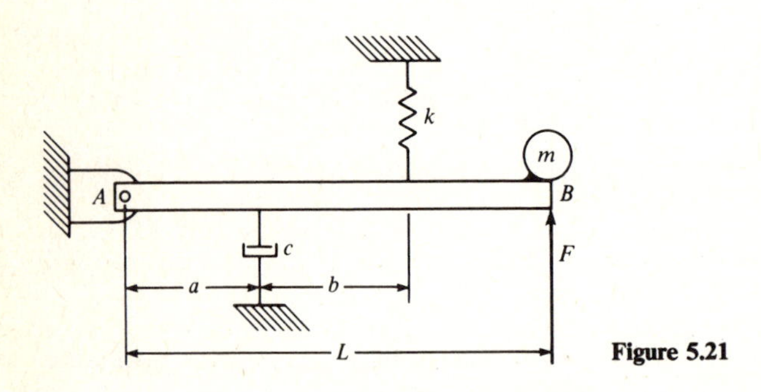

Figure 5.21

5.26 For the system shown in Fig. 5.21, $m = 0.9$ kg, $k = 1.2$ N/mm, $c = 0.15$ N · s/mm, $a = 100$ mm, $b = 250$ mm, and $L = 400$ mm. If $F = 400 \sin \omega_f t$ N, determine the range of values of ω_f for which the assumption of small oscillations is valid.

5.27 The system shown in Fig. 5.22 has $m_1 = 2m_2$, $k = 10$ lb/in, $c = 2$ lb · s/in, $F = 18 \sin 10t$ lb, and a steady-state amplitude of 1.8 in. After removal of m_2, the steady-state amplitude becomes 3.2 in. Determine m_1.

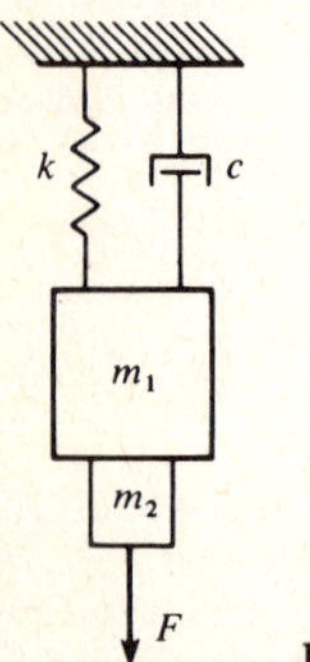

Figure 5.22

5.28 A machine weighing 500 lb is mounted on vibration isolators for which $k = 1200$ lb/in (total). Damping in the isolators is such that $\zeta = 0.15$. If the machine is excited by the harmonic force $F = 800 \sin 32t$ lb, determine, for the steady-state response, (*a*) the vibration amplitude, (*b*) the transmissibility, and (*c*) the maximum force transmitted to the foundation.

5.29 Show that transmissibility can be defined as

$$T = \frac{\sqrt{k^2 - c\omega_f^2}}{\sqrt{\left(k - m\omega_f^2\right)^2 + \left(c\omega_f\right)^2}}$$

5.30 Show that the magnification factor can be written as

$$\mathrm{MF} = \frac{k}{\sqrt{\left(k - m\omega_f^2\right)^2 + \left(c\omega_f\right)^2}}$$

5.31 Using the data of Prob. 5.28, determine the weight which must be added to the machine to reduce the maximum transmitted force by 60 percent. What is the corresponding change in steady-state vibration amplitude?

5.32 If new isolators are installed on the machine of Prob. 5.28, what should be the spring constant if the maximum transmitted force is not to exceed 600 lb? Assume that the damping factor remains the same.

5.33 A plastic grille on a commercial refrigerator has a mass of 0.2 kg, and its attachment has an equivalent spring constant of 0.8 N/mm. The equivalent viscous damping factor is $\zeta = 0.25$. In operation the grille is excited by a harmonic force $F = 25 \sin 75t$ N due to operation of the compressor unit. For the resulting steady-state vibration, determine the (*a*) amplitude, (*b*) phase angle, and (*c*) maximum velocity.

5.34 For the system shown in Fig. 5.23, $k = 45$ lb/in, $c = 3$ lb · s/in, $W = 19.32$ lb, $F_1 = 90 \sin 15t$, and $F_2 = 60 \sin 30t$. Determine (*a*) the differential equation of motion and (*b*) the complete solution for zero initial conditions.

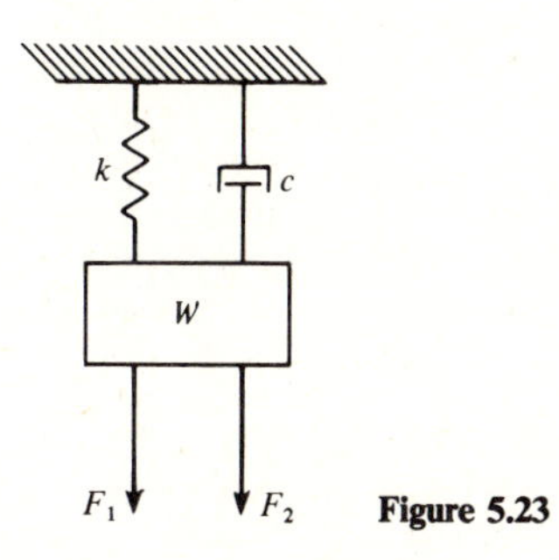

Figure 5.23

5.35 For a damped system excited by a harmonic force, plot the phase angle ψ as a function of frequency for (*a*) $\zeta = 0$, (*b*) $\zeta = 0.5$, (*c*) $\zeta = 0.707$, and (*d*) $\zeta = 1$.

5.36 A homogeneous steel disk contains a hole as shown in Fig. 5.24. The disk is part of a machine having a total mass of 100 kg and mounted such that $k = 200$ N/mm and $c = 15$ N · s/mm. The disk is 30 mm thick and has a density of 78×10^2 kg/m^3. If the disk rotates at 600 r/min, determine (*a*) the steady-state amplitude, (*b*) the phase angle, and (*c*) the time lag between maximum force and maximum displacement.

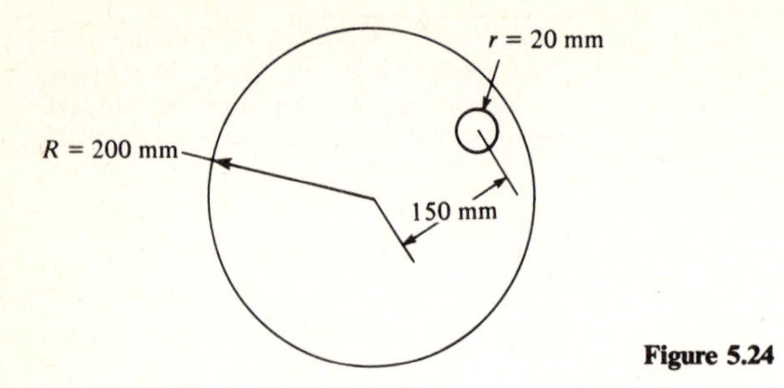

Figure 5.24

5.37 A Pelton-wheel impulse turbine (Fig. 5.25) has 36 buckets arranged symmetrically. While operating at 520 r/min, one of the buckets breaks off. Each bucket weighs 1.5 lb, and the total weight of the machine is 300 lb. The spring constant is 600 lb/in, and damping is such that $\zeta = 0.5$. Determine (*a*) the amplitude of the resulting steady-state vibration and (*b*) the maximum dynamic force transmitted to the support.

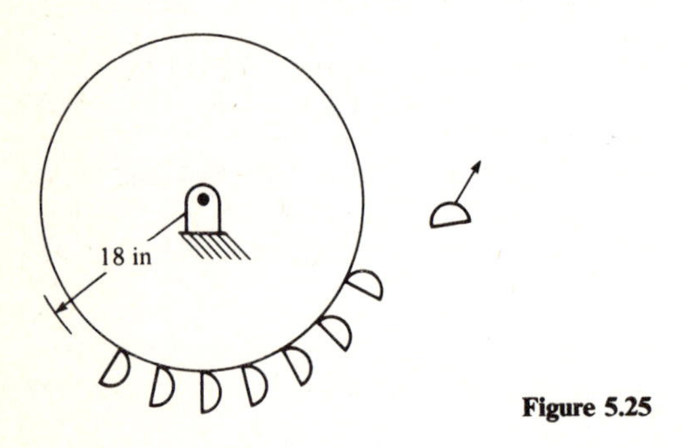

Figure 5.25

5.38 A variable-speed motor having a total mass of 150 kg is mounted on four springs, each having an elastic constant of 130 N/mm. Damping is negligible. The rotor contains an unbalance equivalent to 0.05 kg located 150 mm from the axis of rotation. Determine (*a*) the motor speed in revolutions per minute at which resonance occurs, (*b*) the steady-state amplitude at a motor speed of 900 r/min, and (*c*) the damping constant required to limit the amplitude at resonance to 0.125 mm.

5.39 Figure 5.26 depicts a common type of vibration exciter. It contains two equal eccentric masses m which rotate with identical speeds but in opposite directions. This arrangement produces only vertical force, as the horizontal components cancel. When attached to another machine or structure, this device can be used to obtain the vibrational characteristics of the machine. In the system shown, the total mass is $M = 85$ kg, and the unbalance of *each* exciter disk is $mR = 20$ kg · mm. After the exciter speed is adjusted to 900 r/min, it is observed that a maximum upward displacement of 20 mm occurs at the instant the eccentric masses are in the horizontal position.

(*a*) Utilizing the definition of phase angle, show that this observation corresponds to $r = 1$.
(*b*) Determine the spring constant.
(*c*) Determine the damping constant.
(*d*) Determine the steady-state amplitude if the exciter speed is doubled.

5.40 The vibration exciter described in Prob. 5.39 has a total weight of 20 lb, and the unbalance of *each* exciter disk is 1.5 lb · in. It is rigidly attached to a machine which weighs 250 lb, and the exciter speed is adjusted until the resonant condition is observed at 1800 r/min. For the machine to which the exciter is attached, determine (*a*) the natural frequency, (*b*) the equivalent spring constant, and (*c*) the critical damping constant.

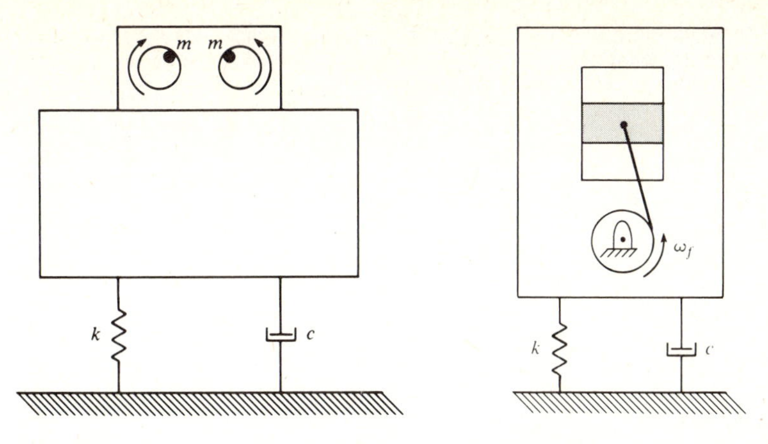

Figure 5.26 **Figure 5.27**

5.41 Figure 5.27 depicts a single-cylinder reciprocating engine. The total mass of the engine is 25 kg, and it is mounted on vibration isolators having $k = 15$ N/mm and $c = 1.25$ N · s/mm. The piston mass is 0.6 kg, and the piston stroke is 100 mm. Assume the piston motion to be harmonic.

(*a*) Determine the differential equation of motion for an engine speed of ω_f rad/s.

(*b*) Determine the engine speed which results in maximum steady-state amplitude.

(*c*) Determine the maximum dynamic force transmitted at an engine speed of 1000 r/min.

5.42 The support of a simple pendulum executes harmonic motion as shown in Fig. 5.28. Determine (*a*) the differential equation of motion for small oscillations and (*b*) the complete solution for small oscillations given the initial conditions $\theta(0) = \theta_o$ and $\dot{\theta}(0) = \dot{\theta}_o$.

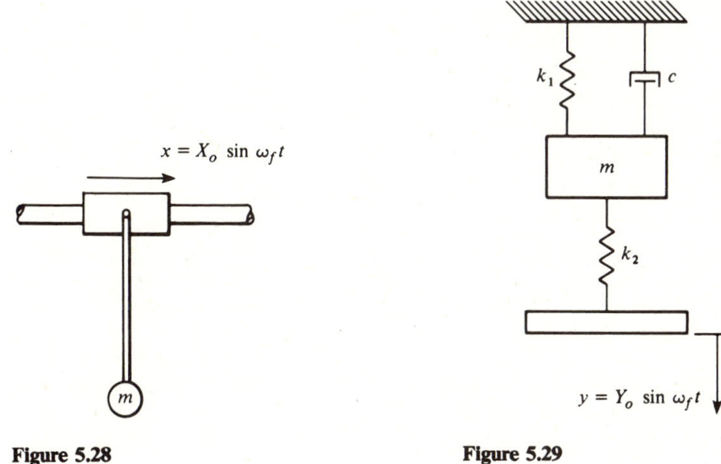

Figure 5.28 **Figure 5.29**

5.43 In the system of Prob. 5.42, $X_o = 2$ in, $L = 12$ in, $m = 0.025$ lb · s²/in, and the initial displacement and velocity and both zero. Determine the range of values of ω_f for which the assumption of small oscillations is valid.

5.44 For the system shown in Fig. 5.29, determine the differential equation of motion and the steady-state solution for the displacement of the mass.

5.45 The system shown in Fig. 5.30 models a vehicle traveling over a rough surface. For a constant vehicle speed v, determine the differential equation of motion of the mass and obtain the steady-state solution.

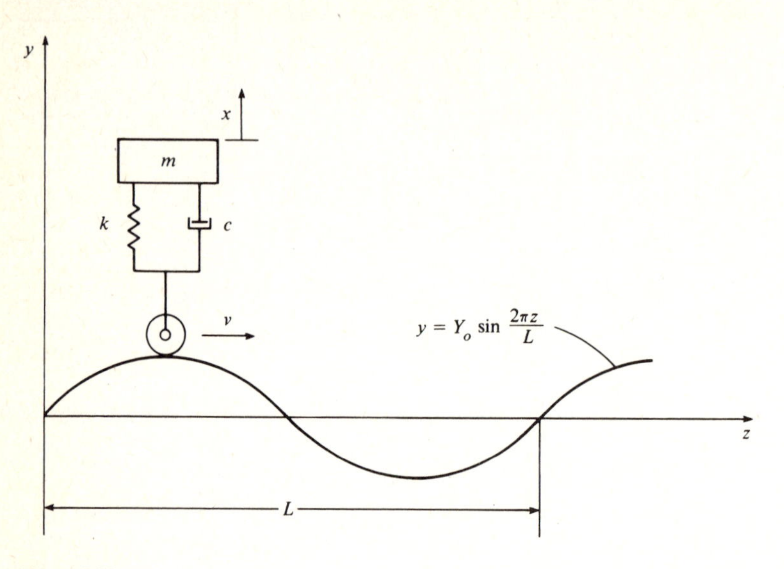

Figure 5.30

5.46 For the system of Fig. 5.30, $m = 12$ kg, $k = 5$ N/mm, $c = 0.2$ N · s/mm, $Y_o = 150$ mm, and $L = 1500$ mm. Determine (*a*) the vehicle speed at resonance, (*b*) the maximum steady-state displacement of the mass and the corresponding vehicle speed, and (*c*) the maximum dynamic force transmitted to the mass at twice the resonant speed.

5.47 A small damped spring-mass system is attached to a moving surface as shown in Fig. 5.31 and is to be used as a displacement-measuring instrument. If damping is such that $\zeta = 0.707$, determine the range of values of the frequency ratio for which the steady-state amplitude of $x(t)$ is at least 95 percent accurate as a measure of Y_o.

5.48 Determine the differential equation of motion for small oscillations of bar AB in the system shown in Fig. 5.32.

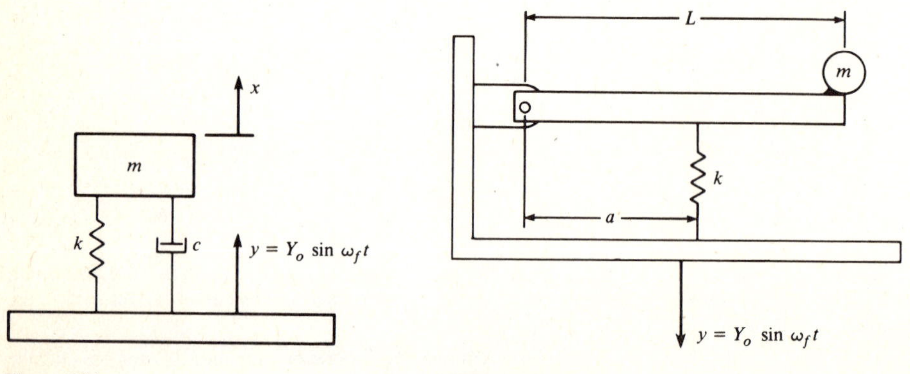

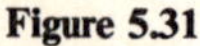
Figure 5.31

Figure 5.32

CHAPTER
SIX
ANALYSIS OF VIBRATIONS IN ROTATING EQUIPMENT

The previous chapters deal with the study of vibrations in simple mechanical systems from a purely mathematical point of view. These simple mechanical systems are no more than mathematical models which reflect in a general way the vibratory response of real mechanical equipment. The purpose of this chapter is to illustrate how knowledge gained from the study of such mathematical models may be coupled with data obtained from actual measurements of vibrations to yield extremely useful and practical information about the condition of mechanical equipment. The discussion will center on rotating mechanical equipment, although the extension of the ideas to reciprocating motion is reasonable straightforward.

6.1 INSTRUMENTATION FOR VIBRATION ANALYSIS

In the majority of cases, the engineer or technologist involved in the design and/or maintenance of rotating mechanical equipment will view vibration as an undesirable but unavoidable phenomenon. The usual effect of excess vibration is premature wear of components such as bearings, gears, and couplings. Rather than attack the impossible problem of eliminating vibration, we propose to show how the proper interpretation of quantitative vibration data may be used to monitor the condition of operating mechanical equipment and lead to reductions in downtime and maintenance costs.

The instrumentation available for the measurement of vibration runs the gamut from the very basic to the very complex and sophisticated, while the expense varies proportionately. The discussion here will center on an instrument which is usually referred to as a *general-purpose vibration analyzer*. Such an instrument is ideally suited to field measurements of vibration, as most available units are compact, portable, and will operate on self-contained rechargeable batteries as well as standard line voltage. The three major components of a portable vibration analyzer are a transducer or vibration pickup, an electronic unit which processes the signal from the transducer, and a stroboscopic light. The purpose and operation of each component will be discussed in turn.

The transducer or pickup is a device which is placed in contact with the machine for which the vibration is to be measured. The vibration pickup is designed so that it produces an electric current or voltage that is directly proportional to the displacement, velocity, or acceleration of the vibration being measured. One commonly used design for the vibration pickup is similar to that shown in Fig. 6.1. A damped spring-mass system is fixed to the base of the instrument but is free to move inside the housing. The mass is wound with a wire coil, and a permanent magnet is fixed to the housing. If the base is placed in contact with a vibrating machine, the mass is set into vibratory motion (this is a case of base excitation), and relative motion occurs between the mass and the housing. As the motion takes place, magnetic flux lines are cut and a voltage is induced, in the coil, which is proportional to the velocity of the mass relative to the base. This type of pickup is velocity-sensitive and is referred to as a *velometer* or simply a *velocity pickup*. The accuracy and sensitivity of such an instrument depend primarily upon the ratio of the frequency of the vibrations being measured to the natural frequency of the instrument, and to a lesser extent on the damping factor. Commercially available velocity pickups have a frequency range on the order of 2 to 12,000 Hz with an accuracy of ± 0.5 percent. Many other types of transducers are available, such as piezoelectric-type accelerometers, but all are characterized by an output current or voltage which directly reflects a vibrational quantity.

The main component of the vibration analyzer is an electronic unit which receives the electric output of the vibration transducer and converts it to useful information. The unit is designed for use with a certain type, or types, of pickup and uses the known characteristics of the pickup to modify the electric signal in

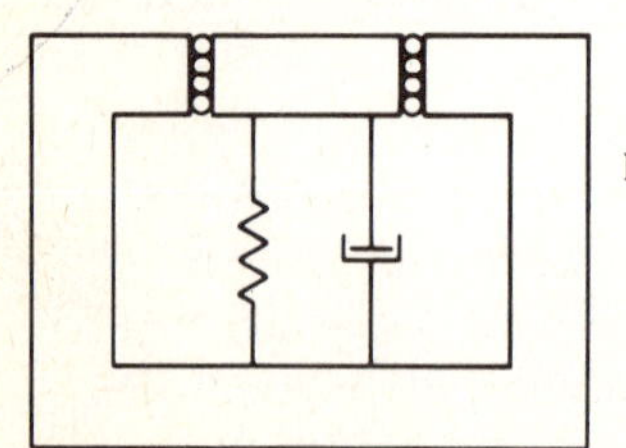

Figure 6.1

an appropriate manner and display it digitally or on a calibrated meter as displacement, velocity, or acceleration. Displacement is usually read in mils (1 mil = 1/1000 in), velocity in inches per second, and acceleration in g's. If an accelerometer is used for measurement, the acceleration is integrated to obtain velocity, and the velocity may in turn be integrated to obtain displacement. The integration feature of the analyzer increases its flexibility as a measuring device.

The vibration measured in rotating machinery is usually composed of several simultaneously occurring vibrations having different frequencies. Thus, the output signal of the transducer will also be composed of the several frequencies. Probably the most powerful feature of the vibration analyzer is that it contains filtering circuitry which allows it to process and display only the portion of the signal, hence the vibration, that occurs at a specific frequency. The frequency filter is manually adjusted and usually operates over the entire frequency range of the transducer. Some analyzers are equipped with automatic scanning filters which scan the signal for a significant component and automatically lock on the frequency of that component. The frequency-filtering capability may be used to great advantage in vibration analysis and is discussed in detail later.

The stroboscopic light is used in conjunction with the filtering circuitry of the analyzer to visually pinpoint mechanical components which are the source of a particular vibration. The stroboscopic light is fired by the analyzer at any vibration frequency and is therefore capable of "freezing" a component, rotating at that frequency, which may be the source of the vibration. The use of the stroboscopic light will be detailed in the discussion of balancing procedures.

6.2 CONDITION MONITORING AND FREQUENCY ANALYSIS

An analysis of the magnitudes of the vibrations occurring at various frequencies in a machine can provide a great deal of information, not only about the mechanical condition of the machine in general but about the condition of specific components of the machine as well. From the discussion of vibration due to unbalance in Sec. 5.9, the circular frequency of such a vibration corresponds to the rotational speed of the unbalanced component. If the overall vibration in a machine exhibits a major component having a frequency the same as the speed of a rotating component, the indication is that the component should be balanced to reduce the vibration level. This is but one example in which the frequency of a vibration can be used to pinpoint the source of that vibration. Essentially, any rotating component will have a specific, identifiable frequency associated with its operational characteristics, and this frequency may be used to identify the vibrations due to that component. Table 6.1 gives some common mechanical components along with the frequency and general vibration characteristics of each. Figure 6.2 is a vibration-severity nomograph which can be used to determine acceptable vibration levels at various operating speeds.

Table 6.1 Vibration Sources

Probable cause	Frequency relative to machine speed	Strobe "picture"	Amplitude	Notes
Unbalance	r/min × 1	One, steady	Radial, steady	Most common cause of vibration
Bent shaft	r/min × 1 or r/min × 2	One, two, or three	Axial, high	Strobe picture depends on machine—usually unsteady
Antifriction bearings	r/min × 20 to r/min × 50	Unstable	Radial, low	Use velocity mode
Sleeve bearings	r/min × 1	One, steady	Shaft = bearing	Compare shaft to bearing readings
Misalignment	r/min × 2	One, two or three	Axial, high	Axial amplitude 0.7 or more of vertical and horizontal
Faulty belts	r/min × 1 to r/min × 5 (belt r/min × 2)	See "Notes" column	Radial, unsteady	Freeze belts with strobe and observe
Oil whip	Less than machine r/min	Unstable	Radial, unsteady, sometimes severe	Frequency is near one-half the r/min
Gears	High (related to number of teeth)	—	Radial; low	Use velocity mode
Looseness	r/min × 2	Two	Proportional to looseness	Frequently coupled with misalignment
Foundation failure	Unsteady	Unstable	Erratic	Shows up when balancing
Resonance	Specific critical speeds	One	High	Increased levels at critical speeds
Beat frequency	Periodically varying	—	Pulsating	Caused by close r/min machines

Source: Courtesy Vitec Incorporated, Cleveland, Ohio.

Vibration data are often used in programs of planned maintenance for mechanical equipment. The technique is generally known as *condition monitoring*, and it uses vibration as an indicator of machinery condition. Applicable to old or new equipment, such a program begins with obtaining a complete record of the vibration data on a given machine while it is operating satisfactorily under standard conditions. This information, called *baseline data*, establishes a basis of comparison for future vibration measurements. The baseline data for a given machine are composed of tabular records or graphs of vibration displacement,

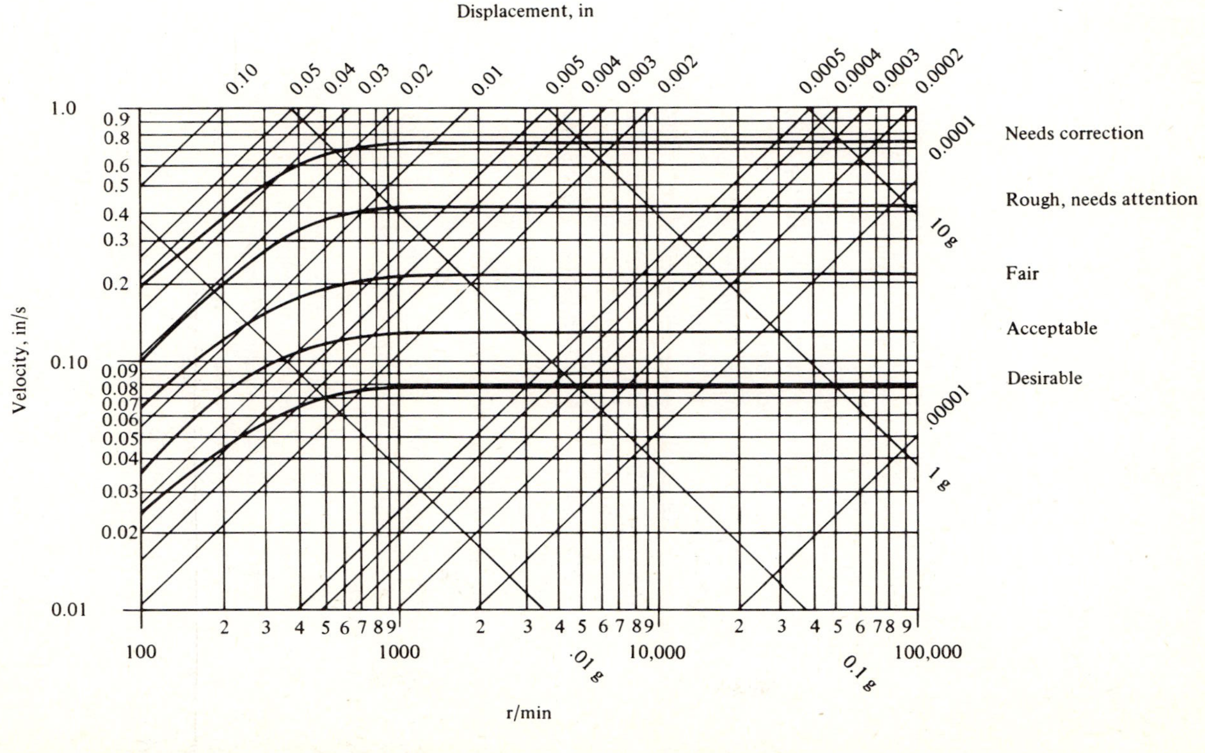

Figure 6.2 Vibration-severity nomograph. *(Courtesy Vitec Incorporated, Cleveland, Ohio.)*

velocity, and acceleration as functions of frequency at specified points on the machine. Usually the vibration measurements are made at each of the machine's bearings and are obtained in horizontal, vertical, and axial directions. Several manufacturers of vibration instrumentation provide equipment capable of producing hard-copy baseline data on XY recorders and automatic frequency-scanning circuits. An example of such baseline data is shown in Fig. 6.3.

Once the baseline data are obtained, periodic vibration measurements are made on the machine at the specified points. The mechanical condition of the machine is thus monitored, and deterioration of any of its components will be indicated by increasing vibration levels. The measurements obtained in the periodic checks need not be as extensive as those made to establish the baseline data. In fact, overall vibration measurements without specific frequency filtering will often be sufficient for monitoring purposes. However, when the periodic measurements indicate a marked increase in vibration levels, a complete check of vibration level versus frequency should be made in a manner identical to that used to obtain the baseline data. A comparison will then reveal the frequency of the vibration which is the source of the increase, and this in turn can be used to identify the deteriorating mechanical component.

A maintenance program based on condition monitoring can reduce overall maintenance costs by extending the maintenance interval for most machinery. This cost reduction is the result of maintenance based on equipment condition as opposed to operating hours or calendar months. In addition, condition monitoring can reduce equipment downtime and costly production outages by giving advance information on the specific repairs and parts required.

6.3 SINGLE-PLANE BALANCING

Earlier we developed a mathematical description of the vibrations due to rotating unbalance by assuming the existence of an eccentric mass m at a known location relative to the center of rotation. Eliminating such vibration requires removal of the eccentric mass or adding an equal mass in a position such as to cancel the effect of the unbalance. This seemingly simple solution is complicated by the fact that in real mechanical equipment the amount and location of the eccentric mass are not visually obtainable. In real equipment, unbalance is due to many factors such as nonuniform material density, machining errors, variation in the sizes of bolts and rivets, and so forth. The unbalance can be best determined from measurements of the vibration resulting from the unbalance. In the following, we shall develop a procedure for single-plane balancing utilizing a general-purpose vibration analyzer.

In balancing terminology, *single plane* refers to any rotating component having its smallest dimension in the direction parallel to the axis of rotation,

Figure 6.3

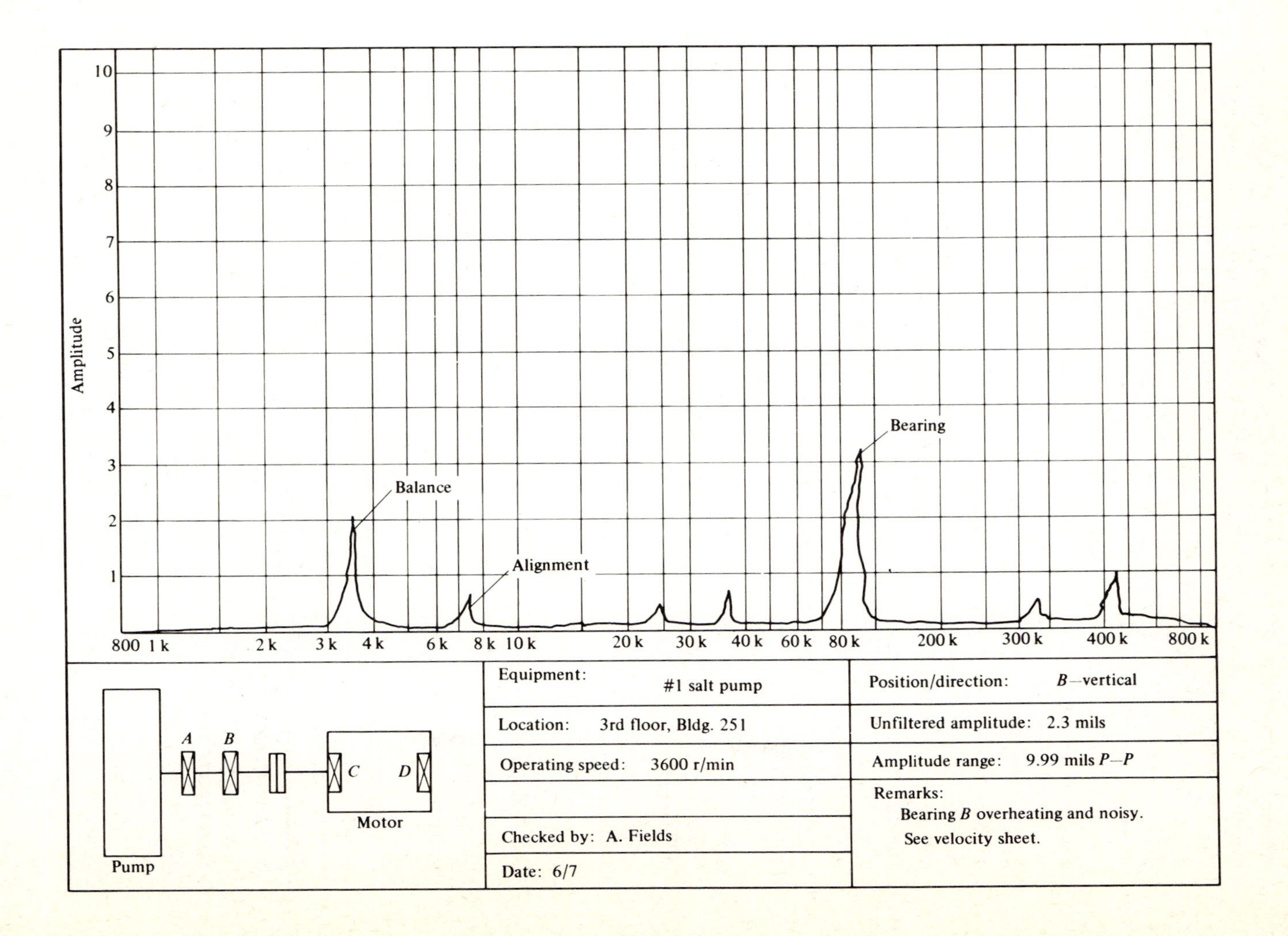

Amplitude
10
9
8
7
6
5
4
3
2
1
800 1 k
2 k
3 k
4 k
6 k
8 k 10 k
20 k
30 k
40 k
60 k
80 k
200 k
300 k
400 k
800 k
Balance
Alignment
Bearing
Pump
A
B
C
D
Motor
Equipment: #1 salt pump
Location: 3rd floor, Bldg. 251
Operating speed: 3600 r/min
Checked by: A. Fields
Date: 6/7
Position/direction: B—vertical
Unfiltered amplitude: 2.3 mils
Amplitude range: 9.99 mils P—P
Remarks:
Bearing B overheating and noisy.
See velocity sheet.

with this dimension small in comparison to the diameter of the component. Actually, any rotating component which is mounted on an overhung shaft may be assumed to fit the definition. The single-plane balancing procedure will be developed in relation to Fig. 6.4, which schematically depicts a centrifugal blower. The impeller is attached to a rotating shaft which has bearings at A and B and is driven by an electric motor through a flexible coupling at C. The shaft and impeller rotate with an angular velocity of ω_f rad/s or $N = (60/2\pi)\omega_f$ r/min. It is assumed that an arbitrary radial line has been scribed on the impeller, and this line will be referred to as the *phase mark*.

To determine if the impeller is unbalanced, the transducer or vibration pickup is held in contact with the housing of the bearing at A, and the amplitude (i.e., maximum displacement) of the vibration having a frequency corresponding to ω_f is measured. To measure this amplitude, the frequency filter of the vibration analyzer must be adjusted to $f_f = \omega_f/2\pi$ Hz or N r/min. The amplitude so obtained is then referred to the nomograph of Fig. 6.2 to determine whether the vibration level is acceptable or if balancing of the impeller is required. We shall assume that balancing is necessary but, before proceeding, one additional and very necessary observation will be made. If the stroboscopic light is connected to the analyzer while the amplitude is measured, the light will flash once per revolution of the impeller owing to the setting of the frequency filter, and the flash will occur at the instant the vibration pickup senses maximum displacement. Owing to phase lag in the response of the structure

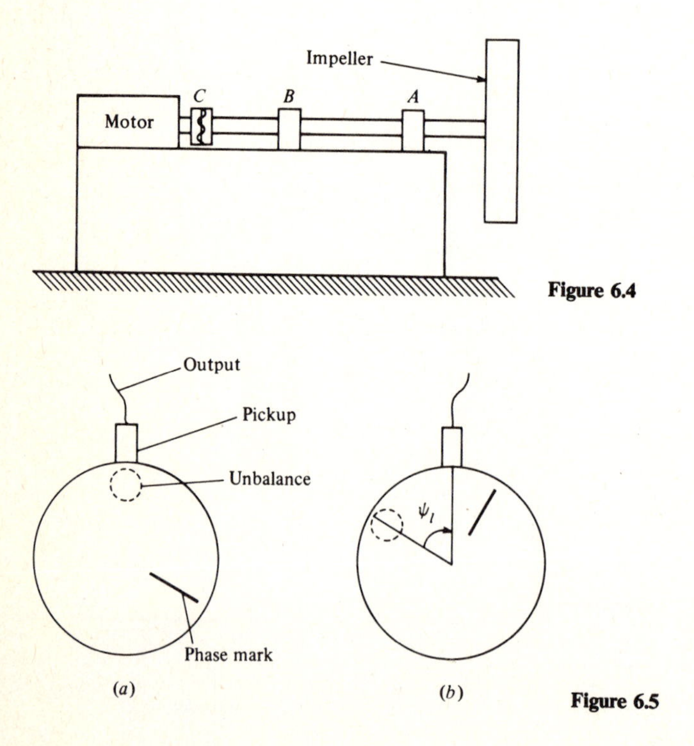

Figure 6.4

Figure 6.5

(Sec. 5.6) and additional phase lag in the response of the pickup, the strobe flash occurs a short time *after* the unbalance mass has passed the position of the pickup. This phase-lag relationship is shown in Fig. 6.5, which shows a front view of the impeller and the orientation of the vibration pickup. In Fig. 6.5*a* the unbalance mass is in the vertical position, and maximum force is exerted in the direction of the pickup. The impeller will rotate through an angle ψ_l corresponding to the time lag between maximum force and maximum displacement as sensed by the pickup. The strobe will then flash when the impeller is in the position shown in Fig. 6.5*b*. The "frozen" position of the phase mark may be observed and is seen to be related to the position of the unbalance.

The single-plane balancing procedure may now be set down. First the machine is run in its original condition, and the amplitude is measured as previously discussed and recorded. In addition, the phase mark is observed using the stroboscopic light, and its position relative to any convenient radial reference line is recorded as ϕ_U as in Fig. 6.6*a*. The amplitude is that due to the original unbalance and is denoted by A_U. We now have an amplitude and phase reference corresponding to the unbalance but still no information as to its amount or location. To quantify the unbalance, we stop the machine, attach a known trial weight T to the impeller, and record its position relative to the phase mark as shown in Fig. 6.6*b*. Note that we are increasing the unbalance of the impeller by a known amount at a known position. The effect of this change in the condition of the impeller will be used to deduce the original unbalance. After the trial weight is added, the machine is run again, and the vibration amplitude and the position of the phase mark are measured and recorded. The amplitude and phase measured in the second run correspond to the combined effects of the original unbalance and the unbalance due to the trial weight; they are denoted by A_{U+T} and ϕ_{U+T}, respectively. The strobe "picture" for this second run is shown in Fig. 6.6*c*.

The two sets of measurements must now be physically interpreted to determine the amount and location of the original unbalance. This is accomplished with the aid of the following observation. If the trial weight is placed in a position such that the net unbalance is shifted in a clockwise direction, the stroboscopically observed position of the phase mark will *shift in the counterclockwise direction by exactly the same amount*, and vice versa. This is illustrated

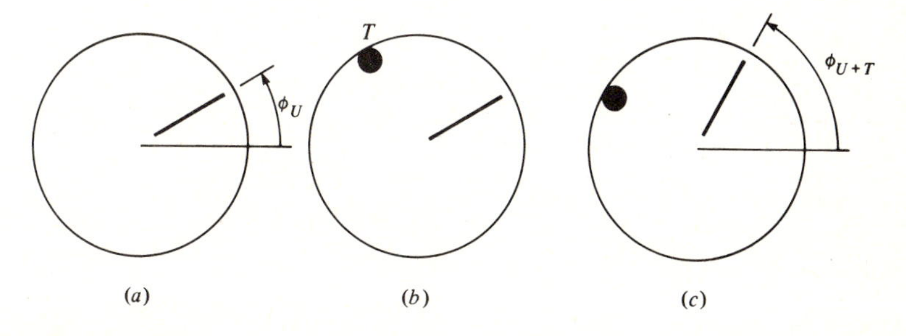

Figure 6.6

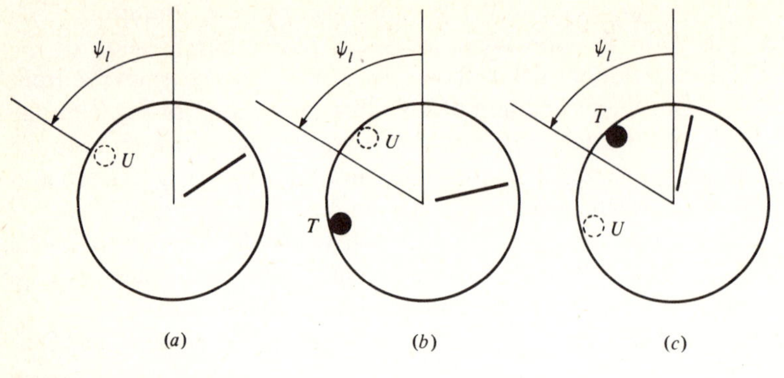

Figure 6.7

in Fig. 6.7, where the impeller is assumed to rotate counterclockwise. Figure 6.7*a* shows the lag angle ψ_l and the position of the phase mark for the original unbalance. In Fig. 6.7*b*, a trial weight has been added such that the net unbalance shifts in the counterclockwise direction as shown. Since ψ_l is constant, the strobe light flashes *sooner*, and the phase mark moves *clockwise*. Figure 6.7*c* shows the case in which the trial weight shifts the net unbalance clockwise. In this case, the light flashes *later*, and the phase mark shifts *counterclockwise*.

The quantitative results are obtained by representing the amplitudes A_U and A_{U+T} as vectors drawn from a common origin and having directions corresponding to the observed positions of the phase mark as shown in Fig. 6.8*a*. The third side of the vector triangle, shown completed in Fig. 6.8*b*, represents the amplitude which would result from the trial weight alone. This amplitude is obtained as

$$A_T = \left[A_U^2 + A_{U+T}^2 - 2A_U A_{U+T}\cos(\phi_{U+T} - \phi_U)\right]^{1/2} \tag{6.1}$$

by applying the law of cosines. Since the amplitude of the vibration due to

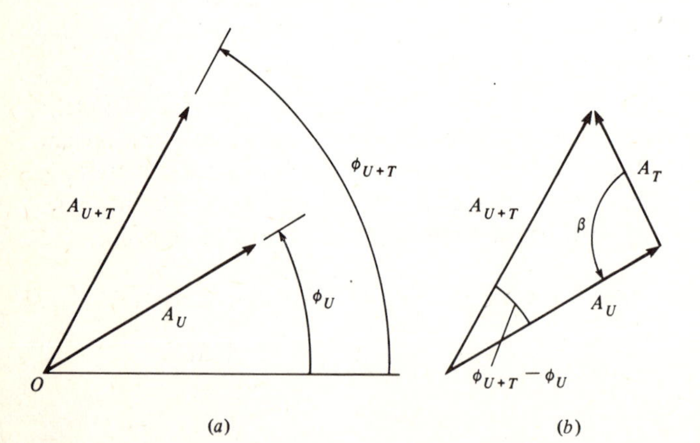

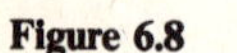
Figure 6.8

unbalance is proportional to the mass, and therefore the weight, of the unbalance, we can write

$$\frac{U}{T} = \frac{A_U}{A_T} \tag{6.2}$$

where U is the weight of the original unbalance. The original unbalance is then given by

$$U = \frac{A_U}{A_T} T \tag{6.3}$$

Now that the amount of unbalance is known, there remains the determination of its position. Again referring to Fig. 6.8*b*, note that if A_T is rotated through the counterclockwise angle β, then A_T and A_U are collinear. Then if A_T and A_U were equal in magnitude, the resultant amplitude A_{U+T} would be zero and the impeller would be balanced. The amplitudes are made equal by replacing the trial weight with a balancing weight B equal to the original unbalance as calculated from Eq. (6.3) and rotating the balance weight through an angle β from the position of the trial weight. The angle is calculated using the law of cosines and is given by

$$\beta = \cos^{-1} \frac{A_U^2 + A_T^2 - A_{U+T}^2}{2A_U A_T} \tag{6.4}$$

It is extremely important to note that since the vector triangle is drawn using angles corresponding to phase-mark positions, any angular change in the position of an unbalance will be in the *opposite* direction to that indicated by the triangle. Since Fig. 6.8*b* indicates that β is a counterclockwise rotation, the balance weight should actually be rotated *clockwise* through the angle β from the trial weight. In practice, application of the single-plane balancing procedure is quite simple, as will be illustrated by the following example.

Example 6.1 The impeller of Fig. 6.4 rotates clockwise at 3600 r/min. The following data were obtained using a portable vibration anlyzer (the phase angles were measured counterclockwise from the right-hand horizontal): Amplitude and phase due to initial unbalance are 5 mils at 40°; trial weight $T = 2$ oz added at a position 30° clockwise from the phase mark; amplitude and phase with the trial weight added are 9 mils at 140°. Determine the magnitude and location of the weight required to balance the impeller.

Solution The vector triangle is drawn as shown in Fig. 6.9, from which

$$A_T = (5^2 + 9^2 - 2 \times 5 \times 9 \cos 100)^{1/2} = 11.03 \text{ mils}$$

The original unbalance is

$$U = \frac{5}{11.03}(2) = 0.906 \text{ oz}$$

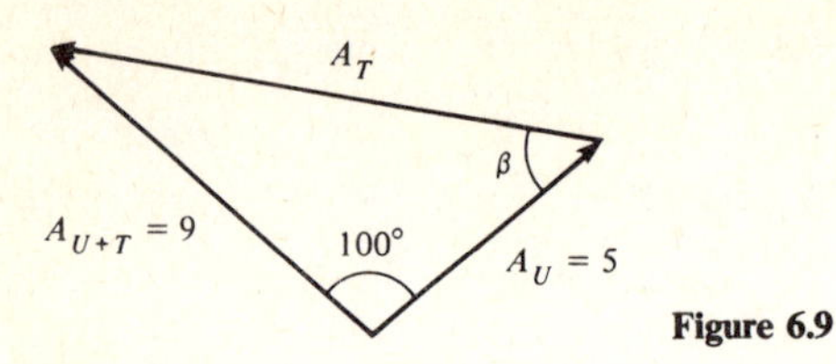

Figure 6.9

and the angle β is

$$\beta = \cos^{-1}\frac{5^2 + (11.03)^2 - 9^2}{2 \times 5 \times 11.03} = 53.5° \text{ counterclockwise}$$

The impeller will be balanced if a weight of 0.906 oz is added 53.5° *clockwise* from the position of the trial weight, or 83.5° clockwise from the phase mark.

6.4 PHASOR CALCULATIONS FOR SINGLE-PLANE BALANCING

As a preface to the more general two-plane balancing procedure to be discussed in Sec. 6.5, an alternative procedure for the single-plane balancing calculations will be developed. Let the unknown original unbalance be represented by the vector

$$\mathbf{U} = U\angle\theta_U \tag{6.5}$$

where U is the magnitude of the unbalance, and θ_U is the angular position of the unbalance relative to the phase mark. As before, the rotor is operated in its original condition, and the displacement and phase of the resulting vibration are measured and recorded. As a vector, the vibration is written as

$$\mathbf{A}_U = A_U\angle - \phi_U \tag{6.6}$$

where A_U is the displacement amplitude, and ϕ_U is the observed phase angle. Note that the direction of the vibration vector is taken as the *negative* of the observed phase angle. This is to ensure that the phase-reversal phenomenon associated with the use of the strobe light is properly accounted for in the calculations. The vibration represented by Eq. (6.6) is due to the unknown unbalance $\mathbf{U}$, and the relation between the two is

$$\mathbf{A}_U = \alpha\mathbf{U} \tag{6.7}$$

where α is a *vector operator* which converts the unbalance vector into the corresponding vibration vector.

Before proceeding, we must define precisely what is meant by the term "vector operator." As used here, a vector operator converts a given vector into a second vector by multiplying the magnitude of the given vector by a constant and shifting its direction through a constant angle. Thus, a vector operator is composed of a magnitude multiplication factor and a shift angle. For example, given the vector

$$\mathbf{A} = 10\angle 20° \tag{6.8}$$

and a vector operator

$$\alpha = 0.4\angle 47° \tag{6.9}$$

the new vector is

$$\mathbf{B} = \alpha\mathbf{A} = 0.4 \times 10\angle(20° + 47°) = 4\angle 67° \tag{6.10}$$

The reader may recognize this simply as phasor algebra as used in electric circuit theory.

If a trial weight is added, a new set of vibration measurements may be obtained and written as

$$\mathbf{A}_{U+T} = A_{U+T}\angle -\phi_{U+T} \tag{6.11}$$

where A_{U+T} and ϕ_{U+T} are as previously defined. The second vibration is due to the combined effects of the original unbalance and the trial weight, and we have

$$\mathbf{A}_{U+T} = \alpha(\mathbf{U} + \mathbf{T}) \tag{6.12}$$

where $\mathbf{T} = T\angle\theta_T$ is the trial-weight vector. Since Eqs. (6.7) and (6.12) comprise a set of two equations in two unknowns α and $\mathbf{U}$, simultaneous solution will yield the magnitude and location of the original unbalance. The results of this operation are

$$\alpha = \frac{\mathbf{A}_{U+T} - \mathbf{A}_U}{\mathbf{T}} \tag{6.13}$$

and

$$\mathbf{U} = \frac{\mathbf{A}_U}{\alpha} \tag{6.14}$$

where it must be understood that the indicated divisions are *phasor* divisions. Once the original unbalance is determined, the required balancing weight is

$$\mathbf{B} = -\mathbf{U} = U\angle(\theta_U + 180°) \tag{6.15}$$

Example 6.2 Solve Example 6.1 using phasor calculations.

SOLUTION Adopting the convention that counterclockwise angles are positive, we have

$$\mathbf{A}_U = 5\angle -40° \qquad \mathbf{T} = 2\angle -30° \qquad \mathbf{A}_{U+T} = 9\angle -140° \tag{6.16}$$

Substituting into Eq. (6.13) gives

$$\alpha = \frac{9\angle -140° - 5\angle -40°}{2\angle -30°}$$

The vector subtraction in the numerator is conveniently handled by resolution into components:

$$\alpha = \frac{-6.89_x - 5.79_y - (3.83_x - 3.21_y)}{2\angle -30°}$$

$$\alpha = \frac{-10.72_x - 2.58_y}{2\angle -30°} = \frac{11.03\angle 193.5°}{2\angle -30°}$$

$$\alpha = 5.515\angle 223.5°$$

The original unbalance is obtained from Eq. (6.14) as

$$\mathbf{U} = \frac{5\angle -40^\circ}{5.515\angle 223.5^\circ} = 0.906\angle -263.5^\circ$$

and the required balancing weight from Eq. (6.15) is

$$\mathbf{B} = 0.906\angle -83.5^\circ$$

which is exactly the same as the previous result. The reader should note that the sign of the phase angle of each of the measured vibrations was changed while the sign of the angle for the trial weight was *not* changed. This is because the angle for the trial weight corresponds to a physical location and is unaffected by phase reversal.

6.5 TWO-PLANE BALANCING

The single-plane balancing procedure will give acceptable results only if the rotating component is thin in comparison to its diameter. In the case of long rotors which do not satisfy this criterion, simultaneous balancing in two planes is required to obtain satisfactory results. The procedure will be referred to as *two-plane balancing* and generally must be applied to a rotating component which is straddle-mounted between bearings.

To illustrate the procedure, consider a long rotor mounted between bearings as shown in Fig. 6.10. Unbalance of the rotor would be indicated by a relatively large vibration displacement at a frequency corresponding to the rotational speed of the rotor. The effective position of the unbalance may be anywhere along the length of the rotor, but it can be shown that balance may be achieved by adding balancing weights in two planes. For accessibility, the two planes chosen for balancing are usually the end planes of the rotor. As an illustration, we shall assume the rotor to have an unbalance of mass m located on the outer surface and at a distance equal to one-fourth the length of the rotor from the left end, as shown in Fig. 6.11a. The amplitude of the force exerted by the

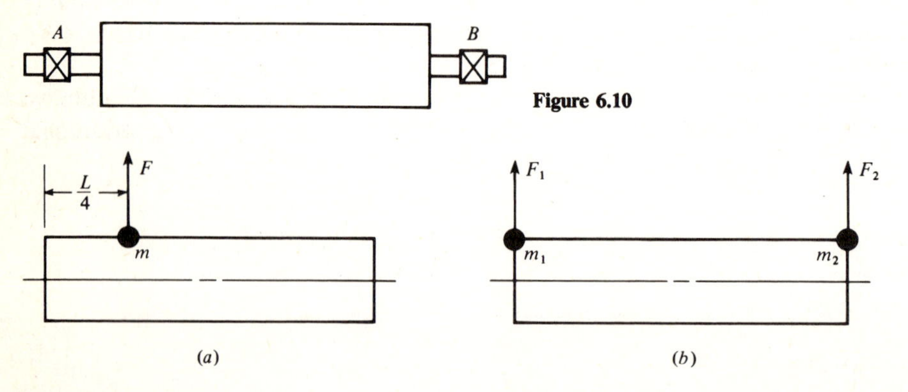

Figure 6.10

Figure 6.11

unbalance mass on the rotor is $F = mR\omega_f^2$, where ω_f is the angular velocity of the rotor in radians per second.

In Fig. 6.11*b*, the original unbalance mass m is replaced by two masses m_1 and m_2 located at the ends of the rotor. The corresponding force amplitudes are $F_1 = m_1 R\omega_f^2$ and $F_2 = m_2 R\omega_f^2$. The two force systems will be equivalent if the resultant force as well as the resultant moment about any point are the same for each system. For force equivalence we have

$$m_1 R\omega_f^2 + m_2 R\omega_f^2 = mR\omega_f^2 \tag{6.16}$$

or

$$m_1 + m_2 = m \tag{6.17}$$

while for moment equivalence about the left end, say, we have

$$Lm_2 R\omega_f^2 = \frac{L}{4} mR\omega_f^2 \tag{6.18}$$

Equations (6.17) and (6.18) give $m_1 = 3/4m$ and $m_2 = 1/4m$. This simple example shows that any unbalance may be considered as two unbalance masses which are located on the end planes of the rotor. Consequently, the rotor may be balanced by adding the appropriate balancing weights to the end planes.

A general condition of unbalance in a long rotor is depicted in Fig. 6.12, where U_L and U_R denote the weights of the original unbalance in the left plane and right plane, respectively. If the rotor is operated in its original unbalance condition and the displacement amplitude and phase of the vibration are measured at bearings A and B, we obtain the original unbalance vibration vectors $\mathbf{V}_A$ and $\mathbf{V}_B$. The magnitude of the vibration vector is its displacement amplitude, and the direction of the vector is taken as the *negative* of the phase angle relative to any convenient reference plane when viewed with the stroboscopic light. Symbolically, the vibration vectors may be written as

$$\mathbf{V}_A = \alpha_{AL}\mathbf{U}_L + \alpha_{AR}\mathbf{U}_R \tag{6.19}$$

and

$$\mathbf{V}_B = \alpha_{BL}\mathbf{U}_L + \alpha_{BR}\mathbf{U}_R \tag{6.20}$$

where α_{AL} is the *vector operator* which reflects the effect of the *left* plane unbalance $\mathbf{U}_L$ on the vibration at bearing A, and α_{AR}, α_{BL}, and α_{BR} are similarly defined. Note that $\mathbf{U}_R$, $\mathbf{U}_L$, and the vector operators are unknown but that $\mathbf{V}_A$ and $\mathbf{V}_B$ are known from actual measurement.

In a manner similar to the single-plane balancing procedure, quantitative results are obtained by adding known trial weights and taking additional

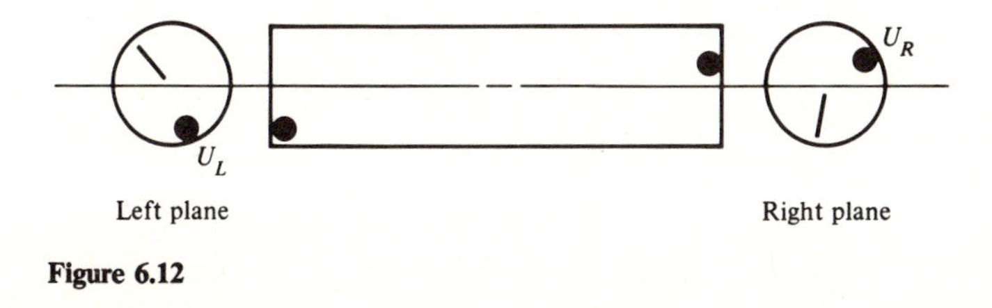

Figure 6.12

measurements. We first add a trial weight $\mathbf{T}_L$ to the left plane at a known position in that plane. We operate the rotor again, and measure the displacement and phase of the vibrations at bearings A and B. This second set of vibrations is represented mathematically by

$$\mathbf{V}'_A = \alpha_{AL}(\mathbf{U}_L + \mathbf{T}_L) + \alpha_{AR}\mathbf{U}_R \tag{6.21}$$

and

$$\mathbf{V}'_B = \alpha_{BL}(\mathbf{U}_L + \mathbf{T}_L) + \alpha_{BR}\mathbf{U}_R \tag{6.22}$$

Subtracting Eq. (6.19) from Eq. (6.21) results in

$$\mathbf{V}'_A - \mathbf{V}_A = \alpha_{AL}\mathbf{T}_L \tag{6.23}$$

which may be solved for α_{AL}. Similarly, subtracting Eq. (6.20) from Eq. (6.22) gives

$$\mathbf{V}'_B - \mathbf{V}_B = \alpha_{BL}\mathbf{T}_L \tag{6.24}$$

which may be solved for α_{BL}.

Next, $\mathbf{T}_L$ is removed, and a trial weight $\mathbf{T}_R$ is added at a known position in the right plane. The rotor is run again, and the vibrations are measured at bearings A and B. This third set of vibrations may be written as

$$\mathbf{V}''_A = \alpha_{AL}\mathbf{U}_L + \alpha_{AR}(\mathbf{U}_R + \mathbf{T}_R) \tag{6.25}$$

and

$$\mathbf{V}''_B = \alpha_{BL}\mathbf{U}_L + \alpha_{BR}(\mathbf{U}_R + \mathbf{T}_R) \tag{6.26}$$

Subtraction of Eq. (6.19) from Eq. (6.25) gives

$$\mathbf{V}''_A - \mathbf{V}_A = \alpha_{AR}\mathbf{T}_R \tag{6.27}$$

while subtracting Eq. (6.20) from Eq. (6.26) yields

$$\mathbf{V}''_B - \mathbf{V}_B = \alpha_{BR}\mathbf{T}_R \tag{6.28}$$

Equations (6.27) and (6.28) may be solved for α_{AR} and α_{BR}, respectively.

With each of the operators known, Eqs. (6.19) and (6.20) may be solved for the unbalance vectors $\mathbf{U}_L$ and $\mathbf{U}_R$. The rotor can then be balanced by adding equal and opposite balancing weights in each plane, Mathematically, the left-plane balance weight is

$$\mathbf{B}_L = -\mathbf{U}_L \tag{6.29}$$

and the right-plane balance weight is

$$\mathbf{B}_R = -\mathbf{U}_R \tag{6.30}$$

Simultaneous solution of Eqs. (6.19) and (6.20) results in

$$\mathbf{U}_L = \frac{\alpha_{BL}\mathbf{V}_A - \alpha_{AR}\mathbf{V}_B}{\alpha_{BR}\alpha_{AL} - \alpha_{AR}\alpha_{BL}} \tag{6.31}$$

and

$$\mathbf{U}_R = \frac{\alpha_{BR}\mathbf{V}_A - \alpha_{AL}\mathbf{V}_B}{\alpha_{BL}\alpha_{AR} - \alpha_{AL}\alpha_{BR}} \tag{6.32}$$

Conceptually the two-plane balancing procedure is straightforward; in practice, however, care must be exercised in performing the calculations, particularly with regard to the angles associated with the various vectors. The following example illustrates the required calculations.

Example 6.3 The data shown below were obtained for the original unbalance, left-plane trial-weight, and right-plane trial-weight measurements in the two-plane balancing procedure. The displacement amplitudes are in mils, and the phase angles were measured relative to the right-hand horizontal. Determine the size and location of the required balance weights.

	Amplitude		Observed phase	
Condition	*A*	*B*	*A*	*B*
Original unbalance	5	7	30°CW	40°CCW
T_L = 2 oz added 20°CW from left-plane phase mark	8	8	5°CW	35°CW
T_R = 1 oz added on the right-plane phase mark (0°)	4	6	5°CCW	15°CW

SOLUTION First it is necessary to write all quantities in vector form. For uniformity, all angles will be specified as counterclockwise. From the data, we have

$$\mathbf{V}_A = 5\angle 30^\circ \qquad \mathbf{V}_B = 7\angle 320^\circ$$

$$\mathbf{V}'_A = 8\angle 5^\circ \qquad \mathbf{V}'_B = 8\angle 35^\circ$$

$$\mathbf{V}''_A = 4\angle 355^\circ \qquad \mathbf{V}''_B = 6\angle 15^\circ$$

$$\mathbf{T}_L = 2\angle 340^\circ \qquad \mathbf{T}_R = 1\angle 0^\circ$$

Note that the angles associated with the trial-weight vectors are measured from the respective phase marks. Substituting into Eq. (6.23) gives

$$8\angle 5^\circ - 5\angle 30^\circ = \alpha_{AL}(2\angle 340^\circ)$$

The vector subtraction indicated on the left of this equation may be performed by drawing the vector triangle as in Fig. 6.13*a*, by resolution into components, or graphically. Solving the triangle using the law of cosines gives

$$8\angle 5^\circ - 5\angle 30^\circ = 4.06\angle 334^\circ = \alpha_{AL}(2\angle 340^\circ)$$

from which

$$\alpha_{AL} = 2.03\angle 354^\circ$$

To find α_{BL}, we substitute into Eq. (6.24) to obtain

$$8\angle 35^\circ - 7\angle 320^\circ = \alpha_{BL}(2\angle 340^\circ)$$

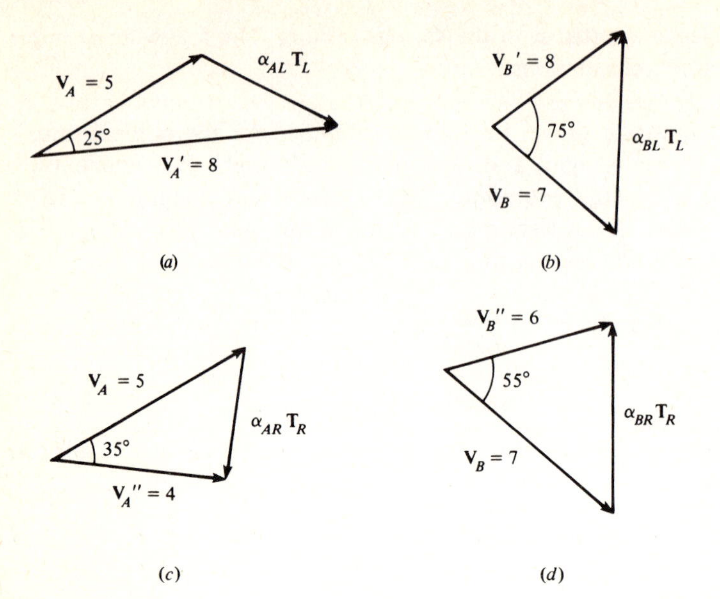

Figure 6.13

which corresponds to the vector triangle in Fig. 6.13*b*. This gives

$$9.17\angle 82.5° = \alpha_{BL}(2\angle 340°)$$

from which
$$\alpha_{BL} = 4.58\angle 102.5°$$

Similarly, Eq. (6.27) becomes

$$4\angle 355° - 5\angle 30° = \alpha_{AR}(1\angle 0°)$$

which is found, by solving the triangle of Fig. 6.13*c*, to be the same as

$$2.87\angle 263° = \alpha_{AR}(1\angle 0°)$$

which gives

$$\alpha_{AR} = 2.87\angle 263°$$

To find α_{BR}, Eq. (6.28) gives

$$6\angle 15° - 7\angle 320° = \alpha_{BR}(1\angle 0°)$$

Again solving the vector triangle (Fig. 6.13*d*), we obtain

$$6.07\angle 85.9° = \alpha_{BR}(1\angle 0°)$$

or
$$\alpha_{BR} = 6.07\angle 85.9°$$

The unbalance weight in the left plane is found by substituting into Eq. (6.31) to obtain

$$\mathbf{U}_L = \frac{6.07\angle 85.9° \times 5\angle 30° - 2.87\angle 263° \times 7\angle 320°}{6.07\angle 85.9° \times 2.03\angle 354° - 2.87\angle 263° \times 4.58\angle 102.5°}$$

or
$$\mathbf{U}_L = \frac{30.35\angle 115.9° - 20.10\angle 233°}{12.32\angle 79.9° - 13.16\angle 5.5°}$$

Performing the vector subtractions indicated in the numerator and denominator (Fig. 6.14*a* and *b*) gives

$$\mathbf{U}_L = \frac{41.04\angle 88°}{15.42\angle 135.2°} = 2.66\angle 312.8°$$

Similarly, Eq. (6.32) gives

$$\mathbf{U}_R = \frac{4.58\angle 102.5° \times 5\angle 30° - 2.03\angle 354° \times 7\angle 320°}{4.58\angle 102.5° \times 2.87\angle 263° - 2.3\angle 354° \times 6.07\angle 85.9°}$$

or
$$\mathbf{U}_R = \frac{22.93\angle 132.5° - 14.21\angle 314°}{13.16\angle 5.5° - 12.32\angle 79.9°}$$

The vector subtractions are as indicated by the triangles shown in Fig. 6.14*c* and *d*. The results are

$$\mathbf{U}_R = \frac{37.14\angle 133.5°}{15.42\angle 315.2°} = 2.41\angle 178.3°$$

Thus, the required balance weights are

$$\mathbf{B}_L = -\mathbf{U}_L = 2.66\angle 132.8° \qquad \text{and} \qquad \mathbf{B}_R = -\mathbf{U}_R = 2.41\angle 358.3°$$

Physically, the rotor will be balanced if a 2.67-oz weight is added to the left plane at an angular position of 187.2° counterclockwise from the left-plane phase mark, and a 2.41-oz weight is added to the right plane at an angular

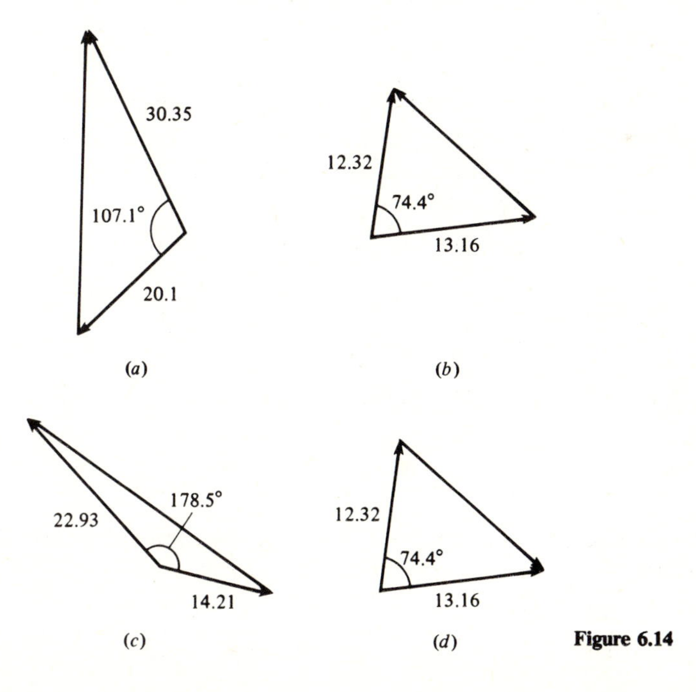

Figure 6.14

position of 2.2° counterclockwise from the right-plane phase mark. The balance weights, as calculated, must be added at the same *radial* position as the trial weights. If a balance weight must be located at a different radial position, the required weight must change in inverse proportion to the radial distance from the center of rotation.

Example 6.4 With the data of Example 6.3, determine α_{AL} using vector addition, by the method of rectangular components.

SOLUTION The defining equation for α_{AL} is

$$\mathbf{V}'_A - \mathbf{V}_A = \alpha_{AL}\mathbf{T}_L$$

which was solved for α_{AL} in the previous example using the vector triangle. This time we resolve $\mathbf{V}_A$ and $\mathbf{V}'_A$ into horizontal and vertical components with the positive directions taken to the right and upward, respectively:

$$\mathbf{V}_A = (5 \cos 30)_x + (5 \sin 30)_y = (4.33)_x + (2.5)_y$$

$$\mathbf{V}'_A = (8 \cos 5)_x + (8 \sin 5)_y = (7.97)_x + (0.7)_y$$

Then $$\mathbf{V}'_A - \mathbf{V}_A = (7.97 - 4.33)_x + (0.7 - 2.5)_y = (3.64)_x - (1.8)_y$$

which, when converted back to polar form, is

$$\mathbf{V}'_A - \mathbf{V}_A = 4.06\angle 333.7°$$

Now $$4.06\angle 333.7° = \alpha_{AL}(2\angle 340°)$$

which gives $$\alpha_{AL} = 2.03\angle 354°$$

as before.

The calculations required for two-plane balancing are tedious when carried out according to the method of Example 6.3. By this point, it may have occurred to the reader that the computations can be greatly simplified by using a calculator capable of performing vector (polar) calculations. Many calculators on the market have this capability, and any such calculator would be a worthwhile addition for anyone frequently involved with two-plane balancing. The calculations may also be handled in tabular form, and a graphical solution technique is available.[1]

PROBLEMS

6.1 Figure 6.15 is a schematic representation of a vibration transducer. Denoting $z = x - y$ as the displacement of the instrument mass relative to the machine displacement, determine (*a*) the differential equation of motion governing $z(t)$; (*b*) the steady-state solution for $z(t)$; and (*c*) the range of values of frequency ratio r for which the magnitude of $z(t)$ is approximately equal to Y_o. Assume $\zeta = 0.707$.

[1] See, for example, *Vector Calculations for Two-Plane Balancing*, Application Rep. 327, IRD Mechanalysis, Inc., Columbus, Ohio 43229.

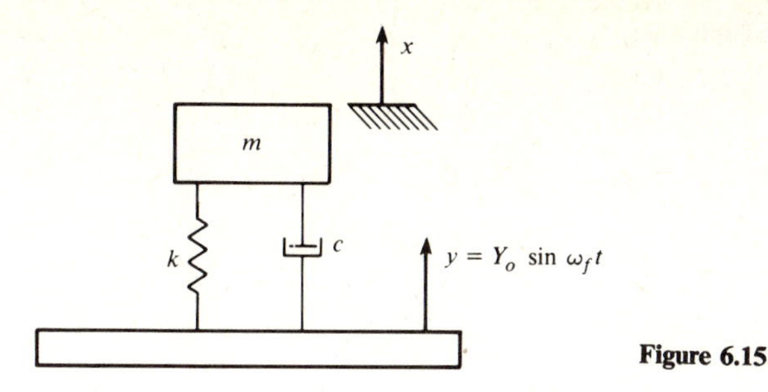

Figure 6.15

6.2 For Fig. 6.15, determine the range of r for which the velocity of the instrument mass approximates the machine velocity. Consider steady-state amplitudes only.

6.3 Repeat Prob. 6.2 for acceleration.

6.4 The following measurements were obtained for an unbalanced flywheel operating at 600 r/min: Amplitude and phase angle due to original unbalance are 8 mils and 35° clockwise from the phase mark; trial weight $T = 5$ oz added 60° clockwise from the phase mark and 3 in radially from the center of rotation; amplitude and phase angle with trial weight added are 7.2 mils and 140° clockwise. Determine the magnitude and angular position of the balancing weight if it is to be located 8 in radially from the center of rotation.

6.5 Determine the required magnitude and angular location of the balancing weight for a single-plane balancing procedure given the following data. Original unbalance: displacement = 8.3 mils, phase angle = 10° clockwise from arbitrary reference. After the addition of a trial weight of 1.5 oz at an angular position 35° counterclockwise from the phase mark: displacement = 12.2 mils, phase angle = 25° counterclockwise.

6.6 A thin rotating component produces a vibration having an amplitude of 14 mils. A strobe light fired at the predominant frequency of the vibration freezes an arbitrary phase mark at 25° clockwise from the vertical. A trial weight of 4 oz is attached at a position 50° clockwise from the phase mark. With the trial weight in place, the displacement is 17 mils with a phase angle of 20° counterclockwise from the vertical. Where should the balancing weight be located relative to the phase mark, and how much should it weigh? Assume that the balancing weight will be located at the same radius as the trial weight.

6.7 Using the data in Table 6.2, calculate the magnitude and angular position of the balancing weights for a two-plane balance. All phase angles are measured from an arbitrary reference, and all weights are added at the same radius.

Table 6.2

Condition	Amplitude, mils		Phase angle	
	A	*B*	*A*	*B*
Original unbalance	4	3	120°CW	220°CW
T_L = 1 oz added 10°CCW from left-plane phase mark	5	3.5	140°CW	150°CW
T_R = 1 oz added 0° from right-plane phase mark	4.5	4	110°CW	60°CCW

6.8 Repeat Prob. 6.7 for the data of Table 6.3.

Table 6.3

Condition	Amplitude, mils		Phase angle	
	A	B	A	B
Original unbalance	8.6	5	63°CW	154°CCW
T_L = 10 oz added 90°CCW from left-plane phase mark	5.9	4.5	123°CW	132°CCW
T_R = 12 oz added 180° from right-plane phase mark	6.2	10.4	36°CW	162°CW

6.9 Repeat Prob. 6.7 for the data of Table 6.4.

Table 6.4

Condition	Amplitude mils		Phase angle	
	A	B	A	B
Original unbalance	12.3	5.5	40°CCW	124°CW
T_L = 3 oz added 90°CCW from left-plane phase mark	8.5	9.4	10°CW	85°CW
T_R = 2 oz added 35°CCW from right-plane phase mark	7.0	5.8	50°CW	48°CCW

CHAPTER

SEVEN

RESPONSE TO NONHARMONIC FORCING FUNCTIONS

In Chaps. 4 and 5, the response of mechanical systems to harmonic forcing functions was discussed. As mentioned earlier, harmonically varying forces commonly occur because of the functional characteristics of various rotating mechanical components. Nonharmonic forces also arise in mechanical systems, and different methods of analysis are required. These more general forcing functions may be periodic or nonperiodic, and mathematical methods for studying each type will be developed in this chapter.

7.1 FOURIER-SERIES EXPANSION OF PERIODIC FUNCTIONS

A periodic function is by definition a function $f(t)$ for which

$$f(t) = f(t + T) \tag{7.1}$$

where T is the period of the function. Figure 7.1 depicts a nonharmonic periodic function. Any periodic function can be represented by a convergent infinite series of harmonic functions known as a Fourier series.[1] Such a series is written as

$$f(t) = \frac{1}{2}a_0 + \sum_{n=1}^{\infty}\left(a_n \cos\frac{2\pi nt}{T} + b_n \sin\frac{2\pi nt}{T}\right) \tag{7.2}$$

[1] Named for Jean Baptiste Joseph Fourier (1768–1830), a French mathematician and physicist who proved the correctness of the procedure.

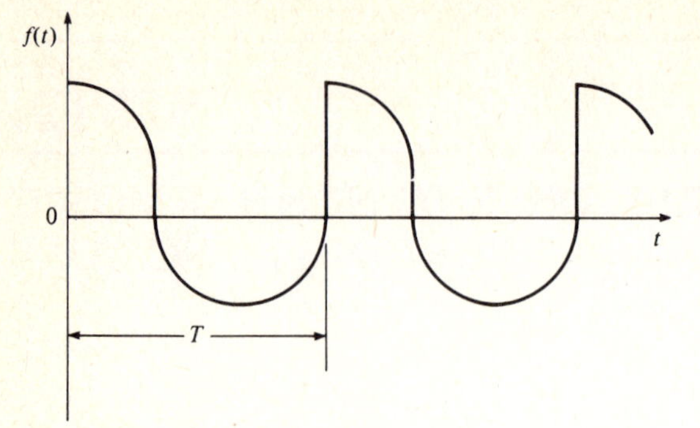

Figure 7.1

where n is the set of positive integers 1, 2, 3, . . . , and T is the period. The coefficients a_n and b_n are defined[2] by

$$a_n = \frac{2}{T}\int_{-T/2}^{T/2} f(t)\cos\frac{2\pi nt}{T}\,dt \qquad n = 0, 1, 2, \ldots \tag{7.3}$$

and

$$b_n = \frac{2}{T}\int_{-T/2}^{T/2} f(t)\sin\frac{2\pi nt}{T}\,dt \qquad n = 1, 2, 3, \ldots \tag{7.4}$$

Note that setting $n = 0$ in Eq. (7.3) gives

$$a_0 = \frac{2}{T}\int_{-T/2}^{T/2} f(t)\,dt \tag{7.5}$$

which shows that a_0 is the average value of $f(t)$ over one period. The significance is that a_0 represents the offset of the harmonic terms above or below the t axis. The coefficients of each of the harmonic terms indicate the degree to which those terms contribute to the representation of $f(t)$.

The Fourier-series expansion of a function is simplified if the function is an odd or even function. A function is said to be *even* if $f(t) = f(-t)$, and *odd* if $f(t) = -f(-t)$. In particular, we note that cosine is an even function, whereas sine is an odd function. Other examples of even and odd functions are shown in Fig. 7.2. If we integrate a periodic function over one period, we obtain

$$\int_{-T/2}^{T/2} f(t)\,dt = \begin{cases} 0 & \text{if } f(t) \text{ is odd} \\ 2\int_0^{T/2} f(t)\,dt & \text{if } f(t) \text{ is even} \end{cases} \tag{7.6}$$

which should be obvious in Fig. 7.2.

Another very important property of even and odd functions is that the product of two functions is even when both functions are even or both are odd, and the product is odd when one of the functions is even and one is odd. This

[2] See App. A.4 for the mathematical details.

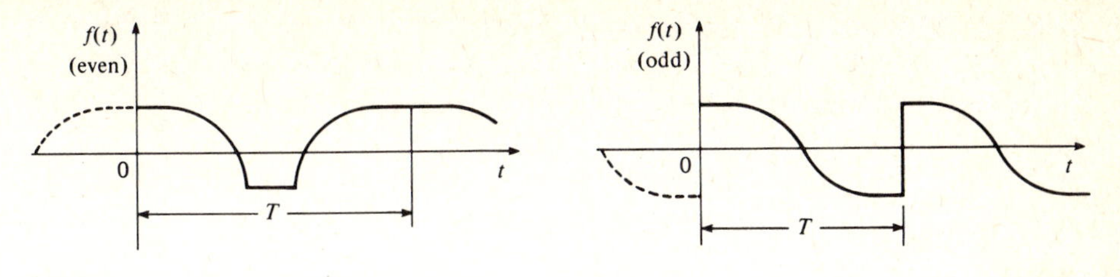

Figure 7.2

property is summarized as

$$(\text{even})(\text{even}) = (\text{odd})(\text{odd}) = \text{even}$$
$$(\text{even})(\text{odd}) = (\text{odd})(\text{even}) = \text{odd}$$

To realize the advantages of these properties, let us consider the Fourier-series expansion of an even function $f(t)$. The Fourier coefficients as defined by Eqs. (7.3) and (7.4) are

$$a_n = \frac{2}{T}\int_{-T/2}^{T/2} f(t)\cos\frac{2\pi nt}{T}\,dt = \frac{4}{T}\int_0^{T/2} f(t)\cos\frac{2\pi nt}{T}\,dt \qquad n = 0, 1, 2, \ldots \tag{7.7}$$

and
$$b_n = \frac{2}{T}\int_{-T/2}^{T/2} f(t)\sin\frac{2\pi nt}{T}\,dt = 0 \qquad n = 1, 2, 3, \ldots \tag{7.8}$$

These results are obtained by observing that since $f(t)$ is even, the integrand of Eq. (7.7) is also an even function whereas the integrand of Eq. (7.8) is an odd function. Thus the Fourier expansion of an even function can be written as

$$f(t) = \frac{1}{2}a_0 + \sum_{n=1}^{\infty} a_n \cos\frac{2\pi nt}{T} \tag{7.9}$$

which is known as a *Fourier cosine series.*

Similarly, we can show that an odd function $f(t)$ has the Fourier coefficients

$$a_n = 0 \qquad n = 0, 1, 2, \ldots \tag{7.10}$$

and
$$b_n = \frac{4}{T}\int_0^{T/2} f(t)\sin\frac{2\pi nt}{T}\,dt \qquad n = 1, 2, 3, \ldots \tag{7.11}$$

and the *Fourier sine series* expansion is

$$f(t) = \sum_{n=1}^{\infty} b_n \sin\frac{2\pi nt}{T} \tag{7.12}$$

7.2 RESPONSE TO NONHARMONIC PERIODIC FORCING FUNCTIONS

Consider the response of a damped spring-mass system subjected to an external forcing function $f(t)$ which is nonharmonic but is known to be periodic with period T. The differential equation of motion is

$$m\ddot{x} + c\dot{x} + kx = f(t) = \frac{1}{2}a_0 + \sum_{n=1}^{\infty} a_n \cos\frac{2\pi nt}{T} + \sum_{n=1}^{\infty} b_n \sin\frac{2\pi nt}{T} \tag{7.13}$$

where we utilize the Fourier-series expansion for $f(t)$. The homogeneous solution giving the free, transient response of the system has been discussed in detail in Chap. 5; since it will decay with time, it will not be considered here. To obtain the particular solution representing the forced steady-state response, we note that the differential equation is linear, and the principle of superposition applies. The particular solution is thus the summation of the individual particular solutions corresponding to each term in the series expansion of $f(t)$.

The particular solution for the constant term $a_0/2$ can be verified by direct substitution as

$$x_p(t) = \frac{a_0}{2k} \tag{7.14}$$

The particular solution corresponding to each harmonic term is found more easily if we introduce $\omega_f = 2\pi/T$, which is referred to as the *fundamental frequency* of the forcing function. We then seek solutions to differential equations of the form

$$m\ddot{x} + c\dot{x} + kx = \begin{cases} a_n \cos n\omega_f t \\ b_n \sin n\omega_f t \end{cases} \qquad n = 1, 2, 3, \ldots \tag{7.15}$$

Following a procedure completely analogous to that used in Sec. 5.6, we can show that the solutions for the $a_n \cos n\omega_f t$ terms are

$$x_p(t) = \frac{a_n/k}{\sqrt{(1 - r_n^2)^2 + (2\zeta r_n)^2}} \cos(n\omega_f t - \psi_n) \tag{7.16}$$

where

$$r_n = nr = n\frac{\omega_f}{\omega} \qquad \text{and} \qquad \psi_n = \tan^{-1}\frac{2\zeta r_n}{1 - r_n^2} \tag{7.17}$$

are the frequency ratio and phase angle for the nth term, respectively. Similarly, for the $b_n \sin n\omega_f t$ terms we have

$$x_p(t) = \frac{b_n/k}{\sqrt{(1 - r_n^2)^2 + (2\zeta r_n)^2}} \sin(n\omega_f t - \psi_n) \tag{7.18}$$

where r_n and ψ_n are as previously defined.

The steady-state response is then found by superposition as

$$x_p(t) = \frac{a_0}{2k} + \sum_{n=1}^{\infty} \frac{a_n/k}{\sqrt{(1 - r_n^2)^2 + (2\zeta r_n)^2}} \cos(n\omega_f t - \psi_n) + \sum_{n=1}^{\infty} \frac{b_n/k}{\sqrt{(1 - r_n^2)^2 + (2\zeta r_n)^2}} \sin(n\omega_f t - \psi_n) \tag{7.19}$$

Because the forcing function is expanded as a Fourier series, the steady-state response given by Eq. (7.19) is also in the form of an infinite series. However, the amplitude coefficients of the cosine and sine terms are essentially in inverse proportion to n so that the contribution of each term decreases rapidly as n increases. In many cases, the first two or three terms of the series are sufficient to describe the system response. Equation (7.19) is very amenable to calculations via digital computer techniques if it is desired to include a large number of terms in the series.

The case of resonance can also occur for nonharmonic periodic forcing functions. If one of the harmonic frequencies $n\omega_f$ is close to or equal to ω, then $r_n \simeq 1$, and the corresponding amplitude ratio in Eq. (7.19) can become large. This is particularly important when there is little damping.

Example 7.1 A damped spring-mass system is subjected to the periodic forcing function shown in Fig. 7.3. (*a*) Obtain the Fourier-series expansion for $f(t)$. (*b*) Determine the steady-state response including three cosine and sine terms.

SOLUTION First we note that $f(t)$ is an odd function, so we can immediately write

$$a_n = 0 \qquad n = 0, 1, 2, \ldots$$

$$b_n = \frac{4}{T}\int_0^{T/2} f(t)\sin n\omega_f t \, dt \qquad n = 1, 2, 3, \ldots$$

Substituting $f(t) = (2F/T)t$ gives

$$b_n = \frac{8F}{T^2}\int_0^{T/2} t \sin n\omega_f t \, dt$$

Integrating by parts and substituting $2\pi/T = \omega_f$ gives the coefficients as

$$b_n = (-1)^{n+1}\frac{2F}{n\pi} \qquad n = 1, 2, 3, \ldots$$

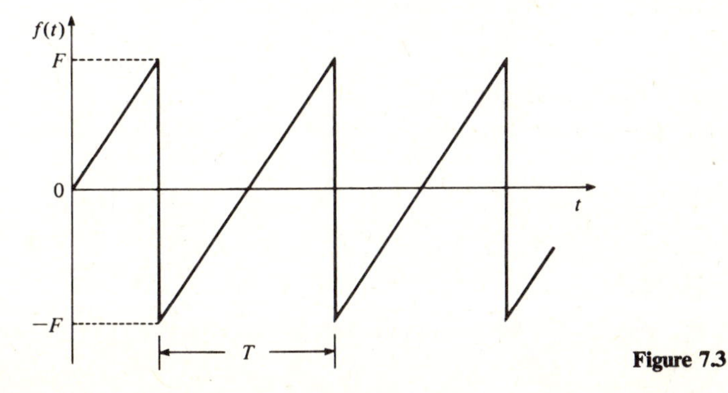

Figure 7.3

The Fourier series for $f(t)$ is a sine series in this case and can be written as

$$f(t) = \sum_{n=1}^{\infty} (-1)^{n+1} \frac{2F}{n\pi} \sin n\omega_f t$$

Substitution in Eq. (7.19) gives the first three terms of the steady-state response as

$$\begin{aligned} x(t) = {} & \frac{2F/\pi k}{\sqrt{(1 - r^2)^2 + (2\zeta r)^2}} \sin(\omega_f t - \psi_1) \\ & - \frac{F/\pi k}{\sqrt{(1 - 4r^2)^2 + (4\zeta r)^2}} \sin(2\omega_f t - \psi_2) \\ & + \frac{2F/3\pi k}{\sqrt{(1 - 9r^2)^2 + (6\zeta r)^2}} \sin(3\omega_f t - \psi_3) \end{aligned}$$

where $r = r_1 = \omega_f/\omega$.

7.3 RESPONSE TO UNIT IMPULSE

A force of large magnitude which acts over a very short time interval is known as an *impulsive force*. The time integral of an impulsive force is finite and is written as

$$\hat{F} = \int F \, dt \tag{7.20}$$

where $\hat{F}$ is the *linear impulse* (in pound seconds or newton seconds) of the force. Figure 7.4 depicts an impulsive force of magnitude $F = \hat{F}/\varepsilon$ acting at $t = a$ over the time interval ε. Note that as $\varepsilon \to 0$ the magnitude of the force becomes infinite but the linear impulse $\hat{F}$ is well defined. The particular case in which $\hat{F}$ is equal to unity is known as a *unit impulse* and is symbolized by the *Dirac delta function* $\delta(t - a)$, which has the following properties:

$$\begin{aligned} \delta(t - a) &= 0 \qquad \text{for } t \neq a \\ \int_0^{\infty} \delta(t - a) \, dt &= 1 \\ \int_0^{\infty} \delta(t - a) f(t) \, dt &= f(a) \end{aligned} \tag{7.21}$$

where $0 < a < \infty$. By utilizing these properties, an impulsive force $F(t)$ acting at $t = a$ to produce a linear impulse $\hat{F}$ of arbitrary magnitude can be expressed as

$$F(t) = \hat{F} \, \delta(t - a) \tag{7.22}$$

Let us now consider the response of a damped spring-mass system to an impulsive force. We shall assume that the system is in equilibrium when at $t = 0$

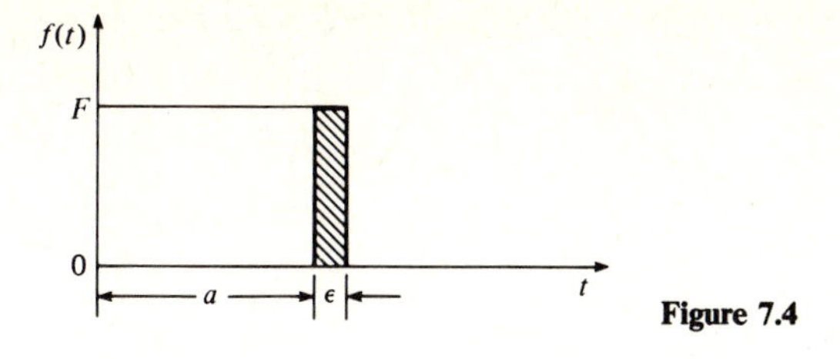

Figure 7.4

it is subjected to an impulse of magnitude $\hat{F}$. With zero initial velocity, application of the principle of impulse and momentum gives

$$\int_0^{\varepsilon} F(t)\, dt = mv(\varepsilon) = m\dot{x}(\varepsilon) \tag{7.23}$$

where $\dot{x}(\varepsilon)$ is the velocity of the mass immediately after application of the impulse. Since ε is very small, we can introduce the notation $\varepsilon = 0+$ and use Eq. (7.22) with $a = 0$ to write

$$\int_0^{0+} \hat{F}\,\delta(t)\, dt = m\dot{x}(0+) \tag{7.24}$$

from which

$$\dot{x}(0+) = \frac{\hat{F}}{m} \tag{7.25}$$

The last result shows that the system is set in motion with an initial velocity $\hat{F}/m$ but no initial displacement, since the time interval is so small. The ensuing motion is the same as that corresponding to free vibration of the system with initial conditions $x_o = 0$ and $\dot{x}_o = \hat{F}/m$. Applying these initial conditions to Eq. (5.21) for an underdamped system gives

$$x(t) = \begin{cases} \dfrac{\hat{F}}{m\omega_d} e^{-\zeta\omega t} \sin \omega_d t & t > 0 \\ 0 & t < 0 \end{cases} \tag{7.26}$$

where $\omega_d = \omega\sqrt{1-\zeta^2}$ is the frequency of damped oscillations.

For convenience in later discussions, we introduce at this point the concept of an *impulsive response function* $g(t)$ defined as the system response to a unit impulse. The system response is then

$$x(t) = \hat{F}g(t) \tag{7.27}$$

and from Eq. (7.26) it is clear that

$$g(t) = \frac{1}{m\omega_d} e^{-\zeta\omega t} \sin \omega_d t \tag{7.28}$$

The impulsive response function will be used extensively in our discussion of general forcing functions.

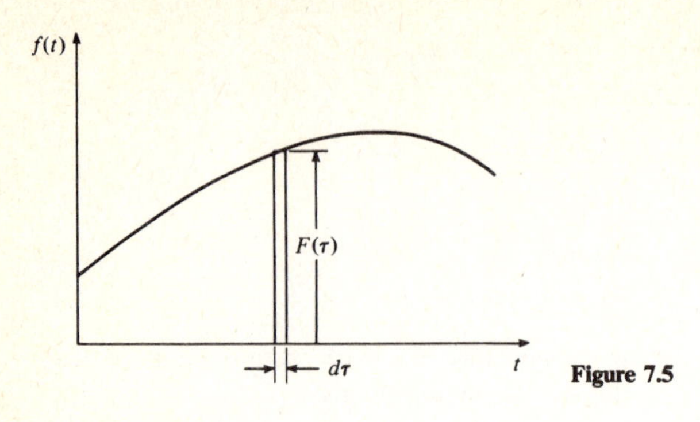

Figure 7.5

7.4 GENERAL FORCING FUNCTIONS; THE CONVOLUTION INTEGRAL

System response to an arbitrary forcing function can be analyzed with the aid of the impulsive response function $g(t)$ defined in the preceding section. Consider the general forcing function $f(t)$, shown in Fig. 7.5, which can be regarded as a series of impulsive forces $F(\tau)$ acting over time interval $d\tau$. The corresponding impulse is $F(\tau)\,d\tau$, and the response to this impulse will apply for all $t > \tau$. Thus the contribution to the response at time t can be expressed as

$$dx = F(\tau)\,d\tau\,g(t-\tau) \tag{7.29}$$

where the argument of the impulsive response function reflects *elapsed time* since the application of the particular impulse. Equation (7.29) gives the *incremental response* to the incremental impulse $F(\tau)\,d\tau$. To determine the total response, we integrate over the entire interval to obtain

$$x(t) = \int_0^t F(\tau)g(t-\tau)\,d\tau \tag{7.30}$$

which is known as the *convolution integral* or *Duhamel's integral*. As a summation process, the convolution integral reflects the application of the principle of superposition, which is valid for the linear systems being studied.

For a damped spring-mass system, substitution of the impulsive response function into Eq. (7.30) yields

$$x(t) = \frac{1}{m\omega_d}\int_0^t F(\tau)e^{-\zeta\omega(t-\tau)}\sin\omega_d(t-\tau)\,d\tau \tag{7.31}$$

as the system response to a general forcing function. We must note, however, that Eq. (7.31) does not account for initial conditions and represents only the forced response.

Example 7.2 Determine the response of a damped spring-mass system to the force shown in Fig. 7.6*a*.

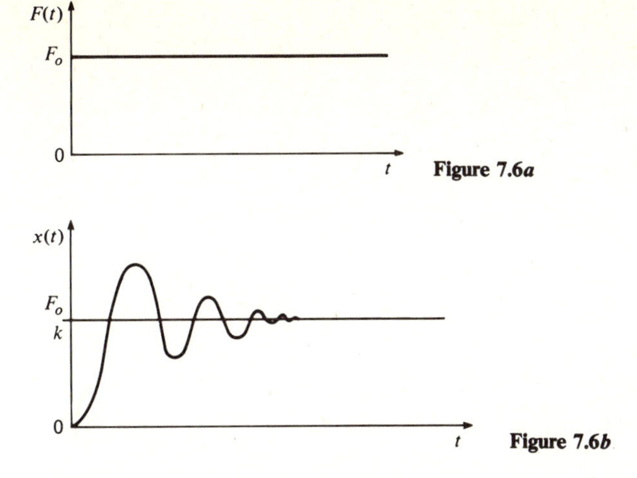

Figure 7.6a

Figure 7.6b

SOLUTION The forcing condition is $F(t) = F_o$ = constant; thus Eq. (7.31) becomes

$$x(t) = \frac{F_o}{m\omega_d}\int_0^t e^{-\zeta\omega(t-\tau)} \sin \omega_d(t - \tau)\, d\tau$$

Using the change of variable $z = t - \tau$, we obtain

$$x(t) = \frac{F_o}{m\omega_d}\int_0^t e^{-\zeta\omega z} \sin \omega_d z\, dz$$

$$= \frac{F_o}{m\omega_d}\left[\frac{e^{-\zeta\omega z}(-\zeta\omega \sin \omega_d z - \omega_d \cos \omega_d z)}{(\zeta\omega)^2 + \omega_d^2}\right]_0^t$$

$$= \frac{F_o}{m\omega_d\left[(\zeta\omega)^2 + \omega_d^2\right]}\left[e^{-\zeta\omega t}(-\zeta\omega \sin \omega_d t - \omega_d \cos \omega_d t + \omega_d\right]$$

Substituting $\omega_d = \omega\sqrt{1 - \zeta^2}$, combining sine and cosine terms, and rearranging result in

$$x(t) = \frac{F_o}{k}\left[1 - \frac{e^{-\zeta\omega t}}{\sqrt{1 - \zeta^2}}\cos(\omega_d t - \psi)\right]$$

where $$\psi = \tan^{-1}\frac{\zeta}{\sqrt{1 - \zeta^2}}$$

The response is as depicted in Fig. 7.6*b*.

Example 7.3 Determine the response of an undamped system to the force shown in Fig. 7.7.

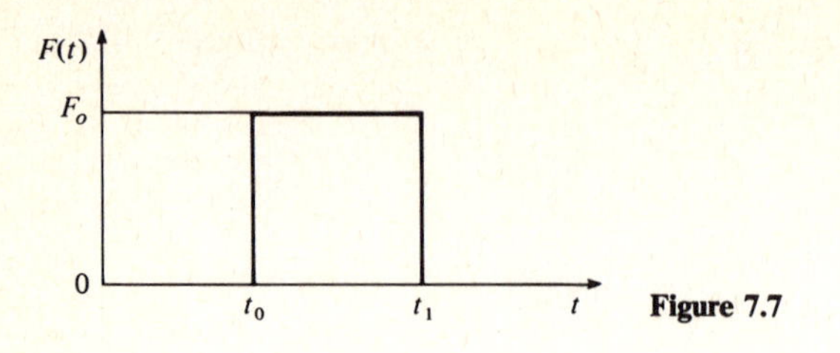

Figure 7.7

SOLUTION The forcing function is described by

$$F(t) = \begin{cases} 0 & \text{for } t < t_o \\ F_o & \text{for } t_o < t < t_1 \\ 0 & \text{for } t_1 < t \end{cases}$$

Clearly, for $t < t_o$ the response is $x(t) = 0$. For $t_o < t < t_1$, we set $F(\tau) = F_o$ in Eq. (7.31) and use $\zeta = 0$ for an undamped system to obtain

$$x(t) = \frac{F_o}{m\omega} \int_{t_o}^{t} \sin \omega(t - \tau)\, d\tau$$

$$= \frac{F_o}{k}\left[1 - \cos \omega(t - t_o)\right]$$

For any time after t_1 (that is, $t_1 < t$) the response is

$$x(t) = \frac{F_o}{m\omega} \int_{t_o}^{t_1} \sin \omega(t - \tau)\, d\tau + [0]$$

$$= \frac{F_o}{k}\left[\cos \omega(t - \tau)\right]_{t_o}^{t_1}$$

$$= \frac{F_o}{k}\left[\cos \omega(t - t_1) - \cos \omega(t - t_o)\right]$$

7.5 LAPLACE-TRANSFORM TECHNIQUES

A very powerful tool for the solution of linear differential equations is the application of the *Laplace transform* or *transformation*. For a given function $f(t)$ specified for $t > 0$, the Laplace transform, denoted by $\mathfrak{L}\{f(t)\}$, is defined by

$$\mathfrak{L}\{f(t)\} = \bar{f}(s) = \int_0^\infty e^{-st} f(t)\, dt \tag{7.32}$$

where the parameter s is real.[3] Note that bar notation is used to indicate that $\bar{f}(s)$ is the Laplace transform of $f(t)$.

[3] In a more general sense mathematically, s may be complex.

To illustrate the application of Laplace transforms to linear differential equations, consider

$$m\ddot{x} + c\dot{x} + kx = F(t) \tag{7.33}$$

which is the differential equation of motion for a damped system subjected to a general forcing function $F(t)$. Multiplying both sides of Eq. (7.33) by $e^{-st}\,dt$ and integrating from zero to infinity gives

$$m\int_0^\infty \ddot{x}(t)e^{-st}\,dt + c\int_0^\infty \dot{x}(t)e^{-st}\,dt + k\int_0^\infty x(t)\, e^{-st}\,dt = \int_0^\infty F(t)e^{-st}\,dt \tag{7.34}$$

By the definition of the Laplace transform, this may be written

$$m\mathcal{L}\{\ddot{x}(t)\} + c\mathcal{L}\{\dot{x}(t)\} + k\mathcal{L}\{x(t)\} = \mathcal{L}\{F(t)\} \tag{7.35}$$

which we shall refer to as the *transformed* equation. Now let us evaluate $\mathcal{L}[\dot{x}(t)]$ according to the definition:

$$\mathcal{L}\{\dot{x}(t)\} = \int_0^\infty \dot{x}(t)e^{-st}\,dt \tag{7.36}$$

Substituting $\dot{x}(t) = dx/dt$ and integrating by parts give

$$\begin{aligned}\int_0^\infty \frac{dx}{dt}e^{-st}\,dt &= x(t)e^{-st}\Big|_0^\infty + \int_0^\infty sx(t)e^{-st}\,dt \\ &= -x(0) + s\int_0^\infty x(t)e^{-st}\,dt \\ &= s\bar{x}(s) - x(0) = \mathcal{L}\{\dot{x}(t)\}\end{aligned} \tag{7.37}$$

Similarly, we can show that

$$\mathcal{L}\{\ddot{x}(t)\} = s^2\bar{x}(s) - sx(0) - \dot{x}(0) \tag{7.38}$$

is the Laplace transform of the second derivative. Substituting these results into Eq. (7.35), we have

$$m\left[s^2\bar{x}(s) - sx(0) - \dot{x}(0)\right] + c\left[\bar{x}(s) - x(0)\right] + k\bar{x}(s) = \bar{F}(s) \tag{7.39}$$

from which

$$\bar{x}(s) = \frac{\bar{F}(s) + (ms + c)x(0) + m\dot{x}(0)}{ms^2 + cs + k} \tag{7.40}$$

Equation (7.40) gives the Laplace transform of the solution $x(t)$. If a function can be found which has a Laplace transform given by the right-hand side of this equation, then we have found the solution. Note particularly that since the initial conditions are included explicitly, the solution so determined will be the complete solution, including both homogeneous and particular parts.

The procedure of determining $x(t)$ given $\bar{x}(s)$ can be regarded as an inverse transformation which can be written symbolically as

$$x(t) = \mathcal{L}^{-1}\{\bar{x}(s)\} \tag{7.41}$$

The task is greatly simplified by the fact that the Laplace transform has been determined for a very large number of functions, and the results tabulated. An abbreviated table of Laplace-transform pairs will be found in App. A.5. The following examples will serve to illustrate the theory as well as the use of the table.

Example 7.4 Determine the Laplace transform of the function $f(t) = a =$ constant.

SOLUTION

$$\mathcal{L}\{f(t)\} = \mathcal{L}\{a\} = \int_0^\infty ae^{-st}\,dt = -a\frac{e^{-st}}{s}\Big|_0^\infty = \frac{a}{s}$$

Example 7.5 Determine the Laplace transform of $\sin \omega t$, where ω is a constant.

SOLUTION

$$\mathcal{L}\{\sin \omega t\} = \int_0^\infty e^{-st} \sin \omega t\,dt = \frac{e^{-st}(-s \sin \omega t - \omega \cos \omega t)}{s^2 + \omega^2}\Bigg|_0^\infty$$

$$= \frac{\omega}{s^2 + \omega^2}$$

Example 7.6 An undamped system having $m = 5$ kg and $k = 3$ N/mm is excited by a force $F(t) = 20(1 - e^{-t})$ N. Determine the displacement $x(t)$ using Laplace transforms if the initial conditions are $x(0) = 5$ mm and $\dot{x}(0) = 0$.

SOLUTION The differential equation of motion is

$$0.005\ddot{x} + 3x = 20(1 - e^{-t})$$

where $m = 5$ kg $= 5$ N $\cdot$ s^2/m has been converted to 0.005 N $\cdot$ s^2/mm for dimensional uniformity. Rewriting the equation as

$$\ddot{x} + 600x = 4000(1 - e^{-t})$$

and transforming give

$$s^2\bar{x}(s) - 5s + 600\bar{x}(s) = \frac{4000}{s} - \frac{4000}{s+1}$$

where the Laplace transform for $F(t)$ was obtained from the table in App. A.5. Solving for $\bar{x}(s)$, we have

$$\bar{x}(s) = \frac{4000}{s(s^2 + 600)} - \frac{4000}{(s+1)(s^2 + 600)} + \frac{5s}{s^2 + 600}$$

Again referring to the table of Laplace-transform pairs, we find

$$\mathcal{L}^{-1}\left\{\frac{4000}{s(s^2+600)}\right\} = \frac{4000}{600}(1 - \cos 24.5t)$$

$$\mathcal{L}^{-1}\left\{\frac{-4000}{(s+1)(s^2+600)}\right\} = -\frac{4000e^{-t}}{1^2+600} - \frac{4000}{24.5\sqrt{1^2+600}}\sin(24.5t - \phi)$$

where $\phi = \tan^{-1}(24.5/1) = 1.57$ rad, and

$$\mathcal{L}^{-1}\left\{\frac{5s}{s^2+600}\right\} = 5 \cos 24.5t$$

Combining similar terms and rearranging give

$$x(t) = 6.67 - 1.67 \cos 24.5t - 6.65e^{-t} - 6.66 \sin(24.5t - 1.57) \text{ mm}$$

Example 7.7 Determine $x(t)$ given its Laplace transform as

$$\bar{x}(s) = \frac{3s+5}{s^2 - 2s - 3}$$

Solution Examination of the table in App. A.5 reveals that $x(t)$ cannot be readily determined by inspection. In such cases, the inverse transform is found only after expanding $\bar{x}(s)$ into partial fractions as follows. First we factor the denominator to obtain

$$\bar{x}(s) = \frac{3s+5}{(s-3)(s+1)}$$

We next seek to determine constants A and B such that

$$\bar{x}(s) = \frac{3s+5}{(s-3)(s+1)} = \frac{A}{s-3} + \frac{B}{s+1}$$

Multiplication by $(s-3)(s+1)$ gives

$$3s + 5 = A(s+1) + B(s-3)$$

or

$$3s + 5 = (A+B)s + A - 3B$$

Equating similar powers of s yields $A = \frac{7}{2}$ and $B = -\frac{1}{2}$. Then

$$\bar{x}(s) = \frac{7/2}{s-3} - \frac{1/2}{s+1}$$

Use of the table now gives

$$x(t) = \tfrac{7}{2}e^{3t} - \tfrac{1}{2}e^{-t}$$

The method of partial fractions utilized in the previous example is discussed in detail in App. A.6. For the more complex functions, the use of this technique is often the only reasonable means of determining inverse Laplace transforms.

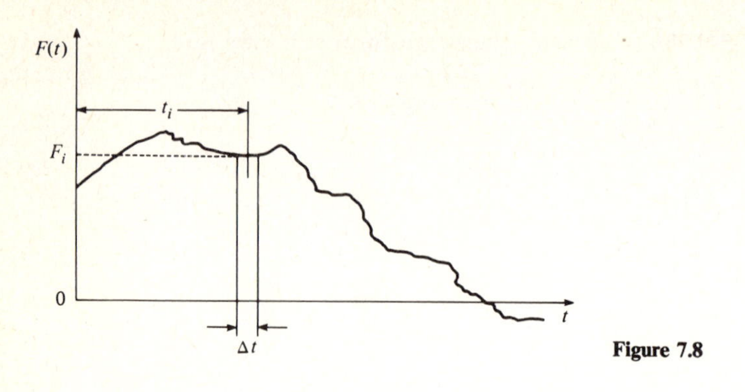

Figure 7.8

7.6 NUMERICAL METHODS

In the preceding sections we discussed several analytical methods of obtaining solutions for the system response to a general forcing function. Each of these methods requires that the forcing function be expressible as an explicit function of time if we are to obtain results mathematically. In many cases the forcing function is determined experimentally and is not amenable to an explicit mathematical expression. Figure 7.8 is an example of such a forcing function. To obtain solutions in these instances, numerical methods are required. A complete discussion of numerical methods is beyond the scope of this text, and the following is intended as an introduction to some basic techniques.

In obtaining a numerical solution for a differential equation of the form

$$\ddot{x} + a_1\dot{x} + a_2 x = F(t) \tag{7.42}$$

with initial conditions $x(0) = x_o$ and $\dot{x}(0) = \dot{x}_o$, we seek numerical values of $x(t)$ at discrete but closely spaced values of t over a range sufficient to ascertain the character of the solution. To facilitate the procedure, we divide $F(t)$ into increments Δt as shown in Fig. 7.8. The increment Δt is arbitrary but in general must be very small to produce accurate results.

Since most numerical schemes are presented on the basis of solving first-order differential equations, it will be convenient to introduce $z = \dot{x}$ into Eq. (7.42) to obtain

$$\dot{z} + a_1 z + a_2 x = F(t) \qquad \dot{x} = z \tag{7.43}$$

as an equivalent first-order system having initial conditions $x(0) = x_o$ and $z(0) = \dot{x}_o$.

Obtaining the numerical solution to the first-order system is a two-step process. A starting solution is used to obtain the first few values, after which a more efficient method is usually used to continue the solution. The starting solution which will be outlined here without proof is the *Runge-Kutta method.*[4]

[4] For the mathematical details, refer to Kaiser S. Kunz, *Numerical Analysis*, McGraw-Hill, New York, 1957, chap. 9.

Rewriting Eqs. (7.43) as

$$\dot{z} = F(t) - a_1 z - a_2 x = h(t, z, x) \tag{7.44}$$

$$\dot{x} = z \tag{7.45}$$

we find successive values of z and x based on known previous values, as follows. Given the values z_i and x_i at time t_i, the values at time t_{i+1} (where $\Delta t = t_{i+1} - t_i$) are calculated from

$$z_{i+1} = z_i + \tfrac{1}{6}(b_1 + 2b_2 + 2b_3 + b_4) \tag{7.46}$$

and

$$x_{i+1} = x_i + \tfrac{1}{6}(c_1 + 2c_2 + 2c_3 + c_4) \tag{7.47}$$

where

$$\begin{aligned} b_1 &= \Delta t\, h(t_i, z_i, x_i) & c_1 &= \Delta t\, z_i \\ b_2 &= \Delta t\, h\left(t_i + \frac{\Delta t}{2}, z_i + \frac{b_1}{2}, x_i + \frac{c_1}{2}\right) & c_2 &= \Delta t\left(z_i + \frac{c_1}{2}\right) \\ b_3 &= \Delta t\, h\left(t_i + \frac{\Delta t}{2}, z_i + \frac{b_2}{2}, x_i + \frac{c_2}{2}\right) & c_3 &= \Delta t\left(z_i + \frac{c_2}{2}\right) \\ b_4 &= \Delta t\, h\left(t_i + \frac{\Delta t}{2}, z_i + b_3, x_i + c_3\right) & c_4 &= \Delta t(z_i + c_3) \end{aligned} \tag{7.48}$$

Note that if the values defined by Eqs. (7.48) are calculated in the order given, only previously calculated values will be needed at each step in the computation. These calculations are well suited to digital computation and, for the problem at hand, require only that we specify the values of $F(t)$ at intervals $\Delta t/2$ to proceed.

With the initial conditions $x(0)$ and $z(0)$ known and $F(t)$ specified at the appropriate intervals, the Runge-Kutta method can be used to calculate as many succeeding values of x and z as desired. Usually, however, only the values for the next few increments are determined and used as starting values in the method chosen for continuing the solution.

The continuation method presented here is *Hamming's modified predictor-corrector method.*[5] To utilize Hamming's method, it is necessary to know the solution at four previous points. Given the solution at the equidistant points t_{i-3}, t_{i-2}, t_{i-1}, and t_i, the solution at t_{i+1} is found as follows: First approximations for z_{i+1} and x_{i+1}, called *predictors*, are calculated using the relation

$$z_{i+1}^{p} = z_{i-3} + \frac{4\,\Delta t}{3}(2\dot{z}_i - \dot{z}_{i-1} + 2\dot{z}_{i-2}) \tag{7.49}$$

and similarly for x_{i+1}. The derivatives are obtained by substituting into Eqs. (7.45) the appropriate values at the previous increments. Next, the predicted values are modified, based on the magnitude of the difference between the

[5] R. W. Hamming, *Numerical Methods for Scientists and Engineers*, McGraw-Hill, New York, 1962.

predicted and corrected values at the previous point, with

$$z_{i+1}^{m} = z_{i+1}^{p} - \frac{112}{121}(z_i^p - z_i^c) \tag{7.50}$$

Finally a corrected value is calculated using the *corrector equation*

$$z_{i+1}^{c} = \frac{1}{8}(9z_i - z_{i-2}) + \frac{3\,\Delta t}{8}(\dot{z}_{i+1}^{m} + 2\dot{z}_i - \dot{z}_{i-1}) \tag{7.51}$$

The modified and corrected values of x are obtained by replacing z with x in the equations. As Hamming's method is not self-starting, the initial conditions together with the solution at three other points from the Runge-Kutta method are required. As both methods are fourth-order approximations [meaning the error allowable is on the order of $(\Delta t)^5$], they are quite compatible and are often used together to obtain digital computer solutions. In fact, commercial programs[6] designed to carry out Hamming's method are available.

PROBLEMS

7.1–7.10 Determine the Fourier-series expansion for the periodic forcing functions shown in Figs. 7.9 through 7.18.

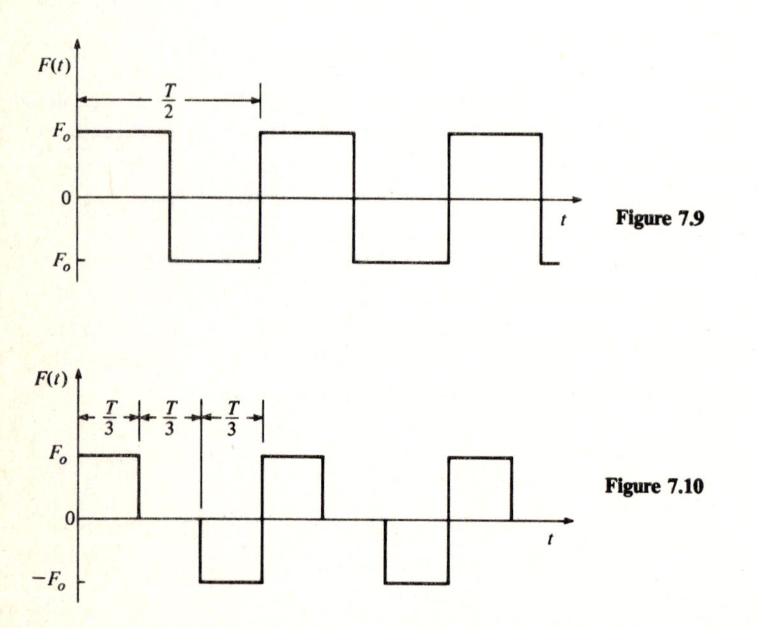

Figure 7.9

Figure 7.10

[6] See for example, *IBM Systems Reference Manual*, Scientific Subroutine Package, Form no. H20-0205-3, p. 337.

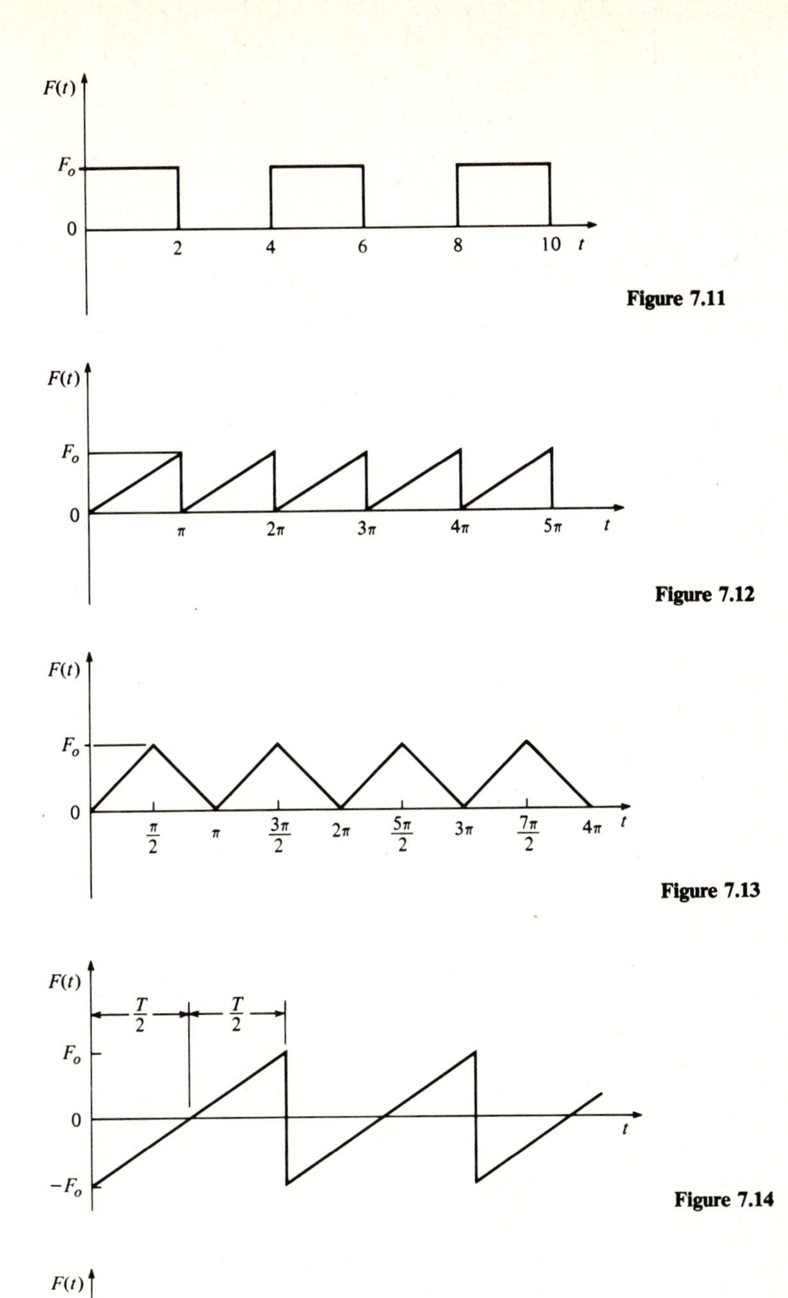

Figure 7.11

Figure 7.12

Figure 7.13

Figure 7.14

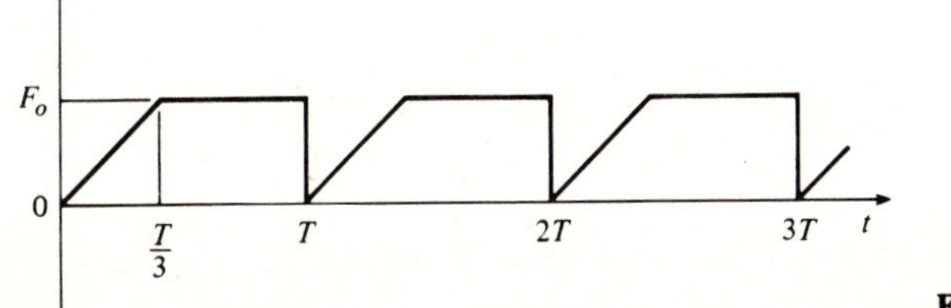

Figure 7.15

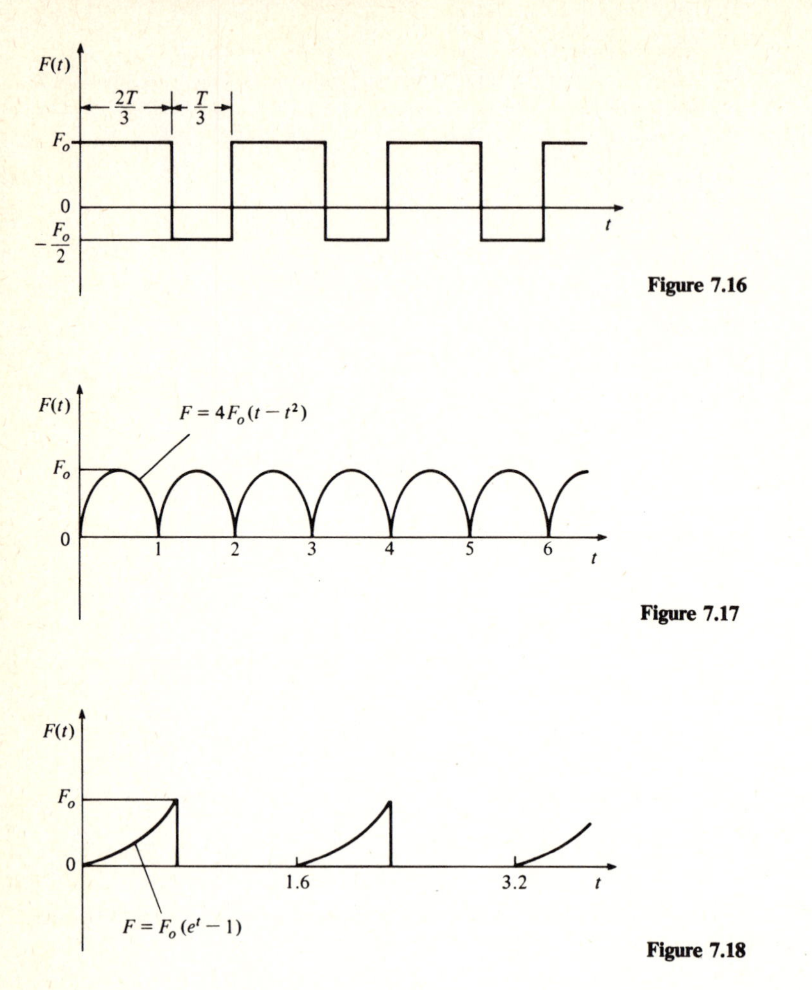

Figure 7.16

Figure 7.17

Figure 7.18

7.11–7.20 For a viscously damped system, write the series representing the steady-state response to the forcing functions shown in Figs. 7.9 through 7.18.

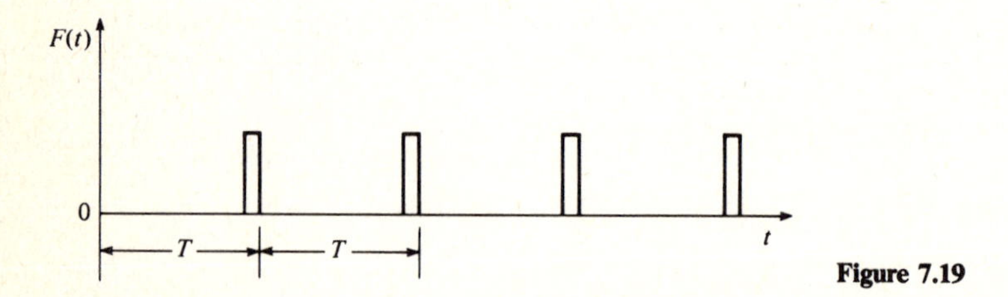

Figure 7.19

7.21 Determine the Fourier-series expansion of a periodic sequence of unit impulses having period T (Fig. 7.19).

7.22 Using the results of Prob. 7.21, write the series representation for the response of a viscously damped system to a periodic sequence of impulses having magnitude $\hat{F}$.

7.23 Determine the impulsive response function $g(t)$ for a critically damped system.

7.24 A damped system having $W = 10$ lb, $k = 15$ lb/in, and $c = 1$ lb · s/in is in equilibrium at $t = 0$. At $t = 2$ s, the system receives an impulse having a magnitude of 8 lb · s. Determine the displacement at $t = 4$ s.

7.25 Determine the response of a damped system to the forcing function shown in Fig. 7.20.

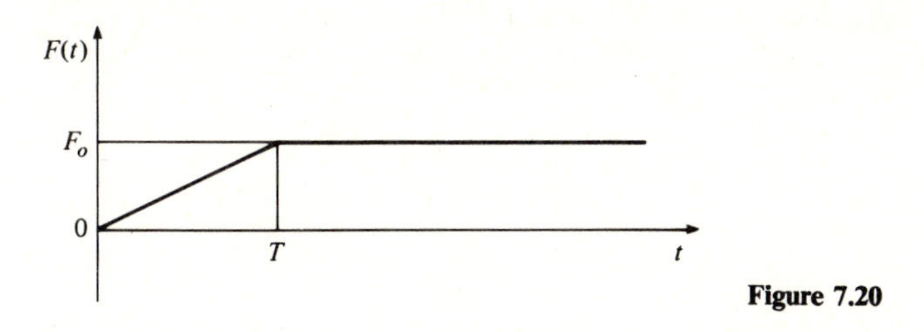

Figure 7.20

7.26 Determine the response of an undamped system to the forcing function shown in Fig. 7.21 for (*a*) $0 < t < T$ and (*b*) $t > T$.

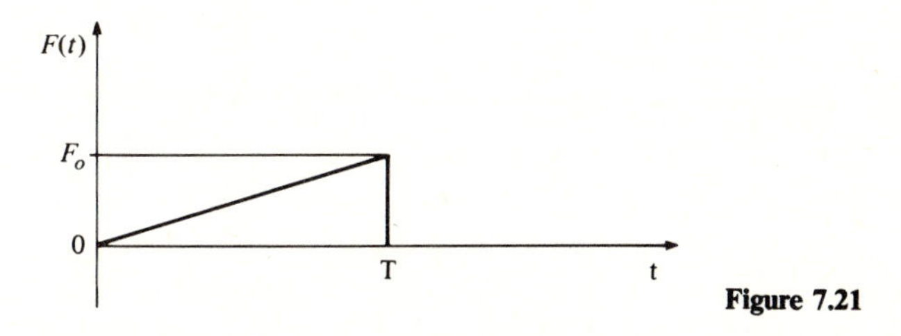

Figure 7.21

7.27 Using Laplace transforms, determine the response of an undamped system to a harmonic force $F_o \sin \omega_f t$. Denote the initial conditions by $x(0) = x_o$ and $\dot{x}(0) = \dot{x}_o$.

7.28 Repeat Prob. 7.27 for a damped system, assuming $\zeta < 1$.

7.29 An undamped system is excited by a force $F = te^{-t} + \sin t$. Assuming a nonresonant condition with $x(0) = \dot{x}(0) = 0$, determine the response using Laplace transforms.

7.30 Given the differential equation $\ddot{x} + 2\dot{x} + x = 4$ with initial conditions $x(0) = 1$ and $\dot{x}(0) = 1$.

(*a*) Determine the complete solution $x(t)$ analytically.

(*b*) Calculate $x(t)$ at 0.1, 0.2, and 0.3 s.

(*c*) Using $\Delta t = 0.1$ s, calculate the same displacements using the Runge-Kutta method.

(*d*) Determine the displacement at $t = 0.4$ s by Hamming's method, and compare with the exact solution. How could the accuracy be improved?

CHAPTER

EIGHT

VIBRATIONS OF SYSTEMS WITH TWO DEGREES OF FREEDOM

To this point, we have considered the vibrations only of systems having one degree of freedom. Recall that the number of degrees of freedom of a system is the same as the number of independent coordinates required to specify the configuration of the system at any time. For a system which executes only one-dimensional motion, the number of degrees of freedom corresponds to the number of inertial elements in the system.

In this chapter, we shall focus our attention on systems having two degrees of freedom. While these are but a special case of the multiple-degree-of-freedom systems to be studied in Chap. 10, two-degree-of-freedom systems are treated separately for convenience. The problem formulation and methods of solution are the same as for more complex systems, but the volume of calculations is greatly reduced. Thus the separate treatment of two-degree-of-freedom systems serves to introduce the concepts in the least encumbered manner.

8.1 FREE UNDAMPED MOTION

As a first example of vibrations having two degrees of freedom, consider the system shown in Fig. 8.1*a*, in which two masses are suspended in a vertical plane. Since the masses are connected elastically through a linear spring, relative motion is possible. Thus, an independent coordinate is required for each mass, and the system has two degrees of freedom. For convenience, the coordinates chosen are the displacements of the masses from the respective equilibrium positions.

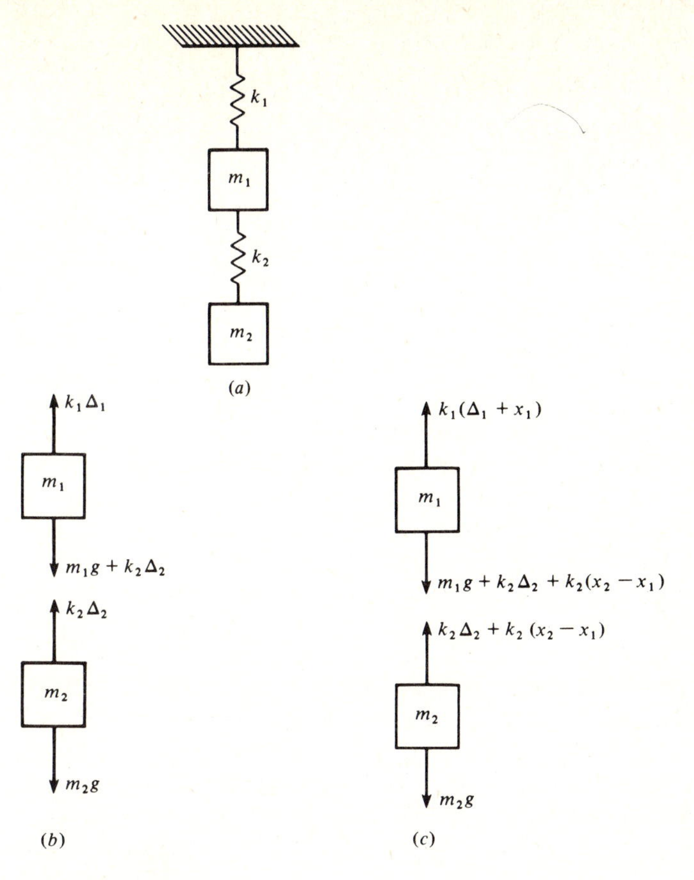

Figure 8.1

Consideration of the static equilibrium results in the free-body diagrams shown in Fig. 8.1*b*, in which Δ_1 and Δ_2 denote the equilibrium deflections of the springs from their free lengths. For equilibrium, we have

$$m_1 g + k_2\Delta_2 - k_1\Delta_1 = 0$$

$$m_2 g - k_2\Delta_2 = 0 \tag{8.1}$$

which will shortly be used to eliminate gravity forces from the analysis, as was done for similar single-degree-of-freedom systems.

To describe the motion of the system mathematically, let us consider the static equilibrium condition to be disturbed in some manner such that each mass is set into motion. Figure 8.1*c* shows the free-body diagrams of the masses in nonequilibrium positions x_1 and x_2. For convenience we have assumed that both displacements are positive and that $x_2 > x_1$ so that spring k_2 is in tension. Applying Newton's second law to each mass gives

$$m_1\ddot{x}_1 = m_1 g + k_2\Delta_2 + k_2(x_2 - x_1) - k_1(\Delta_1 + x_1)$$

$$m_2\ddot{x}_2 = m_2 g - k_2\Delta_2 - k_2(x_2 - x_1) \tag{8.2}$$

Rearranging, we have

$$m_1\ddot{x}_1 + (k_1 + k_2)x_1 - k_2x_2 = m_1g + k_2\Delta_2 - k_1\Delta_1$$
$$m_2\ddot{x}_2 + k_2x_2 - k_2x_1 = m_2g - k_2\Delta_2 \tag{8.3}$$

Comparison of the right-hand sides of Eqs. (8.3) with Eqs. (8.1) allows us to eliminate the equilibrium terms to obtain

$$\ddot{x}_1 + \frac{k_1 + k_2}{m_1}x_1 - \frac{k_2}{m_1}x_2 = 0$$
$$\ddot{x}_2 + \frac{k_2}{m_2}x_2 - \frac{k_2}{m_2}x_1 = 0 \tag{8.4}$$

as the differential equations of motion for the system. These are simultaneous, second-order, linear differential equations with constant coefficients. Further, the equations are homogeneous, since the system is free of all external forces and thus executes free oscillations about equilibrium.

To obtain the solution for the motion of each mass as a function of time, Eqs. (8.4) must be solved simultaneously. This is necessary because the equations are coupled by the appearance of both x_1 and x_2 in each equation. Physically, this results from the spring connection between the masses; the system is said to be *elastically coupled*. Methods of obtaining solutions for the two-degree-of-freedom system represented by Eqs. (8.4) are relatively straightforward, as will be discussed in the following sections.

Example 8.1 Determine the differential equations of motion for the two-degree-of-freedom torsional system shown in Fig. 8.2*a*.

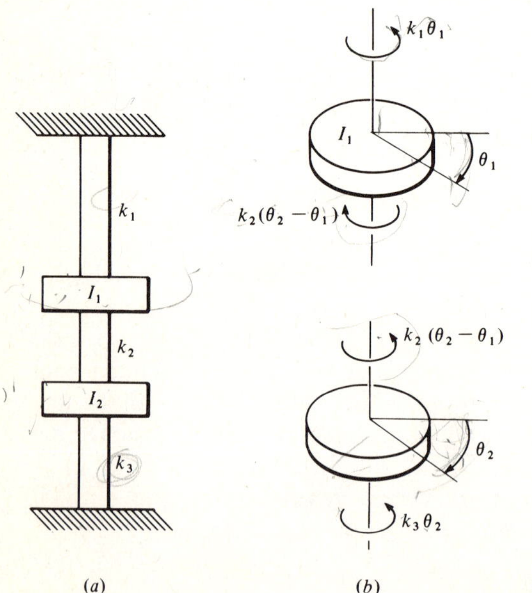

Figure 8.2

SOLUTION If we denote by θ_1 and θ_2 the angles of rotation of the inertial disks, the free-body diagrams are as shown in Fig. 8.2*b*, for $\theta_2 > \theta_1$. From the free-body diagrams, the equations of motion are

$$I_1\ddot{\theta}_1 = -k_1\theta_1 + k_2(\theta_2 - \theta_1)$$

and

$$I_2\ddot{\theta}_2 = -k_2(\theta_2 - \theta_1) - k_3\theta_2$$

These can be rearranged to give

$$\ddot{\theta}_1 + \frac{k_1 + k_2}{I_1}\theta_1 - k_2\theta_2 = 0$$

and

$$\ddot{\theta}_2 + \frac{k_2 + k_3}{I_2}\theta_2 - k_2\theta_1 = 0$$

which are elastically coupled through k_2.

8.2 PRINCIPAL MODES

Referring again to Fig. 8.1*a*, we shall now seek solutions $x_1(t)$ and $x_2(t)$ which describe the displacement of m_1 and m_2, respectively. The solutions must satisfy Eqs. (8.4) for all values of time and satisfy any specified initial conditions. Since the form of the differential equations is similar to that of the equations governing single-degree-of-freedom systems, it seems reasonable to seek solutions having similar form as well. With this observation as a starting point, let us assume solutions of the form

$$\begin{aligned} x_1(t) &= A_1 \sin(\omega t + \phi) \\ x_2(t) &= A_2 \sin(\omega t + \phi) \end{aligned} \tag{8.5}$$

where A_1 and A_2 are the respective amplitudes, ω is the circular frequency, and ϕ is the phase angle. Our task is to find values for A_1, A_2, ω, and ϕ which satisfy Eqs. (8.4) identically. Performing the necessary differentiation and substituting into the differential equations give

$$\begin{aligned} \left[\left(-\omega^2 + \frac{k_1 + k_2}{m_1}\right)A_1 - \frac{k_2}{m_1}A_2\right] \sin(\omega t + \phi) &= 0 \\ \left[\left(-\omega^2 + \frac{k_2}{m_2}\right)A_2 - \frac{k_2}{m_2}A_1\right] \sin(\omega t + \phi) &= 0 \end{aligned} \tag{8.6}$$

Since we seek solutions which are independent of time, we can drop the sine terms to obtain

$$\begin{aligned} \left(-\omega^2 + \frac{k_1 + k_2}{m_1}\right)A_1 - \frac{k_2}{m_1}A_2 &= 0 \\ -\frac{k_2}{m_2}A_1 + \left(-\omega^2 + \frac{k_2}{m_2}\right)A_2 &= 0 \end{aligned} \tag{8.7}$$

Thus the simultaneous differential equations are reduced to a pair of simultaneous linear algebraic equations. Examination of Eqs. (8.7) reveals that we have only *two* equations from which *three* unknowns ω, A_1, and A_2 must be determined. Our first impulse may be that such a system cannot be solved. However, by noting that the equations are homogeneous, we may utilize certain facts from linear algebra to facilitate the solution process.

For convenience and as an introduction to methods to be used in Chap. 10, we can rewrite Eqs. (8.7) in the equivalent matrix form

$$\begin{bmatrix} -\omega^2 + \dfrac{k_1 + k_2}{m_1} & -\dfrac{k_2}{m_1} \\ -\dfrac{k_2}{m_2} & -\omega^2 + \dfrac{k_2}{m_2} \end{bmatrix} \begin{bmatrix} A_1 \\ A_2 \end{bmatrix} = [0] \tag{8.8}$$

where [0] denotes the 2×1 column matrix having all zero terms. This system has the *trivial solution* $A_1 = A_2 = 0$, but this solution will be ignored, as it simply represents static equilibrium. The system represented by Eq. (8.8) will have a nontrivial solution if and only if the determinant of the coefficient matrix is identically equal to zero.[1] Symbolically, this determinant is written as

$$\begin{vmatrix} -\omega^2 + \dfrac{k_1 + k_2}{m_1} & -\dfrac{k_2}{m_1} \\ -\dfrac{k_2}{m_2} & -\omega^2 + \dfrac{k_2}{m_2} \end{vmatrix} = 0 \tag{8.9}$$

and is known as the *characteristic determinant* of the system. This terminology is indicative of the fact that this determinant characterizes the conditions under which nontrivial solutions exist. Expanding the determinant and rearranging give

$$\omega^4 - \left(\frac{k_2}{m_2} + \frac{k_1 + k_2}{m_1}\right)\omega^2 + \frac{k_1 k_2}{m_1 m_2} = 0 \tag{8.10}$$

Equation (8.10) is known in general as the *characteristic equation* and is also referred to as the *frequency equation* of the vibration system. Formally, there exist four values of ω which will satisfy the frequency equation, but two of these can be shown to be negative for mechanical systems, since the m's and k's are real positive constants. Since negative circular frequencies have no physical significance, we shall consider only the two positive roots of Eq. (8.10).

To simplify the analysis and provide a clearer understanding of the solution, we shall now examine the special case in which $m_1 = m_2 = m$, $k_1 = 3k$, and

[1] See, for example, D. L. Kreider, R. G. Kuller, D. R. Ostberg, and F. W. Perkins, *An Introduction to Linear Analysis*, Addison-Wesley, Reading, Mass., 1966, p. 467.

$k_2 = 2k$. If we substitute these values, the frequency equation becomes

$$\omega^4 - 7\frac{k}{m}\omega^2 + 6\frac{k^2}{m^2} = 0 \tag{8.11}$$

Since only even powers of ω appear in Eq. (8.11) it can be solved as a quadratic equation in ω^2 to obtain

$$\omega^2 = \frac{7}{2}\frac{k}{m} \pm \frac{5}{2}\frac{k}{m} \tag{8.12}$$

from which the positive roots are obtained as

$$\omega_1 = \sqrt{\frac{k}{m}} \qquad \omega_2 = \sqrt{6\frac{k}{m}} \tag{8.13}$$

which are known as the *natural circular frequencies* of the vibrating system. Both ω_1 and ω_2 satisfy the frequency equation, and they represent two distinct circular frequencies at which the system can oscillate. It is customary to denote the smaller of the two values as ω_1, in which case it is called the *fundamental* frequency of the system.

Having determined the natural frequencies, we must now return to the task of determining the amplitudes A_1 and A_2. Rewriting Eqs. (8.7) using the assumed values of mass and spring constants results in

$$\left(-\omega^2 + 5\frac{k}{m}\right)A_1 - 2\frac{k}{m}A_2 = 0$$

$$-2\frac{k}{m}A_1 + \left(-\omega^2 + 2\frac{k}{m}\right)A_2 = 0 \tag{8.14}$$

from which we must attempt to determine A_1 and A_2. We cannot, of course, find absolute values of the amplitudes from this set of homogeneous equations, but we can gain valuable information as follows. If we let $\omega^2 = \omega_1^2 = k/m$ in either of Eqs. (8.14), we obtain the relation $A_2 = 2A_1$. This relation establishes the *fundamental* or *first mode* and represents the pattern of the harmonic vibration. The corresponding displacement equations for the fundamental mode are

$$x_1(t) = A_1 \sin(\omega_1 t + \phi_1)$$

$$x_2(t) = 2A_1 \sin(\omega_1 t + \phi_1) \tag{8.15}$$

where $\omega_1 = \sqrt{k/m}$, and ϕ_1 is the phase angle of the first mode. The constants A_1 and ϕ_1 are determined from the initial conditions. We observe from Eqs. (8.15) that the masses move together but that the amplitude of m_2 is twice that of m_1.

If we next substitute $\omega_2 = \sqrt{6k/m}$ into either of Eqs. (8.14), we obtain $A_2 = -0.5A_1$. This defines the second mode of vibration of the system. The displacements for the second mode are

$$x_1(t) = A_1 \sin(\omega_2 t + \phi_2)$$

$$x_2(t) = -0.5A_1 \sin(\omega_2 t + \phi_2) \tag{8.16}$$

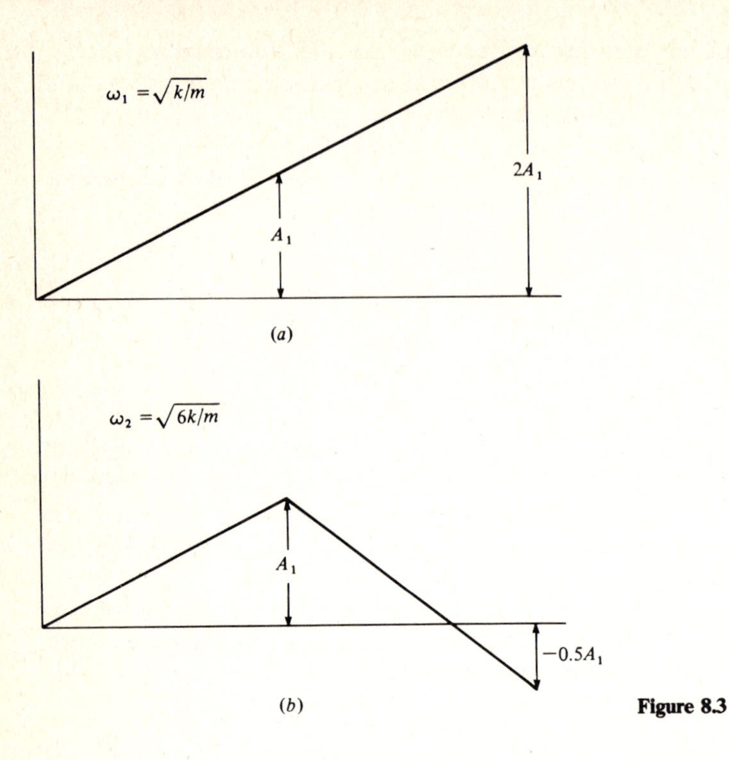

Figure 8.3

with A_1 and ϕ_2 determined by the initial conditions of the motion. In this mode, the masses move in opposite directions with the displacement of m_2 only half that of m_1.

The solutions given by Eqs. (8.15) and (8.16) are known as the *principal-mode* vibrations. The principal modes are depicted graphically by the *mode shapes* of Fig. 8.3. A mode shape is simply a plot of the amplitudes of each point in the system. Figure 8.3*b* shows that the second principal-mode vibration exhibits a point for which the displacement is zero at all times. Such a point is called a *node*.

8.3 GENERAL SOLUTION AND INITIAL CONDITIONS

While the principal-mode solutions obtained by the previous analysis may exist independently, they do not represent the general solution for vibrations of the two-degree-of-freedom system. To verify this statement, we need only note that the governing differential equations are linear and thus subject to the principle of superposition. Since the principal modes are independent solutions, the sum of the principal modes is also a solution. The general solutions for the displacement of each mass are then given by

$$x_1(t) = A_1^{(1)} \sin(\omega_1 t + \phi_1) + A_1^{(2)} \sin(\omega_2 t + \phi_2)$$

$$x_2(t) = A_2^{(1)} \sin(\omega_1 t + \phi_1) + A_2^{(2)} \sin(\omega_2 t + \phi_2) \tag{8.17}$$

Here the superscript in parentheses is used to indicate the frequency of the harmonic component. Thus $A_1^{(1)}$ is the amplitude of the harmonic component of x_1 having circular frequency ω_1, and so forth.

Close scrutiny in Eqs. (8.17) reveals the presence of six unknown constants: four amplitudes and two phase angles. From the principal-mode analysis, the amplitude equations can be used to relate $A_1^{(1)}$ to $A_2^{(1)}$ and $A_1^{(2)}$ to $A_2^{(2)}$. If we define the *amplitude ratios* as

$$\beta_1 = \frac{A_2^{(1)}}{A_1^{(1)}} \qquad \beta_2 = \frac{A_2^{(2)}}{A_1^{(2)}} \tag{8.18}$$

the general solution becomes

$$x_1(t) = A_1^{(1)} \sin(\omega_1 t + \phi_1) + A_1^{(2)} \sin(\omega_2 t + \phi_2)$$
$$x_2(t) = \beta_1 A_1^{(1)} \sin(\omega_1 t + \phi_1) + \beta_2 A_1^{(2)} \sin(\omega_2 t + \phi_2) \tag{8.19}$$

In this form, the solution contains four arbitrary constants $A_1^{(1)}$, $A_1^{(2)}$, ϕ_1, and ϕ_2, which can be evaluated by applying the initial conditions of displacement and velocity for each mass. Since the sum of harmonic components having different frequencies is not generally harmonic, the displacements given by Eqs. (8.19) represent complex waveforms, the shapes of which depend specifically upon the initial conditions. Thus, as opposed to one degree of freedom, simple harmonic motion is not likely to occur for a system with two degrees of freedom.

Example 8.2 Given the system of Fig. 8.1 with $m_1 = m_2 = m$, $k_1 = 3k$, and $k_2 = 2k$, determine the initial conditions necessary for the system to vibrate in its fundamental mode.

SOLUTION From Sec. 8.2, $\omega_1 = \sqrt{k/m}$, $\omega_2 = \sqrt{6k/m}$, $\beta_1 = 2$, and $\beta_2 = -0.5$, and the general solution is

$$x_1(t) = A_1^{(1)} \sin(\omega_1 t + \phi_1) + A_1^{(2)} \sin(\omega_2 t + \phi_2)$$
$$x_2(t) = 2A_1^{(1)} \sin(\omega_1 t + \phi_1) - 0.5A_1^{(2)} \sin(\omega_2 t + \phi_2)$$

For the fundamental mode to exist, we must have $A_1^{(2)} = 0$. The initial conditions can then be written

$$x_1(0) = A_1^{(1)} \sin \phi_1$$
$$x_2(0) = 2A_1^{(1)} \sin \phi_1$$
$$\dot{x}_1(0) = \omega_1 A_1^{(1)} \cos \phi_1$$
$$\dot{x}_2(0) = 2\omega_1 A_1^{(1)} \cos \phi_1$$

From the last equations, we observe that the fundamental mode will exist only if

$$x_2(0) = \beta_1 x_1(0)$$
$$\dot{x}_2(0) = \beta_1 \dot{x}_1(0)$$

Thus, for a principal mode to exist, the initial conditions for the masses must be related by the amplitude ratios. For this example, vibration will correspond to the second mode if

$$x_2(0) = -0.5x_1(0)$$
$$\dot{x}_2(0) = -0.5\dot{x}_1(0)$$

Example 8.3 Obtain the principal-mode solutions for the system shown in Fig. 8.4*a*.

SOLUTION Referring to the free-body diagrams in Fig. 8.4*b* and assuming $x_2 > x_1$ give

$$m\ddot{x}_1 = -kx_1 + k(x_2 - x_1)$$
$$m\ddot{x}_2 = -k(x_2 - x_1) - kx_2$$

as the differential equations of motion. Rearranging, we obtain

$$\ddot{x}_1 + \frac{2k}{m}x_1 - \frac{k}{m}x_2 = 0$$
$$\ddot{x}_2 + \frac{2k}{m}x_2 - \frac{k}{m}x_1 = 0$$

Assuming the solutions as $x_1(t) = A_1 \sin(\omega t + \phi)$ and $x_2(t) = A_2 \sin(\omega t + \phi)$ and substitution into the differential equations give

$$\left[\left(-\omega^2 + \frac{2k}{m}\right)A_1 - \frac{k}{m}A_2\right] \sin(\omega t + \phi) = 0$$
$$\left[\left(-\frac{k}{m}A_1 + \left(-\omega^2 + \frac{2k}{m}\right)A_2\right] \sin(\omega t + \phi) = 0$$

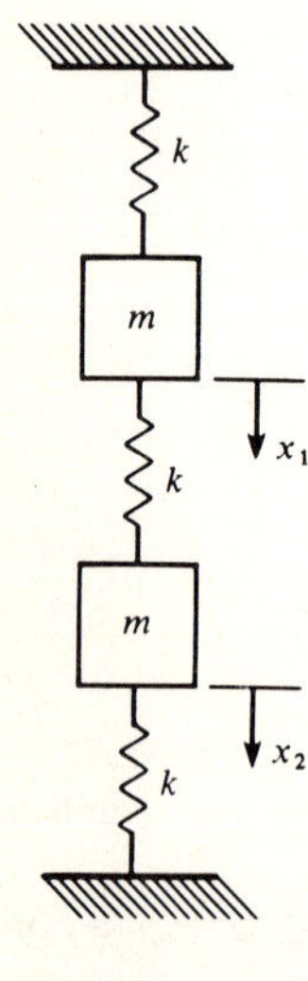

(*a*)

kx_1

m

$k(x_2 - x_1)$

$k(x_2 - x_1)$

m

kx_2

(*b*)

Figure 8.4

The characteristic determinant is then

$$\begin{vmatrix} -\omega^2 + \dfrac{2k}{m} & -\dfrac{k}{m} \\ -\dfrac{k}{m} & -\omega^2 + \dfrac{2k}{m} \end{vmatrix} = 0$$

which, when expanded and rearranged, results in the frequency equation

$$\omega^4 - 4\frac{k}{m}\omega^2 + 3\frac{k^2}{m^2} = 0$$

The frequency equation can be factored to obtain the natural circular frequencies as

$$\omega_1 = \sqrt{\frac{k}{m}} \qquad \omega_2 = \sqrt{3\frac{k}{m}}$$

The amplitude equations are

$$\left(-\omega^2 + 2\frac{k}{m}\right)A_1 - \frac{k}{m}A_2 = 0$$

$$-\frac{k}{m}A_1 + \left(-\omega^2 + 2\frac{k}{m}\right)A_2 = 0$$

Substituting the natural frequencies one at a time into the amplitude equations yields the amplitude ratios

$$\beta_1 = \frac{A_2}{A_1} = 1 \qquad \text{for } \omega_1 = \sqrt{\frac{k}{m}}$$

$$\beta_2 = \frac{A_2}{A_1} = -1 \qquad \text{for } \omega_2 = \sqrt{3\frac{k}{m}}$$

The fundamental mode is described by

$$x_1(t) = A_1 \sin(\omega_1 t + \phi_1)$$

$$x_2(t) = A_1 \sin(\omega_1 t + \phi_1)$$

while the second principal mode is

$$x_1(t) = A_1 \sin(\omega_2 t + \phi_2)$$

$$x_2(t) = -A_1 \sin(\omega_2 t + \phi_2)$$

Example 8.4 Using the data of Example 8.3 with $m = 0.05\ \text{lb} \cdot \text{s}^2/\text{in}$ and $k = 20\ \text{lb/in}$, determine $x_1(t)$ and $x_2(t)$ if the initial conditions are $x_1(0) = 1$ in, $x_2(0) = -2$ in, and $\dot{x}_1(0) = \dot{x}_2(0) = 0$.

SOLUTION For the principal-mode solutions obtained in Example 8.3, the natural circular frequencies are

$$\omega_1 = \sqrt{\frac{20}{0.05}} = 20 \text{ rad/s} \qquad \omega_2 = \sqrt{3\frac{20}{0.05}} = 34.64 \text{ rad/s}$$

The general solution is

$$x_1(t) = A_1^{(1)} \sin(20t + \phi_1) + A_1^{(2)} \sin(34.64t + \phi_2)$$
$$x_2(t) = A_1^{(1)} \sin(20t + \phi_1) - A_1^{(2)} \sin(34.64t + \phi_2)$$

Applying the given initial conditions, we obtain

$$x_1(0) = 1 = A_1^{(1)} \sin \phi_1 + A_1^{(2)} \sin \phi_2$$
$$x_2(0) = -2 = A_1^{(1)} \sin \phi_1 - A_1^{(2)} \sin \phi_2$$
$$\dot{x}_1(0) = 0 = 20A_1^{(1)} \cos \phi_1 + 34.64A_1^{(2)} \cos \phi_2$$
$$\dot{x}_2(0) = 0 = 20A_1^{(1)} \cos \phi_1 - 34.64A_1^{(2)} \cos \phi_2$$

Simultaneous solution gives $A_1^{(1)} = -0.5$, $A_1^{(2)} = 1.5$, and $\phi_1 = \phi_2 = \pi/2$. Using the relation $\sin(\omega t + \pi/2) = \cos \omega t$, we write the displacements as

$$x_1(t) = -0.5 \cos 20t + 1.5 \cos 34.64t$$
$$x_2(t) = -0.5 \cos 20t - 1.5 \cos 34.64t$$

The displacements are as shown in Figs. 8.5, which illustrate the complex nature of the waveforms.

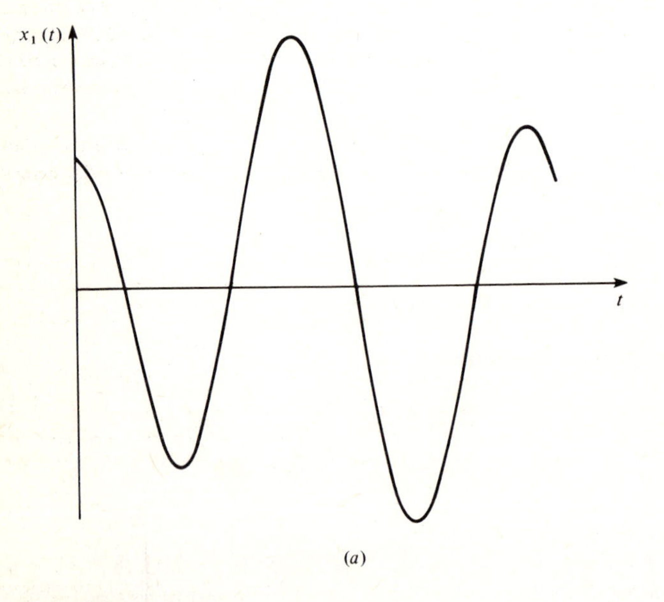

(a)

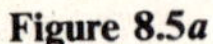

Figure 8.5a

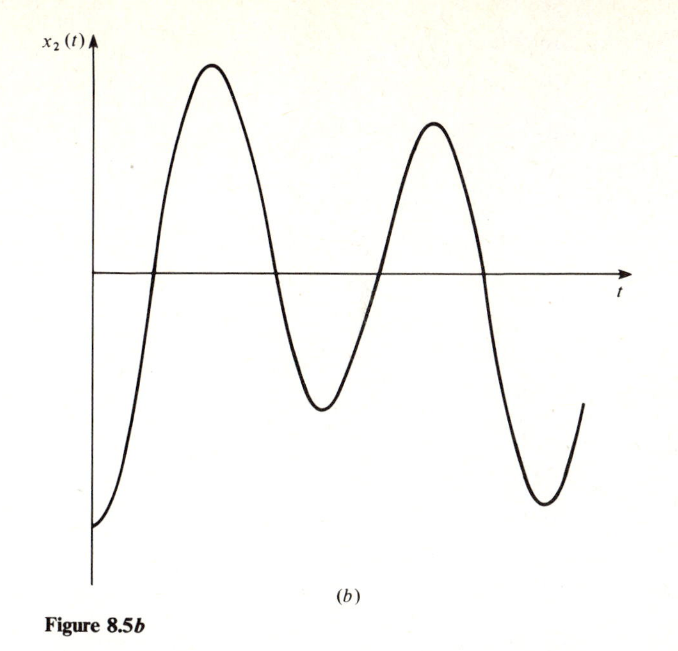

Figure 8.5b

8.4 COUPLING; PRINCIPAL COORDINATES

Let us refer again to the differential equations of motion for the system of Fig. 8.1, which are rewritten here for convenience as

$$\ddot{x}_1 + \frac{k_1 + k_2}{m_1} x_1 - \frac{k_2}{m_1} x_2 = 0$$

$$\ddot{x}_2 + \frac{k_2}{m_2} x_2 - \frac{k_2}{m_2} x_1 = 0 \tag{8.20}$$

We have already observed that Eqs. (8.20) are elastically coupled through k_2. We have also seen the additional complexity introduced by the corresponding requirement that we solve the differential equations simultaneously.

We shall now examine the possibility of *uncoupling* the differential equations by some suitable coordinate transformation. To uncouple Eqs. (8.20), we seek new coordinates

$$q_1 = q_1(x_1, x_2)$$

$$q_2 = q_2(x_1, x_2) \tag{8.21}$$

such that Eqs. (8.20) can be transformed into

$$\ddot{q}_1 + \omega_1^2 q_1 = 0$$

$$\ddot{q}_2 + \omega_2^2 q_2 = 0 \tag{8.22}$$

The advantage afforded by such a procedure should be obvious, since Eqs.

(8.22) are independent and the solutions can be written immediately as

$$q_1(t) = A_1 \sin(\omega_1 t + \phi_1)$$
$$q_2(t) = A_2 \sin(\omega_2 t + \phi_2) \tag{8.23}$$

Although the mathematical details of the proof are too complex for inclusion here, we can state that a *linear* transformation

$$q_1 = a_{11}x_1 + a_{12}x_2$$
$$q_2 = a_{21}x_1 + a_{22}x_2 \tag{8.24}$$

does indeed exist and can be used to uncouple Eqs. (8.20). To illustrate, let us again consider the system for which $m_1 = m_2 = m$, $k_1 = 3k$, and $k_2 = 2k$. Rewriting Eqs. (8.24) in the inverted form

$$x_1 = b_{11}q_1 + b_{12}q_2$$
$$x_2 = b_{21}q_1 + b_{22}q_2 \tag{8.25}$$

and differentiating twice, we obtain

$$\ddot{x}_1 = b_{11}\ddot{q}_1 + b_{12}\ddot{q}_2$$
$$\ddot{x}_2 = b_{21}\ddot{q}_1 + b_{22}\ddot{q}_2 \tag{8.26}$$

Substitution of Eqs. (8.25) and (8.26) as well as the assumed values of mass and spring stiffness into Eqs. (8.20) gives the transformed differential equations as

$$b_{11}\ddot{q}_1 + b_{12}\ddot{q}_2 + \frac{5k}{m}(b_{11}q_1 + b_{12}q_2) - \frac{2k}{m}(b_{21}q_1 + b_{22}q_2) = 0$$
$$b_{21}\ddot{q}_1 + b_{22}\ddot{q}_2 + \frac{2k}{m}(b_{21}q_1 + b_{22}q_2) - \frac{2k}{m}(b_{11}q_1 + b_{12}q_2) = 0 \tag{8.27}$$

Combining Eqs. (8.27) to eliminate $\ddot{q}_2$ gives

$$(b_{11}b_{22} - b_{21}b_{12})\ddot{q}_1 + \frac{k}{m}(5b_{11}b_{22} - 2b_{21}b_{22} - 2b_{21}b_{12} + 2b_{11}b_{12})q_1$$
$$+ \frac{k}{m}(3b_{12}b_{22} - 2b_{22}^2 + 2b_{12}^2)q_2 = 0 \tag{8.28}$$

while elimination of $\ddot{q}_1$ yields

$$(b_{12}b_{21} - b_{22}b_{11})\ddot{q}_2 + \frac{k}{m}(5b_{12}b_{21} - 2b_{22}b_{21} - 2b_{22}b_{11} + 2b_{12}b_{11})q_2$$
$$+ \frac{k}{m}(3b_{11}b_{21} - 2b_{21}^2 + 2b_{11}^2)q_1 = 0 \tag{8.29}$$

Comparison of Eqs. (8.28) and (8.29) with Eqs. (8.22) shows that the desired form will be obtained if

$$3b_{12}b_{22} - 2b_{22}^2 + 2b_{12}^2 = 0$$
$$3b_{11}b_{21} - 2b_{21}^1 + 2b_{11}^2 = 0 \tag{8.30}$$

and
$$b_{11}b_{22} - b_{21}b_{12} \neq 0 \tag{8.31}$$

Equations (8.30) contain too many unknowns for explicit solution; we can, however, treat them as quadratic equations in the unknowns b_{22}/b_{12} and b_{21}/b_{11}, respectively, and solve by the quadratic formula. Proceeding in this manner, we find that the only values which satisfy Eqs. (8.30) *and* the inequality (8.31) are

$$\frac{b_{21}}{b_{11}} = 2 \qquad \frac{b_{22}}{b_{12}} = -0.5 \tag{8.32}$$

Thus, two of the coefficients are arbitrary. Setting $b_{11} = b_{12} = 1$, we obtain $b_{21} = 2$ and $b_{22} = -0.5$, and the desired transformation is given by

$$x_1 = q_1 + q_2$$
$$x_2 = 2q_1 - 0.5q_2 \tag{8.33}$$

Substituting the b values into Eqs. (8.28) and (8.29) gives the transformed differential equations of motion as

$$\ddot{q}_1 + \frac{k}{m}q_1 = 0$$
$$\ddot{q}_2 + 6\frac{k}{m}q_2 = 0 \tag{8.34}$$

which by inspection give $\omega_1 = \sqrt{k/m}$ and $\omega_2 = \sqrt{6k/m}$ as the natural circular frequencies of the system.

The coordinates q_1 and q_2 for which the equations of motion of the system are independent are known as *principal coordinates* or *natural coordinates*. While the mathematics involved in determining the principal coordinates is somewhat tedious, principal coordinates will prove quite useful in obtaining the forced response of systems with several degrees of freedom. In this context, we note that the ratios b_{21}/b_{11} and b_{22}/b_{12} obtained for the previous system are exactly the same as the amplitude ratios β_1 and β_2 obtained by using the direct solution procedure. This is not coincidental and will be used to advantage later on.

In addition to elastic coupling, a system may possess *inertial coupling*. This exists when the equations of motion are coupled through the acceleration terms. Such systems may also be uncoupled via the principal coordinates.

Example 8.5 A nonuniform bar having mass moment of inertia I_G about its center of mass G is mounted on linear springs k_1 and k_2 as shown in Fig. 8.6*a*. Determine (*a*) the equations of motion for the system, (*b*) the natural frequencies, (*c*) the amplitude ratios, and (*d*) the principal coordinates.

SOLUTION The free-body diagram of the bar in a displaced position is shown in Fig. 8.6*b*. If we eliminate the weight of the bar and the equilibrium spring deflections, the motion of the center of mass is given by Newton's second law as

$$m\ddot{x} = -k_1(x - a\theta) - k_2(x + b\theta)$$

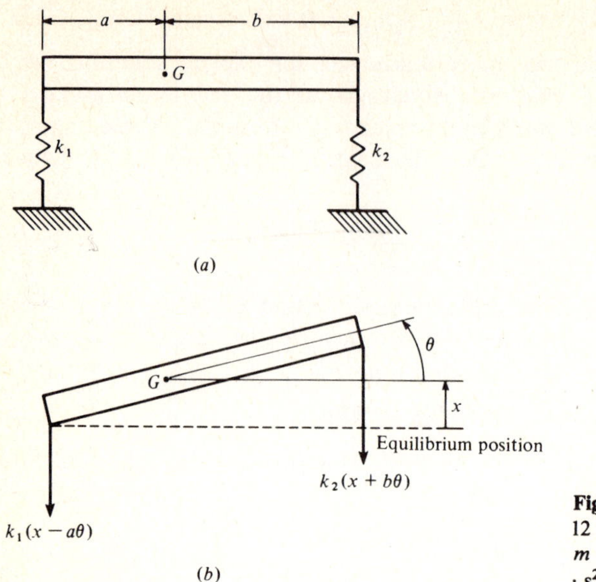

Figure 8.6 $k_1 = 10$ lb/in, $k_2 = 12$ lb/in, $a = 12$ in, $b = 8$ in, $m = 0.16$ lb · s²/in, $I_G = 20$ lb · s² · in

The equation governing rotation is obtained by summing moments about the mass center and is given by

$$I_G\ddot{\theta} = k_1(x - a\theta)a - k_2(x + b\theta)b$$

where we have assumed small angular oscillations such that $\sin\theta \simeq \theta$. After rearranging, the differential equations of motion become

$$\ddot{x} + \frac{k_1 + k_2}{m}x + \frac{k_2 b - k_1 a}{m}\theta = 0$$

$$\ddot{\theta} + \frac{k_1 a^2 + k_2 b^2}{I_G}\theta + \frac{k_2 b - k_1 a}{I_G}x = 0$$

These equations are seen to be elastically coupled.

We assume solutions of the form

$$x(t) = A_1 \sin(\omega t + \phi)$$
$$\theta(t) = A_2 \sin(\omega t + \phi)$$

Differentiation and substitution into the equations of motion yield

$$\left(\frac{k_1 + k_2}{m} - \omega^2\right)A_1 + \frac{k_2 b - k_1 a}{m}A_2 = 0$$

$$\frac{k_2 b - k_1 a}{I_G}A_1 + \left(\frac{k_1 a^2 + k_2 b^2}{I_G} - \omega^2\right)A_2 = 0$$

as the amplitude equations. With numerical values substituted, these become

$$(137.5 - \omega^2)A_1 - 150A_2 = 0$$
$$-1.2A_1 + (110.4 - \omega^2)A_2 = 0$$

The natural frequencies are given by the characteristic determinant

$$\begin{vmatrix} 137.5 - \omega^2 & -150 \\ -1.2 & 110.4 - \omega^2 \end{vmatrix} = 0$$

Expanding the determinant, we obtain the frequency equation

$$\omega^4 - 247.9\omega^2 + 15{,}000 = 0$$

which has positive roots $\omega_1 = 10.24$ rad/s and $\omega_2 = 11.96$ rad/s. Substitution of ω_1 into either amplitude equation gives

$$\beta_1 = \frac{A_2}{A_1} = 0.218$$

while substitution of ω_2 gives

$$\beta_2 = \frac{A_2}{A_1} = -0.037$$

The general solutions are then

$$x(t) = A_1^{(1)} \sin(10.24t + \phi_1) + A_1^{(2)} \sin(11.96t + \phi_2)$$
$$\theta(t) = 0.218A_1^{(1)} \sin(10.24t + \phi_1) - 0.037A_1^{(2)} \sin(11.96t + \phi_2)$$

To determine the principal coordinates, we assume the relations

$$x = q_1 + q_2$$
$$\theta = \beta_1 q_1 + \beta_2 q_2 = 0.218q_1 - 0.037q_2$$

and substitute into the original differential equations to obtain

$$\ddot{q}_1 + \ddot{q}_2 + 137.5q_1 + 137.5q_2 - 32.7q_1 + 5.55q_2 = 0$$
$$0.218\ddot{q}_1 - 0.037\ddot{q}_2 + 24.07q_1 - 3.97q_2 - 1.2q_1 - 1.2q_2 = 0$$

or

$$\ddot{q}_1 + \ddot{q}_2 + 104.8q_1 + 143.05q_2 = 0$$
$$0.218\ddot{q}_1 - 0.037\ddot{q}_2 + 22.87q_1 - 5.17q_2 = 0$$

Successive elimination of $\ddot{q}_2$ and $\ddot{q}_1$ yields

$$\ddot{q}_1 + 104.86q_1 = 0$$

and

$$\ddot{q}_2 + 143.04q_2 = 0$$

Since these equations are uncoupled, the assumed relations are correct, and the principal coordinates are

$$q_1 = 0.145x + 3.92\theta$$
$$q_2 = 0.855x - 3.92\theta$$

8.5 DAMPED FREE VIBRATION

Let us now consider the free-vibration response of the viscously damped system shown in Fig. 8.7*a*. Choosing to measure the displacement of each mass from its equilibrium position, we draw free-body diagrams as in Fig. 8.7*b*. For convenience, $x_1(t)$ and $x_2(t)$ are assumed positive, with $x_2 > x_1$ and $\dot{x}_2 > \dot{x}_1$. Writing Newton's second law for each mass, we obtain

$$\begin{aligned} m_1\ddot{x}_1 &= -k_1x_1 - c_1\dot{x}_1 + k_2(x_2 - x_1) + c_2(\dot{x}_2 - \dot{x}_1) \\ m_2\ddot{x}_2 &= -k_2(x_2 - x_1) - c_2(\dot{x}_2 - \dot{x}_1) \end{aligned} \tag{8.35}$$

which can be rewritten as

$$\begin{aligned} m\ddot{x}_1 + (c_1 + c_2)\dot{x}_1 + (k_1 + k_2)x_1 - k_2x_2 - c_2\dot{x}_2 &= 0 \\ m\ddot{x}_2 + c_2\dot{x}_2 + k_2x_2 - k_2x_1 - c_2\dot{x}_1 &= 0 \end{aligned} \tag{8.36}$$

Equations (8.36) are the differential equations of motion for the system and are observed to be elastically coupled and *viscously coupled* through the damping terms.

At this point we introduce the matrix equivalent of Eqs. (8.36) for mathematical convenience and as an introduction to methods used later for digital computation purposes. We define the *mass matrix*

$$[m] = \begin{bmatrix} m_1 & 0 \\ 0 & m_2 \end{bmatrix} = \begin{bmatrix} m_{11} & m_{12} \\ m_{21} & m_{22} \end{bmatrix} \tag{8.37}$$

the *damping matrix*

$$[c] = \begin{bmatrix} c_1 + c_2 & -c_2 \\ -c_2 & c_2 \end{bmatrix} = \begin{bmatrix} c_{11} & c_{12} \\ c_{21} & c_{22} \end{bmatrix} \tag{8.38}$$

the *stiffness matrix*

$$[k] = \begin{bmatrix} k_1 + k_2 & -k_2 \\ -k_2 & k_2 \end{bmatrix} = \begin{bmatrix} k_{11} & k_{12} \\ k_{21} & k_{22} \end{bmatrix} \tag{8.39}$$

and the *displacement matrix* or *displacement vector*

$$\{x(t)\} = \begin{Bmatrix} x_1(t) \\ x_2(t) \end{Bmatrix} \tag{8.40}$$

Equations (8.36) can now be written in the compact form

$$[m]\{\ddot{x}\} + [c]\{\dot{x}\} + [k]\{x\} = 0 \tag{8.41}$$

Note that, for the example at hand, the mass, damping, and stiffness matrices are each symmetric about the main diagonal.

Considering our experience with damped single-degree-of-freedom systems, we assume solutions of the form

$$\begin{aligned} x_1(t) &= C_1e^{st} \\ x_2(t) &= C_2e^{st} \end{aligned} \tag{8.42}$$

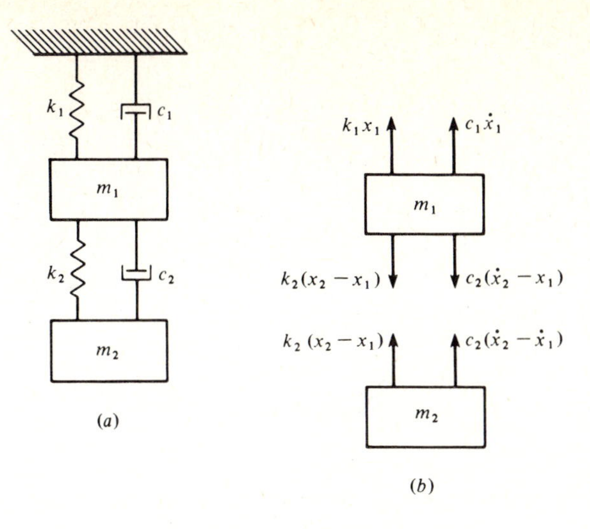

Figure 8.7

or

$$\{x(t)\} = \begin{Bmatrix} C_1 \\ C_2 \end{Bmatrix} e^{st} \tag{8.43}$$

Differentiating gives the velocity and acceleration vectors as

$$\{\dot{x}(t)\} = \{sC\}e^{st} \tag{8.44}$$

and

$$\{\ddot{x}(t)\} = \{s^2C\}e^{st} \tag{8.45}$$

respectively. Substituting Eqs. (8.43) through (8.45) into Eq. (8.41), we obtain

$$[m]\{s^2C\}e^{st} + [c]\{sC\}e^{st} + [k]\{C\}e^{st} = 0 \tag{8.46}$$

Since e^{st} is never zero, we must solve the matrix equation

$$[m]\{s^2C\} + [c]\{sC\} + [k]\{C\} = 0 \tag{8.47}$$

which is equivalent to two linear algebraic equations in the two unknowns C_1 and C_2. As Eq. (8.47) is homogeneous, nontrivial solutions exist only if the determinant of the coefficient matrix vanishes. Equation (8.47) can be written as

$$[D]\{C\} = \begin{bmatrix} d_{11} & d_{12} \\ d_{21} & d_{22} \end{bmatrix} \begin{Bmatrix} C_1 \\ C_2 \end{Bmatrix} = 0 \tag{8.48}$$

where

$$d_{11} = m_1 s^2 + (c_1 + c_2)s + k_1 + k_2$$

$$d_{12} = c_2 s - k_2$$

$$d_{21} = c_2 s - k_2$$

$$d_{22} = m_2 s^2 + c_2 s + k_2$$

Since the determinant of $[D]$ must be zero, this is perfectly analogous with the characteristic determinant discussed in the solution for the undamped system.

Expanding the determinant gives

$$m_1 m_2 s^4 + [m_1 c_2 + m_2(c_1 + c_2)]s^3 + [m_1 k_2 + m_2(k_1 + k_2) + c_1 c_2]s^2 + (k_1 c_2 + k_2 c_1)s + k_1 k_2 = 0 \tag{8.49}$$

Equation (8.49) is of fourth degree in s and is analogous to the frequency equation obtained for the undamped system. Given their physical nature, all the coefficients in the equation are positive; an analysis of sign changes reveals that there are three possibilities regarding the four roots:

1. All four roots can be real and negative.
2. The four roots can occur as two pairs of complex conjugates having negative real parts.
3. Two roots can be real and negative while the other two roots are complex conjugates.

Before discussing each possibility, let us note that corresponding to each of the four roots there exists a solution to Eq. (8.48) in the form $\beta_i = C_2/C_1$ for $i = 1, 2, 3, 4$. Formally, we shall obtain four amplitude ratios β_i and four solutions for each of $x_1(t)$ and $x_2(t)$. The total of eight solutions will contain four arbitrary constants which can be evaluated by applying the initial conditions of the system.

If the roots of Eq. (8.49) are all real and negative, each displacement solution obtained is of decaying exponential form. As in the undamped case, the complete solution is the sum of the independent solutions, so the displacements will be given by

$$x_1(t) = C_1^{(1)} e^{s_1 t} + C_1^{(2)} e^{s_2 t} + C_1^{(3)} e^{s_3 t} + C_1^{(4)} e^{s_4 t}$$

$$x_2(t) = \beta_1 C_1^{(1)} e^{s_1 t} + \beta_2 C_1^{(2)} e^{s_2 t} + \beta_3 C_1^{(3)} e^{s_3 t} + \beta_4 C_1^{(4)} e^{s_4 t} \tag{8.50}$$

where $\beta_i = C_2^{(i)}/C_1^{(i)}$ is obtained by substituting root s_i into either of the equations represented by (8.48). The displacements exhibit no oscillation, and the response is similar to an overdamped simple harmonic oscillator.

If the roots occur as pairs of complex conjugates, they can be expressed as

$$s_1 = -a_1 + ib_1$$

$$s_2 = -a_1 - ib_1$$

$$s_3 = -a_2 + ib_2$$

$$s_4 = -a_2 - ib_2 \tag{8.51}$$

If s_1 is substituted into either amplitude equation, we obtain the amplitude ratio $\beta_1 = C_2^{(1)}/C_1^{(1)}$, and the corresponding displacements can be written as

$$x_1(t) = C_1^{(1)} e^{(-a_1 + ib_1)t}$$

$$x_2(t) = \beta_1 C_1^{(1)} e^{(-a_1 + ib_1)t} \tag{8.52}$$

Similarly, substituting s_2 gives

$$x_1(t) = C_1^{(2)} e^{(-a_1 - ib_1)t}$$
$$x_2(t) = \beta_2 C_1^{(2)} e^{(-a_1 - ib_1)t} \tag{8.53}$$

Since Eqs. (8.52) and (8.53) represent independent solutions to the governing differential equations (which are linear), the sum of these is also a solution. Summing and applying the Euler equations as in Sec. 2.1, we obtain

$$x_1(t) = e^{-a_1 t}\left[\left(C_1^{(1)} + C_1^{(2)}\right)\cos b_1 t + i\left(C_1^{(1)} - C_1^{(2)}\right)\sin b_1 t\right] \tag{8.54}$$
$$x_2(t) = e^{-a_1 t}\left[\left(\beta_1 C_1^{(1)} + \beta_2 C_1^{(2)}\right)\cos b_1 t + i\left(\beta_1 C_1^{(1)} - \beta_2 C_1^{(2)}\right)\sin b_1 t\right]$$

It can be shown that all coefficients of sine and cosine terms in Eqs. (8.54) are real, so that the displacements represented by the solutions are exponentially decaying harmonic oscillations. The response is similar to that of an underdamped single-degree-of-freedom system. The remaining roots s_3 and s_4 will result in displacement solutions similar to Eqs. (8.54) and will contain two unknown constants $C_1^{(3)}$ and $C_1^{(4)}$. Thus the complete solution contains four constants, which are again determined by the initial conditions.

In the final case, the roots will be of the form

$$s_1 = -a$$
$$s_2 = -b$$
$$s_3 = -c + id$$
$$s_4 = -c - id \tag{8.55}$$

The displacement solutions are obtained by a combination of the methods used for the first two cases and will be of the form

$$x_1(t) = C_1^{(1)} e^{-at} + C_1^{(2)} e^{-bt}$$
$$+ e^{-ct}\left[\left(C_1^{(3)} + C_1^{(4)}\right)\cos dt + i\left(C_1^{(3)} - C_1^{(4)}\right)\sin dt\right] \tag{8.56}$$
$$x_2(t) = \beta_1 C_1^{(1)} e^{-at} + \beta_2 C_1^{(2)} e^{-bt}$$
$$+ e^{-ct}\left[\left(\beta_3 C_1^{(3)} + \beta_4 C_1^{(4)}\right)\cos dt + i\left(\beta_3 C_1^{(3)} - \beta_4 C_1^{(4)}\right)\sin dt\right]$$

Thus the displacements are composed of decaying exponentials and decaying oscillations. The exact composition of the solutions depends upon the initial conditions. In any case, the motion eventually ceases as is expected for a damped system.

We have discussed the possible forms of the displacement solutions for the damped two-degree-of-freedom system, but we have not mentioned methods of finding the roots of Eq. (8.49). This is a quartic equation with all coefficients real and positive. The solution method is by no means simple, but it is straightforward, and the details are given in App. A.8. In addition, a FORTRAN program for finding the roots is included in App. B.

Example 8.6 Determine the displacements $x_1(t)$ and $x_2(t)$ for the system of Fig. 8.7*a* given the following data: $m_1 = m_2 = 0.1 \text{ lb} \cdot \text{s}^2/\text{in}$, $k_1 = k_2 = 20$

lb/in, $c_1 = 1.0$ lb · s/in, $c_2 = 0$. The initial conditions are $x_1(0) = 1$ in, $x_2(0) = 2$ in, $\dot{x}_1(0) = \dot{x}_2(0) = 0$.

SOLUTION Substituting the given values into Eq. (8.49), we obtain

$$0.01s^4 + 0.1s^3 + 6s^2 + 20s + 400 = 0$$

or

$$s^4 + 10s^3 + 600s^2 + 2000s + 40{,}000 = 0$$

Utilizing the procedure given in App. A.8, we find that the roots are approximately

$$s_1 = -3.495 + 8.245i$$
$$s_2 = -3.495 - 8.245i$$
$$s_3 = -1.505 + 22.28i$$
$$s_4 = -1.505 - 22.28i$$

To find the amplitude ratios, we substitute the known physical constants into Eqs. (8.48) to obtain the amplitude equations

$$(0.1s^2 + s + 40)C_1 - 20C_2 = 0$$

and

$$-20C_1 + (0.1s^2 + 20)C_2 = 0$$

From the first of these equations, the amplitude ratios are given by

$$\beta_i = \frac{C_2^{(i)}}{C_1^{(i)}} = \frac{0.1s_i^2 + s_i + 40}{20} \qquad i = 1, 2, 3, 4$$

Substituting each of the four roots in turn gives

$$\beta_1 = 1.546 + 0.124i$$
$$\beta_2 = 1.546 - 0.124i$$
$$\beta_3 = -10.92 + 15.57i$$
$$\beta_4 = -10.92 - 15.57i$$

from which we note that, in this case, the amplitude ratios occur as pairs of complex conjugates also. To illustrate the importance of this fact, let us write

$$\beta_1 = a + bi \qquad \beta_2 = a - bi$$

and

$$C_2^{(1)} = \beta_1 C_1^{(1)} = (a + bi)C_1^{(1)}$$
$$C_2^{(2)} = \beta_2 C_1^{(2)} = (a - bi)C_1^{(2)}$$

The last two equations will hold for real, nonzero values of a and b only if $C_1^{(1)}$ and $C_1^{(2)}$ are complex conjugates, as are $C_2^{(1)}$ and $C_2^{(2)}$.

The displacement solutions are of the form of Eqs. (8.54) and can be written as

$$x_1(t) = e^{-3.495t}\left[\left(C_1^{(1)} + C_1^{(2)}\right)\cos 8.245t + i\left(C_1^{(1)} - C_1^{(2)}\right)\sin 8.245t\right]$$
$$+ e^{-1.505t}\left[\left(C_1^{(3)} + C_1^{(4)}\right)\cos 22.28t + i\left(C_1^{(3)} - C_1^{(4)}\right)\sin 22.28t\right]$$

$$x_2(t) = e^{-3.495t}\left[\left(\beta_1 C_1^{(1)} + \beta_2 C_1^{(2)}\right)\cos 8.245t\right.$$
$$\left. + i\left(\beta_1 C_1^{(1)} - \beta_2 C_1^{(2)}\right)\sin 8.245t\right]$$
$$+ e^{-1.505t}\left[\left(\beta_3 C_1^{(3)} + \beta_4 C_1^{(4)}\right)\cos 22.28t\right.$$
$$\left. + i\left(\beta_3 C_1^{(3)} - \beta_4 C_1^{(4)}\right)\sin 22.28t\right]$$

Before we attempt to apply the initial conditions, some additional simplification is required. Using the known conjugate relations, we have

$$C_1^{(1)} + C_1^{(2)} = A_1$$
$$i(C_1^{(1)} - C_1^{(2)}) = B_1$$
$$C_1^{(3)} + C_1^{(4)} = A_2$$
$$i(C_1^{(3)} - C_1^{(4)}) = B_2$$
$$\beta_1 C_1^{(1)} + \beta_2 C_1^{(2)} = 1.546A_1 - 0.124B_1$$
$$i(\beta_1 C_1^{(1)} - \beta_2 C_1^{(2)}) = -0.124A_1 - 1.546B_1$$
$$\beta_3 C_1^{(3)} + \beta_4 C_1^{(4)} = -10.92A_2 - 15.57B_2$$
$$i(\beta_3 C_1^{(3)} - \beta_4 C_1^{(4)}) = -15.57A_2 + 10.92B_2$$

where $C_1^{(1)} = \frac{1}{2}(A_1 + iB_1)$ and $C_1^{(3)} = \frac{1}{2}(A_2 + iB_2)$. The initial conditions are now applied to obtain

$$x_1(0) = 1 = A_1 + A_2$$
$$x_2(0) = 2 = 1.546A_1 - 0.124B_1 - 10.92A_2 - 15.57B_2$$
$$\dot{x}_1(0) = 0 = -3.495A_1 + 8.245B_1 - 1.505A_2 + 22.28B_2$$
$$\dot{x}_2(0) = 0 = -6.426A_1 - 12.31B_1 - 330.5A_2 + 266.7B_2$$

Simultaneous solution gives

$$A_1 = 1.038 \qquad A_2 = -0.038$$
$$B_1 = 0.439 \qquad B_2 = -0.002$$

and the displacements are then described by

$$x_1(t) = e^{-3.495t}(1.038\cos 8.245t + 0.439\sin 8.245t)$$
$$- e^{-1.505t}(0.038\cos 22.28t + 0.002\sin 22.28t)$$
$$x_2(t) = e^{-3.495t}(1.550\cos 8.245t - 0.807\sin 8.245t)$$
$$+ e^{-1.505t}(0.450\cos 22.28t + 0.570\sin 22.28t)$$

It may be a gross understatement to say that the preceding example involves considerable calculation. The procedures discussed in Chaps. 9 and 10 for solving such problems via analog and digital computer, respectively, will reduce the computations enormously; the power of these methods will be much more appreciated after similar problems have been solved by direct calculation.

8.6 FORCED VIBRATION OF UNDAMPED SYSTEMS

Having examined the free response of undamped and damped two-degree-of-freedom systems, we now turn our attention to the response of undamped systems to harmonic forcing functions. The method of analysis will be discussed with reference to the system shown in Fig. 8.8*a*, in which the harmonic forces

$$\begin{aligned} F_1(t) &= F_1 \sin \omega_f t \\ F_2(t) &= F_2 \sin \omega_f t \end{aligned} \tag{8.57}$$

are applied to masses m_1 and m_2, respectively. From the dynamic free-body diagrams of Fig. 8.8*b*, the equations of motion are

$$\begin{aligned} m_1\ddot{x}_1 + (k_1 + k_2)x_1 - k_2x_2 &= F_1 \sin \omega_f t \\ m_2\ddot{x}_2 + k_2x_2 - k_2x_1 &= F_2 \sin \omega_f t \end{aligned} \tag{8.58}$$

which, as expected, are coupled and inhomogeneous. Recalling our experience with the harmonic response of undamped systems with one degree of freedom, we assume that the response of each mass will have a form similar to that of the forcing functions, and we write

$$\begin{aligned} x_1(t) &= X_1 \sin \omega_f t \\ x_2(t) &= X_2 \sin \omega_f t \end{aligned} \tag{8.59}$$

Differentiating and substituting into Eqs. (8.58), we obtain

$$\begin{aligned} \left(-m_1\omega_f^2 + k_1 + k_2\right)X_1 \sin \omega_f t - k_2X_2 \sin \omega_f t &= F_1 \sin \omega_f t \\ \left(-m_2\omega_f^2 + k_2\right)X_2 \sin \omega_f t - k_2X_1 \sin \omega_f t &= F_2 \sin \omega_f t \end{aligned} \tag{8.60}$$

which can be written in matrix form as

$$\begin{bmatrix} k_1 + k_2 - m_1\omega_f^2 & -k_2 \\ -k_2 & k_2 - m_2\omega_f^2 \end{bmatrix} \begin{Bmatrix} X_1 \\ X_2 \end{Bmatrix} = \begin{Bmatrix} F_1 \\ F_2 \end{Bmatrix} \tag{8.61}$$

The assumed solutions will be verified as correct if Eq. (8.61) can be solved for the amplitudes X_1 and X_2. Rewriting the last equation as

$$\begin{bmatrix} d_{11} & d_{12} \\ d_{21} & d_{22} \end{bmatrix} \begin{Bmatrix} X_1 \\ X_2 \end{Bmatrix} = \begin{Bmatrix} F_1 \\ F_2 \end{Bmatrix} \tag{8.62}$$

we apply Cramer's rule to obtain

$$X_1 = \frac{d_{22}F_1 - d_{12}F_2}{d_{11}d_{22} - d_{12}d_{21}} \qquad X_2 = \frac{d_{11}F_2 - d_{21}F_1}{d_{11}d_{22} - d_{12}d_{21}} \tag{8.63}$$

which are the desired amplitudes provided $d_{11}d_{22} - d_{12}d_{21} \neq 0$. Recall from Sec. 8.1 that the condition $d_{11}d_{22} - d_{12}d_{21} = 0$ defines the natural frequencies of the system ω_1 and ω_2. Thus the amplitudes given by Eqs. (8.63) are valid only if the frequency of the forcing function is different from both ω_1 and ω_2. If $\omega_f = \omega_1$ or $\omega_f = \omega_2$, the denominator vanishes, and we observe that the two-degree-of-freedom system possesses two resonant frequencies. We also note that each mass will exhibit resonance even if the resonant force acts on only one mass. This is clear from Eqs. (8.63) and occurs because the motions are coupled.

Example 8.7 Determine the steady-state response of the system shown in Fig. 8.8 if $m_1 = m_2 = m$, $k_1 = 3k$, $k_2 = 2k$, and $F_2 = 0$.

SOLUTION The equations of motion are

$$m\ddot{x}_1 + 5kx_1 - 2kx_2 = F_1 \sin \omega_f t$$

$$m\ddot{x}_2 + 2kx_2 - 2kx_1 = 0$$

If we assume harmonic solutions as per Eqs. (8.59), the matrix form is

$$\begin{bmatrix} 5k - m\omega_f^2 & -2k \\ -2k & 2k - m\omega_f^2 \end{bmatrix} \begin{Bmatrix} X_1 \\ X_2 \end{Bmatrix} = \begin{Bmatrix} F_1 \\ 0 \end{Bmatrix}$$

After expanding and simplifying, the determinant of the coefficient matrix is

$$d_{11}d_{22} - d_{12}d_{21} = \left(\omega_f^4 - 7\omega_f^2 \frac{k}{m} + 6\frac{k^2}{m^2}\right)m^2$$

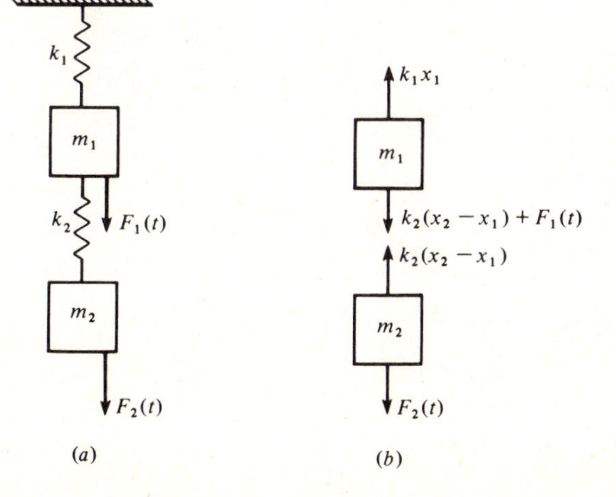

Figure 8.8

which can be factored to obtain

$$d_{11}d_{22} - d_{12}d_{21} = \left(\omega_f^2 - \frac{k}{m}\right)\left(\omega_f^2 - 6\frac{k}{m}\right)m^2$$

$$= \left(\omega_f^2 - \omega_1^2\right)\left(\omega_f^2 - \omega_2^2\right)m^2$$

where $\omega_1^2 = k/m$ and $\omega_2^2 = 6k/m$ are the squares of the system natural frequencies.

The steady-state amplitudes are found, using Eqs. (8.63), as

$$X_1 = \frac{\left(2k - m\omega_f^2\right)F_1}{\left(\omega_f^2 - \omega_1^2\right)\left(\omega_f^2 - \omega_2^2\right)m^2}$$

$$X_2 = \frac{2kF_1}{\left(\omega_f^2 - \omega_1^2\right)\left(\omega_f^2 - \omega_2^2\right)m^2}$$

and the steady-state response is then

$$x_1(t) = X_1 \sin \omega_f t$$

$$x_2(t) = X_2 \sin \omega_f t$$

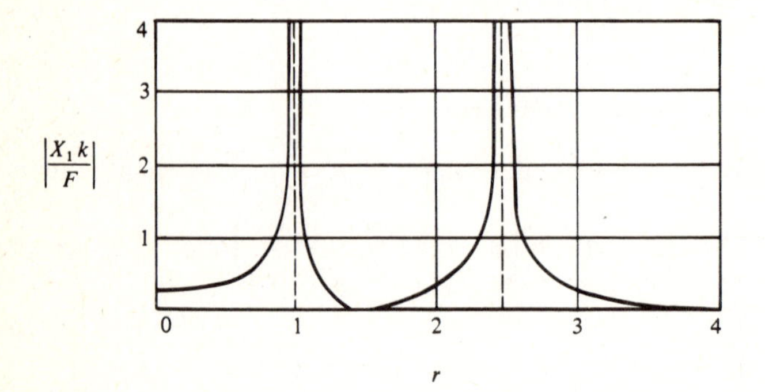

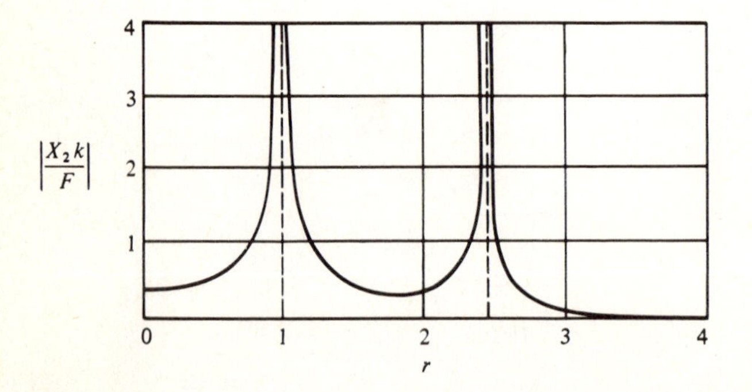

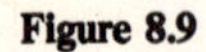
Figure 8.9

To examine the behavior of the system in more detail, it is convenient to rewrite the expressions for the amplitudes as

$$X_1 = \frac{F_1}{6k} \frac{2 - \omega_f^2/\omega_1^2}{(1 - \omega_f^2/\omega_1^2)(1 - \omega_f^2/\omega_2^2)}$$

$$X_2 = \frac{F_1}{3k} \frac{1}{(1 - \omega_f^2/\omega_1^2)(1 - \omega_f^2/\omega_2^2)}$$

and plot the steady-state amplitudes as functions of the frequency ratio ω_f/ω_1. The frequency-response curves are shown in Fig. 8.9.

8.7 UNDAMPED VIBRATION ABSORBER

In the discussion of resonance for a single-degree-of-freedom system, we observed that the frequency of such a system could be changed and the resonant condition alleviated by changing either the mass or the spring constant. When this is not possible, a resonant condition can also be eliminated by adding a second spring and mass to produce a two-degree-of-freedom system which exhibits no response to the excitation frequency. Consider Fig. 8.10, in which mass m_2 and spring k_2 have been added to the original forced system. From the analysis of the previous section, the steady-state amplitudes are governed by the matrix equation

$$\begin{bmatrix} k_1 + k_2 - m_1\omega_f^2 & -k_2 \\ -k_2 & k_2 - m_2\omega_f^2 \end{bmatrix} \begin{Bmatrix} X_1 \\ X_2 \end{Bmatrix} = \begin{Bmatrix} F_1 \\ 0 \end{Bmatrix} \tag{8.64}$$

from which

$$X_1 = \frac{(k_2 - m_2\omega_f^2)F_1}{(k_1 + k_2 - m_1\omega_f^2)(k_2 - m_2\omega_f^2) - k_2^2} \tag{8.65}$$

$$X_2 = \frac{k_2 F_1}{(k_1 + k_2 - m_1\omega_f^2)(k_2 - m_2\omega_f^2) - k_2^2}$$

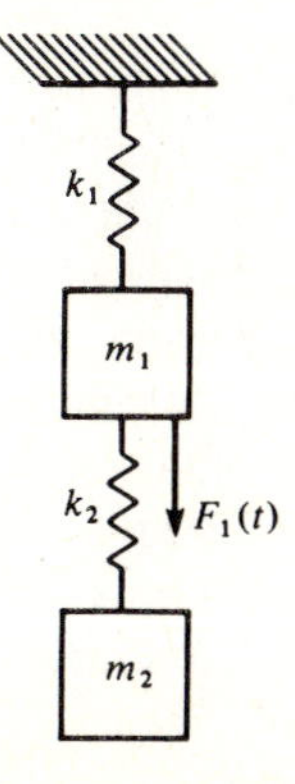

Figure 8.10

From the first of Eqs. (8.65), we observe that the steady-state amplitude of m_1 will become *zero* if the added mass and spring are chosen such that $\omega_f^2 = k_2/m_2$. The corresponding amplitude of m_2 is then $X_2 = -F_1/k_2$, giving the steady-state displacement as

$$x_2(t) = -\frac{F_1}{k_2}\sin \omega_f t \tag{8.66}$$

The force exerted by spring k_2 on mass m_1 is then

$$k_2 x_2(t) = -F_1 \sin \omega_f t \tag{8.67}$$

which exactly balances the applied force $F_1 \sin \omega_f t$ at all times.

A second spring and mass added in this manner is known as a *vibration absorber*. It is especially useful for rotating machinery that operates at a constant speed which very nearly corresponds to the resonant frequency. To consider this phenomenon in more detail, let us introduce the notation

$$\omega_m = \sqrt{\frac{k_1}{m_1}} = \text{natural frequency of main system alone}$$

$$\omega_a = \sqrt{\frac{k_2}{m_2}} = \text{natural frequency of absorber system alone}$$

$$\mu = \frac{m_2}{m_1} = \text{ratio of absorber mass to main mass}$$

$$X_0 = \frac{F_1}{k_1} = \text{equivalent static deflection of main mass}$$

With this notation, the amplitudes may be written as

$$X_1 = \frac{(1 - \omega_f^2/\omega_a^2)X_0}{\left[1 + \mu(\omega_a/\omega_m)^2 - (\omega_f/\omega_m)^2\right]\left[1 - (\omega_f/\omega_a)^2\right] - \mu(\omega_a/\omega_m)^2}$$

$$X_2 = \frac{X_0}{\left[1 + \mu(\omega_a/\omega_m)^2 - (\omega_f/\omega_m)^2\right]\left[1 - (\omega_f/\omega_a)^2\right] - \mu(\omega_a/\omega_m)^2} \tag{8.68}$$

In this form, we can readily plot the amplitude X_1 (as compared to X_0) as a function of ω_f/ω_a for a given mass ratio μ, as shown in Fig. 8.11. The figure also shows that X_1 is identically zero for $\omega_f/\omega_a = 1$. Two other important facts can be discerned from the figure, First, even though the absorber is designed to eliminate vibration of the main mass at one frequency only, this vibration may be greatly reduced over a finite range of frequencies as indicated by the shaded region. Second, two new resonant frequencies ω_1 and ω_2 exist such that $\omega_1 < \omega_f$ and $\omega_2 > \omega_f$. The lower of the two new resonant frequencies is significant, since a rotating machine would pass through that frequency when accelerating to its operating speed from rest. Even so, this is more desirable than continuous operation at resonance.

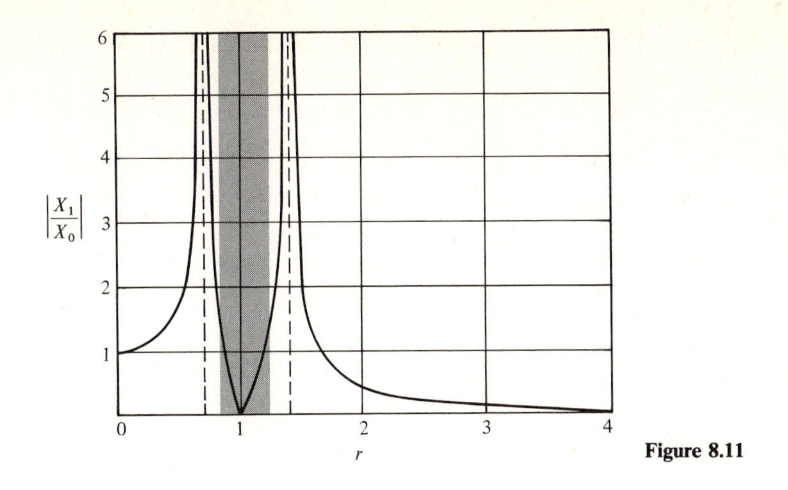

Figure 8.11

8.8 HARMONICALLY FORCED VIBRATIONS OF DAMPED TWO-DEGREES-OF-FREEDOM SYSTEMS

To discuss the methods of analysis used to obtain the displacement response of systems with two degrees of freedom, we shall consider the system shown in Fig. 8.12*a*. The lower mass m_2 is subjected to a harmonic forcing function F as shown. No loss of generality occurs if we assume m_1 to be free of any direct external force, but the analysis is simplified considerably.

With reference to the free-body diagrams shown in Fig. 8.12*b*, the equations of motion can be written as

$$m_1\ddot{x}_1 + (c_1 + c_2)\dot{x}_1 + (k_1 + k_2)x_1 - c_2\dot{x}_2 - k_2x_2 = 0$$

$$m_2\ddot{x}_2 + c_2\dot{x}_2 + k_2x_2 - c_2\dot{x}_1 - k_2x_1 = F \tag{8.69}$$

As expected, the equations of motion are coupled viscously and elastically, and one is inhomogeneous. Since we have already discussed methods of obtaining the homogeneous solutions (free response), we shall concentrate only on obtaining the particular solutions of Eqs. (8.69). Note that although only one of the equations is inhomogeneous, we must find particular solutions for both $x_1(t)$ and $x_2(t)$ because of the coupling. The manipulations required to obtain the solution will be reduced, however, if we express the harmonic force as[2]

$$F = F_o e^{i\omega_f t} \tag{8.70}$$

where the actual applied force is given by the real part of this expression.

In Sec. 5.6 we obtained the forced response to harmonic forces of damped systems having one degree of freedom, and those results can be used to advantage in analyzing this problem. For those systems, we found that the

[2] Recall the Euler relation $e^{i\omega_f t} = \cos \omega_f t + i \sin \omega_f t$; thus, Eq. (8.70) represents $F = F_o \cos \omega_f t$. For $F = F_o \sin \omega_f t$, let $F = iF_o e^{-i\omega_f t}$.

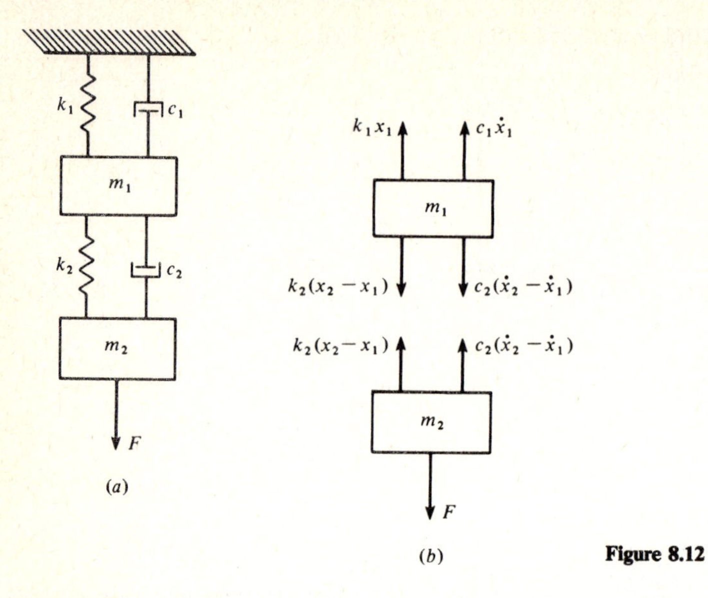

Figure 8.12

steady-state response to a harmonic force was a harmonic function of time having the same frequency as the force but lagging the force by a phase angle ψ. Physically, the phase lag exists because damping retards the displacement response in time. Mathematically, it exists because the presence of the first derivative in the equation of motion requires that both sine and cosine terms appear in the particular solution. Following this experience, we are led to assume that the displacements of the two-degree-of-freedom system are of the form

$$x_1(t) = \bar{X}_1 e^{i\omega_f t} \qquad x_2(t) = \bar{X}_2 e^{i\omega_f t} \tag{8.71}$$

where $\bar{X}_1$ and $\bar{X}_2$ are complex constants. It is to be understood that the actual displacements are given by the real parts of Eqs. (8.71).

To show that the assumed solutions exhibit phase lag, we note that any complex constant ($\bar{X}_1$, for example) can be written in the form

$$\bar{X}_1 = U + iV = X_1 e^{-i\psi_1} \tag{8.72}$$

where
$$X_1 = \sqrt{U^2 + V^2} \qquad \psi_1 = \tan^{-1}\left(-\frac{V}{U}\right) \tag{8.73}$$

are the amplitude and phase angle. We then have

$$\begin{aligned} x_1(t) &= \bar{X}_1 e^{i\omega_f t} = X_1 e^{-i\psi_1} e^{i\omega_f t} = X_1 e^{i(\omega_f t - \psi_1)} \\ &= X_1 \cos(\omega_f t - \psi_1) + iX_1 \sin(\omega_f t - \psi_1) \end{aligned} \tag{8.74}$$

and the real part of Eq. (8.74) shows that the assumed displacement has the desired form.

Differentiating the assumed solutions as required and substituting into the equations of motion give

$$\left[k_1 + k_2 - m_1\omega_f^2 + i\omega_f(c_1 + c_2)\right]\bar{X}_1 e^{i\omega_f t} - (k_2 + i\omega_f c_2)\bar{X}_2 e^{i\omega_f t} = 0$$
$$(k_2 - m_2\omega_f^2 + i\omega_f c_2)\bar{X}_2 e^{i\omega_f t} - (k_2 + i\omega_f c_2)\bar{X}_1 e^{i\omega_f t} = F_0 e^{i\omega_f t} \tag{8.75}$$

Since the exponential function is common to each term, Eqs. (8.75) reduce to a system of two complex algebraic equations in the complex unknowns $\bar{X}_1$ and $\bar{X}_2$. In matrix form, we have

$$\begin{bmatrix} k_1 + k_2 - m_1\omega_f^2 + i\omega_f(c_1 + c_2) & -(k_2 + i\omega_f c_2) \\ -(k_2 + i\omega_f c_2) & k_2 - m_2\omega_f^2 + i\omega_f c_2 \end{bmatrix} \begin{Bmatrix} \bar{X}_1 \\ \bar{X}_2 \end{Bmatrix} = \begin{Bmatrix} 0 \\ F_o \end{Bmatrix} \tag{8.76}$$

Solving Eq. (8.76) by Cramer's rule gives, as the complex amplitudes,

$$\bar{X}_1 = \frac{(k_2 + ic_2\omega_f)F_0}{\left[k_1 + k_2 - m_1\omega_f^2 + i\omega_f(c_1 + c_2)\right](k_2 - m_2\omega_f^2 + i\omega_f c_2) - (k_2 + i\omega_f c_2)^2}$$
$$\bar{X}_2 = \frac{\left[k_1 + k_2 - m_1\omega_f^2 + i\omega_f(c_1 + c_2)\right]F_o}{\left[k_1 + k_2 - m_1\omega_f^2 + i\omega_f(c_1 + c_2)\right](k_2 - m_2\omega_f^2 + i\omega_f c_2) - (k_2 + i\omega_f c_2)^2} \tag{8.77}$$

To continue the solution in terms of general variables becomes very unwieldy. Thus, let us at this point consider the case in which $m_1 = m_2 = m$, $c_1 = c_2 = c$, and $k_1 = k_2 = k$ for simplicity. Substitution of these values gives

$$\bar{X}_1 = \frac{(k + ic\omega_f)F_o}{(2k - m\omega_f^2 + 2ic\omega_f)(k - m\omega_f^2 + i\omega_f c) - (k + i\omega_f c)^2}$$
$$\bar{X}_2 = \frac{(2k - m\omega_f^2 + 2ic\omega_f)F_o}{(2k - m\omega_f^2 + 2ic\omega_f)(k - m\omega_f^2 + i\omega_f c) - (k + i\omega_f c)^2} \tag{8.78}$$

which, after expansion and simplification of the denominator terms, become

$$\bar{X}_1 = \frac{(k + ic\omega_f)F_o}{m^2\omega_f^4 - 3km\omega_f^2 + k^2 - c^2\omega_f^2 + ic\omega_f(2k - 3m\omega_f^2)}$$
$$\bar{X}_2 = \frac{(2k - m\omega_f^2 + 2ic\omega_f)F_o}{m^2\omega_f^4 - 3km\omega_f^2 + k^2 - c^2\omega_f^2 + ic\omega_f(2k - 3m\omega_f^2)} \tag{8.79}$$

Note that if $c = 0$, Eqs. (8.79) define the amplitudes of the forced response of the undamped system, and that the first three terms of the denominator correspond to the frequency equation. Thus we can write

$$\bar{X}_1 = \frac{(k + ic\omega_f)F_o}{m^2(\omega_f^2 - \omega_1^2)(\omega_f^2 - \omega_2^2) - c^2\omega_f^2 + ic\omega_f(2k - m\omega_f^2)}$$

$$\bar{X}_2 = \frac{(2k - m\omega_f^2 + 2ic\omega_f)F_0}{m^2(\omega_f^2 - \omega_1^2)(\omega_f^2 - \omega_2^2) - c^2\omega_f^2 + ic\omega_f(2k - m\omega_f^2)} \tag{8.80}$$

where ω_1 and ω_2 are the undamped natural frequencies.

Since we are primarily interested in finding the amplitudes of the forced response as given by the real parts of Eqs. (8.71), the following can be used to advantage. If a complex constant $\bar{A}$ can be written as

$$\bar{A} = \frac{B + iC}{D + iE} \tag{8.81}$$

we can multiply numerator and denominator by $D - iE$ to obtain

$$A = \frac{B + iC}{D + iE}\,\frac{D - iE}{D - iE} = \frac{BD - CE + i(CD - BE)}{D^2 + E^2} \tag{8.82}$$

In polar form, $\bar{A}$ can be written

$$\bar{A} = Ae^{i\theta} = \frac{\sqrt{(BD - CE)^2 + (CD - BE)^2}}{D^2 + E^2}\,e^{i\theta} \tag{8.83}$$

where

$$\theta = \tan^{-1}\frac{CD - BE}{BD - CE} \tag{8.84}$$

By expanding and simplifying the real part of Eq. (8.83), we find the magnitude of $\bar{A}$ as

$$A = \sqrt{\frac{B^2 + C^2}{D^2 + E^2}} \tag{8.85}$$

Applying this procedure to Eqs. (8.80) gives

$$X_1 = \frac{F_o\sqrt{k^2 + (c\omega_f)^2}}{\sqrt{\left[m^2(\omega_f^2 - \omega_1^2)(\omega_f^2 - \omega_2^2) - c^2\omega_f^2\right]^2 + c^2\omega_f^2(2k - m\omega_f)^2}}$$

$$X_2 = \frac{F_o\sqrt{(2k - m\omega_f^2)^2 + (2c\omega_f)^2}}{\sqrt{\left[m^2(\omega_f^2 - \omega_1^2)(\omega_f^2 - \omega_2^2) - c^2\omega_f^2\right]^2 + c^2\omega_f^2(2k - m\omega_f)^2}} \tag{8.86}$$

as the amplitudes of the forced response. To study these amplitudes in more detail, Eqs. (8.86) can be made nondimensional by introducing

$$r = \frac{\omega_f}{\omega_1} \qquad X_o = \frac{F_o}{k} \qquad \zeta = \frac{c}{m\omega_1} \tag{8.87}$$

where $\omega_1^2 = \gamma_1 k/m = 0.382k/m$ is the square of the fundamental natural frequency. Also $\omega_2^2 = \gamma_2 k/m = 2.62k/m$ for this system. Introducing these variables gives

$$X_1 = \frac{X_o\sqrt{1 + (\gamma_1 \zeta r)^2}}{\sqrt{\left[\gamma_1^2(r^2 - 1)(r^2 - \gamma_2^2/\gamma_1^2) - (\gamma_1 \zeta r)^2\right]^2 + (\gamma_1 \zeta r)^2(2 - \gamma_1 r^2)^2}}$$

$$X_2 = \frac{X_o\sqrt{(2 - \gamma_1 r^2)^2 + (\gamma_1 \zeta r)^2}}{\sqrt{\left[\gamma_1^2(r^2 - 1)(r^2 - \gamma_2^2/\gamma_1^2) - (\gamma_1 \zeta r)^2\right]^2 + (\gamma_1 \zeta r)^2(2 - \gamma_1 r^2)^2}} \tag{8.88}$$

In this form the steady-state amplitudes can be plotted against the frequency ratio $r = \omega_f/\omega_1$ for various values of the damping variable ζ as in Fig. 8.13, which shows X_2/X_o. Note that in this context ζ is not the same as the damping factor defined for systems with one degree of freedom. Rather, it is a conveniently chosen variable which indicates the amount of damping present in the system. The curve shows that the behavior is very similar to that of a single-degree-of-freedom system except that relatively large amplitudes occur near *two* values of r, corresponding to $\omega_f = \omega_1$ and $\omega_f = \omega_2$. This is expected, as two resonant frequencies exist. We also note that as damping increases, the amplitude decreases generally, and the decrease is marked at higher frequency ratios. A similar plot of X_1/X_o would exhibit the same phenomena except that lower values result, since much of the force applied to m_2 is dissipated by the dashpot between the masses.

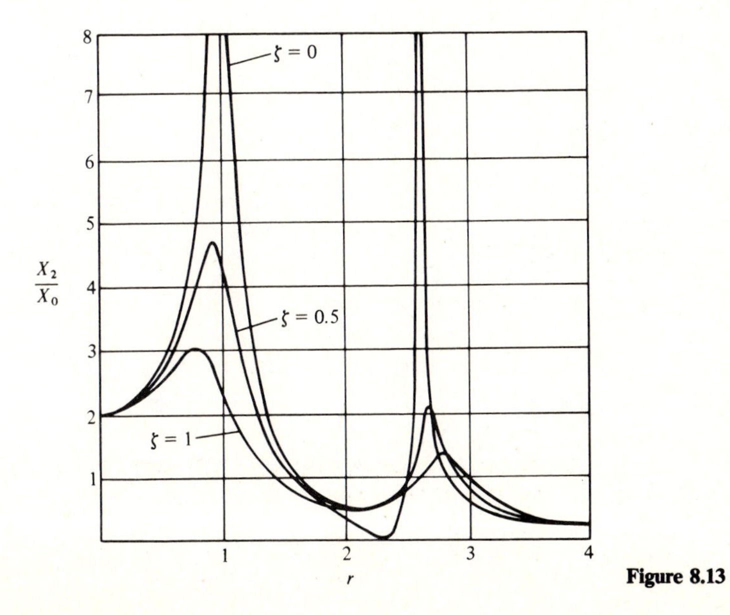

Figure 8.13

8.9 VISCOUS VIBRATION ABSORBER

The undamped vibration absorber discussed in Sec. 8.7 is said to be *tuned* because it is designed to eliminate vibration at a given frequency. Thus, its application is restricted to equipment which operates at a single constant speed. For machinery which exhibits vibrations at different frequencies because of variable operating speeds, an *untuned viscous vibration absorber* can be used to reduce the vibration levels. As shown in Fig. 8.14, this device is composed of a mass m_2 which is attached through a viscous damper c to the main system composed of mass m_1 and spring constant k_1.

The differential equations of motion for the system are

$$m_1\ddot{x}_1 + c\dot{x}_1 + k_1x_1 - c\dot{x}_2 = F$$

$$m_2\ddot{x}_2 + c\dot{x}_2 - c\dot{x}_1 = 0 \tag{8.89}$$

For a harmonic force expressed as $F = F_o e^{i\omega_f t}$, the method of Sec. 8.9 can be applied to obtain the steady-state amplitudes

$$X_1 = \frac{F_o\sqrt{(m_2\omega_f^2)^2 + (c\omega_f)^2}}{\sqrt{[m_2\omega_f^2(k_1 - m_1\omega_f^2)]^2 + (c\omega_f)^2[m_2\omega_f^2 - (k_1 - m_1\omega_f^2)]^2}}$$

$$X_2 = \frac{-F_o(c\omega_f)^2}{\sqrt{[m_2\omega_f^2(k_1 - m_1\omega_f^2)]^2 + (c\omega_f)^2[m_2\omega_f^2 - (k_1 - m_1\omega_f^2)]^2}} \tag{8.90}$$

As we are primarily interested in the effect of the absorber on the vibration of the main mass m_1, we shall concentrate on the first of Eqs. (8.90) and make it nondimensional by introducing

$$X_o = \frac{F_o}{k_1} \qquad \omega_1^2 = \frac{k_1}{m_1} \qquad \zeta = \frac{c}{2m_1\omega_1} \qquad \mu = \frac{m_2}{m_1} \qquad r = \frac{\omega_f}{\omega_1} \tag{8.91}$$

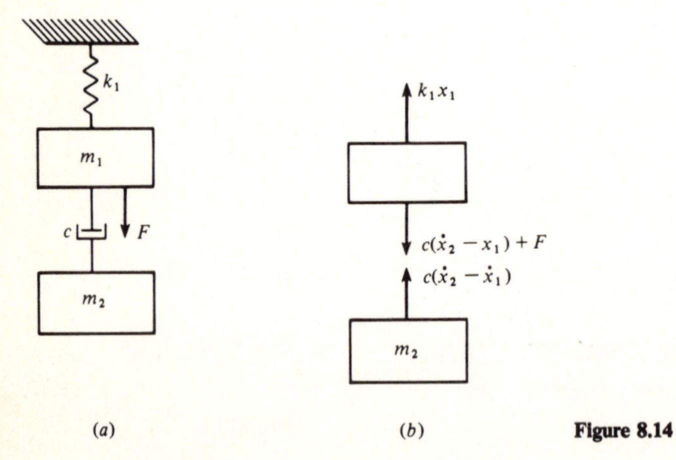

Figure 8.14

In terms of these variables, we have

$$X_1 = \frac{X_o\sqrt{\mu^2 r^2 + 4\zeta^2}}{\sqrt{\mu^2(1 - r^2)^2 + 4\zeta^2[\mu r^2 - (1 - r^2)]^2}} \tag{8.92}$$

which shows that for a given $X_o = F_o/k_1$, the amplitude of the main mass is a function of the three parameters μ, ζ, and r. The response is most conveniently studied by plotting curves of X_1/X_o against frequency ratio r for various damping factors ζ while holding the mass ratio constant as shown in Fig. 8.15. Of particular interest is the fact that each curve passes through the common point P as shown. If the value of ζ can be determined for which the curve has a horizontal tangent at P, the peak amplitude will be minimized. For a given mass ratio μ, it can be shown that this optimum damping value is given by

$$\zeta_o = \frac{1}{\sqrt{2(1 + \mu)(2 + \mu)}} \tag{8.93}$$

and the peak amplitude occurs for the frequency ratio

$$r = \sqrt{\frac{2}{2 + \mu}} \tag{8.94}$$

Equation (8.94) is obtained by equating the values of X_1/X_o for any two values of ζ and solving the resulting expression for r. Setting the slope of the curve equal to zero at this value of r then yields ζ_o.

As the viscous vibration absorber is effective over a range of frequencies, its torsional counterpart is often used to reduce the torsional crankshaft oscillations which occur in reciprocating engines. Such a device is depicted in Fig. 8.16. A disk having moment of inertia I_d is mounted inside a housing which is attached

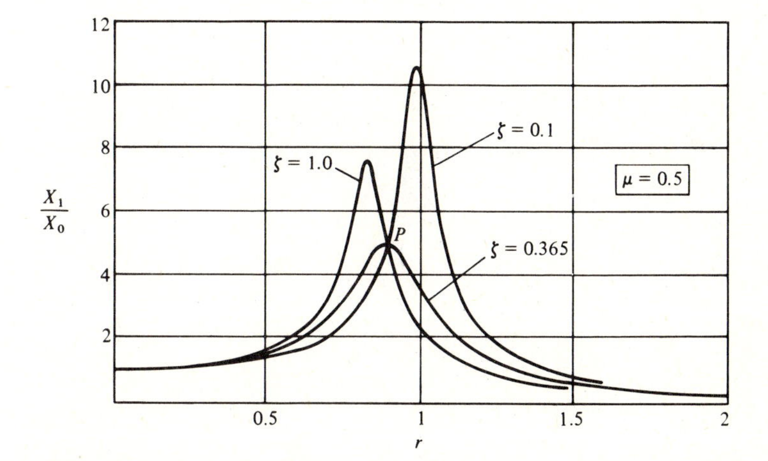

Figure 8.15

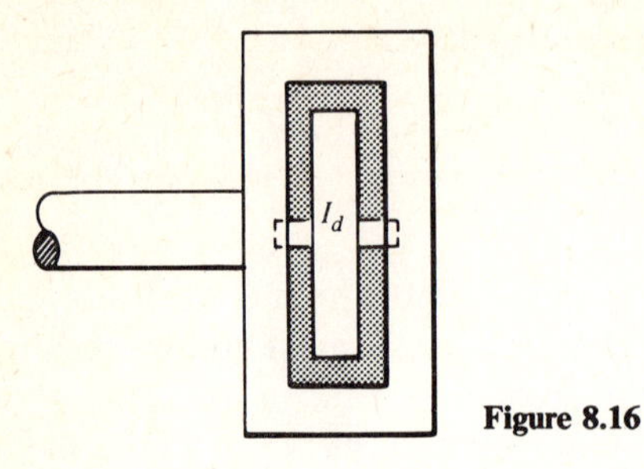

Figure 8.16

to the crankshaft. The disk is free to rotate inside and relative to the housing. Damping is introduced by filling the space between housing and disk with a silicone oil. Damping torque is produced by the viscosity of the oil and is proportional to the relative angular velocity of housing and disk. This type of torsional vibration absorber is known as a *Houdaille damper* or *viscous Lanchester damper*.

PROBLEMS

8.1 For the system shown in Fig. 8.17:

(*a*) Write the differential equations of motion for each mass after first drawing the appropriate free-body diagrams.

(*b*) Obtain the frequency equation.

(*c*) Assuming equal masses and spring constants, determine the natural frequencies and the corresponding amplitude ratios.

(*d*) Sketch the mode shapes.

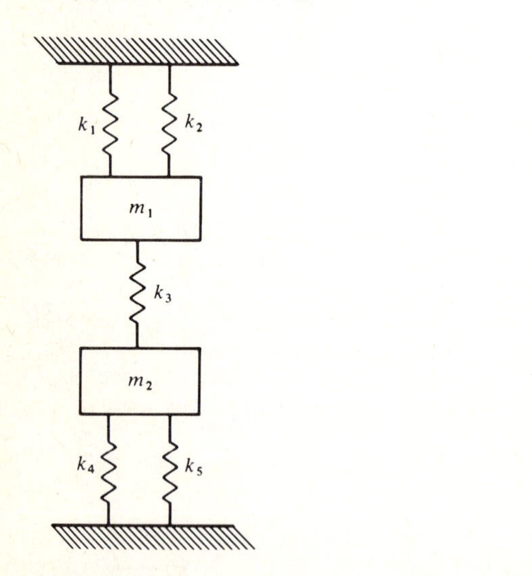

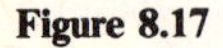
Figure 8.17

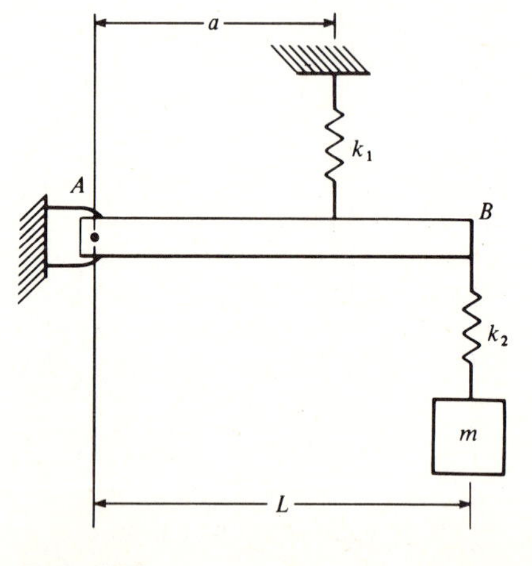

Figure 8.18

8.2 In the system shown in Fig. 8.18, rigid bar AB has total mass M and pivots freely about a pin connection at A. At equilibrium, bar AB is in the horizontal position. Obtain the differential equations of motion for this system, assuming the bar executes small oscillations about equilibrium. How would you classify the coupling exhibited?

8.3 The system of Fig. 8.18 has $L = 10$ in, $a = 6$ in, $M = 0.2$ lb · s^2/in, $m = 0.05$ lb · s^2/in, $k_1 = 40$ lb/in, and $k_2 = 20$ lb/in. If the small mass is displaced downward 1 in from its equilibrium position and released, determine the response of the system.

8.4 Determine the principal coordinates for the system shown in Fig. 8.18.

8.5 Obtain the differential equations of motion for the double pendulum shown in Fig. 8.19, (*a*) using angular coordinates θ_1 and θ_2 and (*b*) using displacement coordinates x_1 and x_2. (*c*) Compare the coupling in the equations obtained by the two methods.

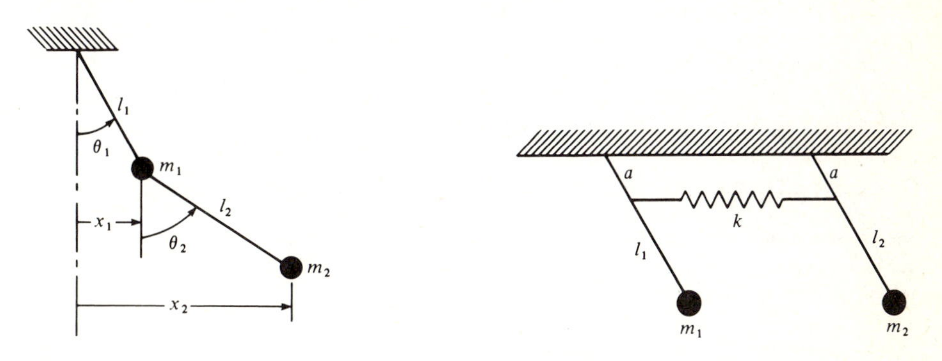

Figure 8.19

Figure 8.20

8.6 Two simple pendulums are connected by a linear spring as shown in Fig. 8.20. Obtain the principal-mode solutions for small oscillations about the vertical.

8.7 Two disks having polar mass moments of inertia I_1 and I_2 are attached to a shaft as shown in Fig. 8.21. The ends of the shaft are fixed, and the three sections of the shaft have torsional spring constants as shown. Determine the differential equations governing the rotational motion of the disks.

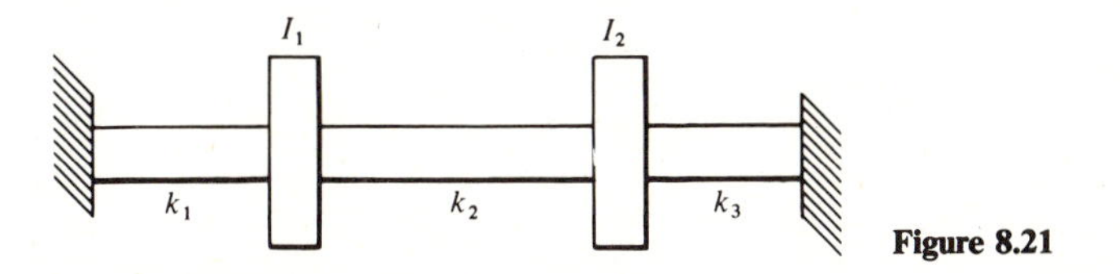

Figure 8.21

8.8 In Fig. 8.21, let $I_1 = I_2$ and $k_1 = k_2 = k_3$, and obtain the natural frequencies and amplitude ratios for free oscillation of the disks.

8.9 A uniform bar of length L and mass m is pinned to the center of a solid cylinder which rolls without slipping on a horizontal surface as shown in Fig. 8.22. The radius and mass of the cylinder are $L/4$ and m, respectively. Determine the natural frequencies of small oscillations of the system.

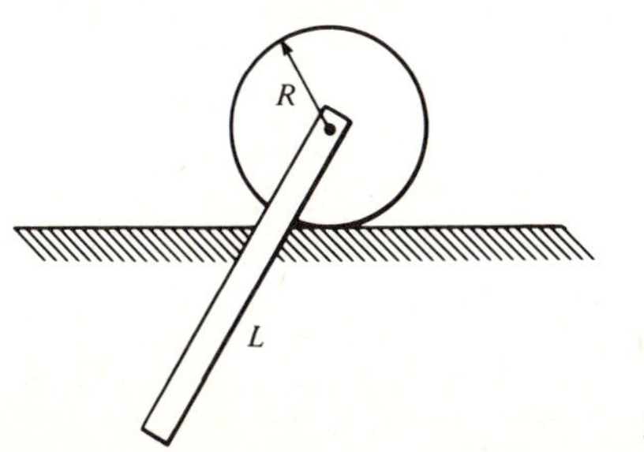

Figure 8.22

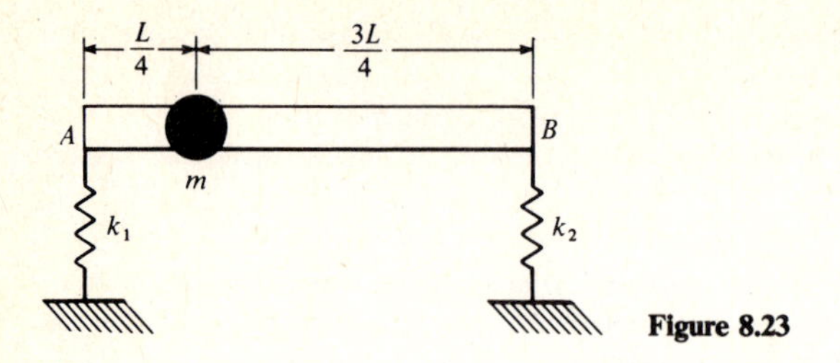

Figure 8.23

8.10 Rigid bar AB having mass M and length L is supported on springs and carries concentrated mass m as shown in Fig. 8.23. Using coordinates corresponding to translation of the center of mass and rotation about the center of mass, determine the differential equations of motion.

8.11 Repeat Prob. 8.10 using coordinates representing the vertical displacements of ends A and B.

8.12 For the system of Fig. 8.23, let $M = 2m$ and $k_2 = \frac{5}{7}k_1$, and determine the natural frequencies and amplitude ratios for principal-mode vibrations.

8.13 Figure 8.24 is a model of an overhead crane. The crane support is modeled as a simply supported beam having cross-sectional moment of inertia I and modulus of elasticity E. The mass of the beam is negligible in comparison to the mass of the crane. Load mass m is lifted through a cable for which the equivalent spring constant is k. Obtain the differential equations of motion for free vibration of the system.

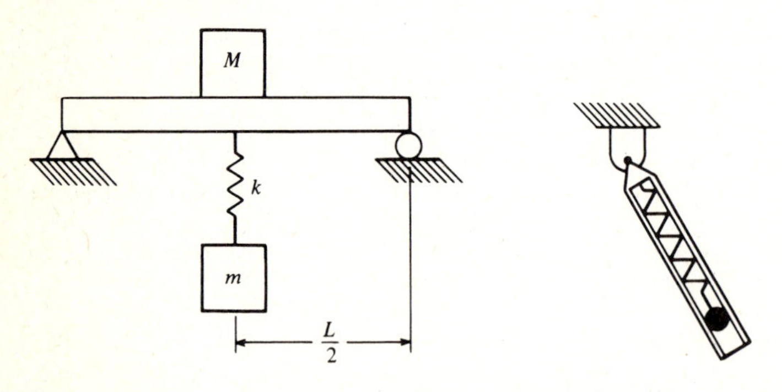

Figure 8.24

Figure 8.25

8.14 A spring-mass system is enclosed in a thin tube, and the assembly is supported in a vertical plane through a pin connection as shown in Fig. 8.25. Assuming the mass of the tube to be negligible, determine the differential equations of motion.

8.15 Obtain the differential equations of motion and natural frequencies for free vibration of the system in Fig. 8.26.

8.16 The differential equations of motion for a linear system are written in the matrix form

$$\begin{bmatrix} m_1 & 0 \\ 0 & m_2 \end{bmatrix} \begin{Bmatrix} \ddot{x}_1 \\ \ddot{x}_2 \end{Bmatrix} + \begin{bmatrix} k_{11} & k_{12} \\ k_{21} & k_{22} \end{bmatrix} \begin{Bmatrix} x_1 \\ x_2 \end{Bmatrix} = 0$$

Show that the amplitude ratios are given by

$$\beta_1 = -\frac{k_{11} - \omega_1^2 m_1}{k_{12}} = -\frac{k_{12}}{k_{22} - \omega_1^2 m_2}$$

$$\beta_2 = -\frac{k_{11} - \omega_2^2 m_2}{k_{12}} = -\frac{k_{12}}{k_{22} - \omega_2^2 m_2}$$

where ω_1 and ω_2 are the system natural frequencies.

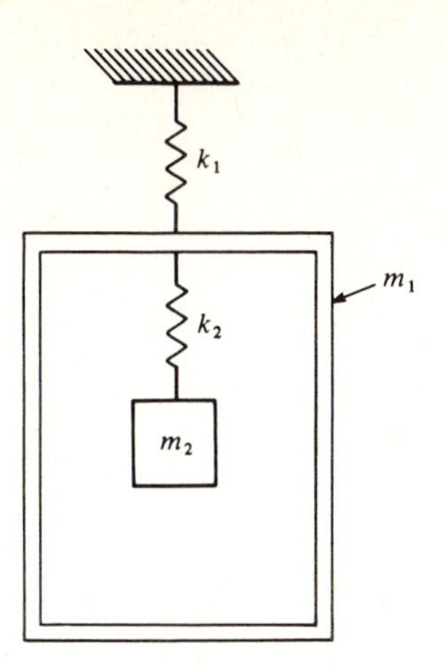

Figure 8.26

8.17 Show that the equations of motion given in Prob. 8.16 can be uncoupled by introducing the coordinate transformation

$$\begin{Bmatrix} x_1 \\ x_2 \end{Bmatrix} = \begin{bmatrix} 1 & 1 \\ \beta_1 & \beta_2 \end{bmatrix} \begin{Bmatrix} q_1 \\ q_2 \end{Bmatrix}$$

and multiplying the result by the matrix

$$\begin{bmatrix} 1 & \beta_1 \\ 1 & \beta_2 \end{bmatrix}$$

where β_1 and β_2 are the amplitude ratios of the principal-mode solutions.

8.18 The motion of the system shown in Fig. 8.27 is to be expressed in terms of coordinates expressing translation of point A and rotation about point A.

(*a*) Write the differential equations of motion, assuming small angular oscillations.
(*b*) Obtain the natural frequencies and the principal-mode solutions.
(*c*) Draw the mode shapes and locate the nodes.

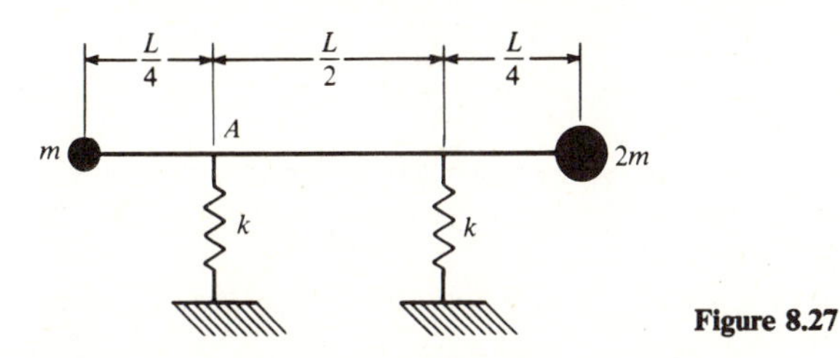

Figure 8.27

8.19 Repeat Prob. 8.18 using principal coordinates.

8.20 For the damped system shown in Fig. 8.28, determine $x_1(t)$ and $x_2(t)$ if $m = 0.5$ kg, $k = 1750$ N/m, and $c = 100$ N · s/m. The initial conditions are zero except $x_1(0) = 30$ mm.

8.21 Derive the differential equations of motion for the system of Fig. 8.29.

8.22 Determine the forced response of the system shown in Fig. 8.30 if $F_1 = F_1 \sin \omega_f t$, $F_2 = 0$.

8.23 Repeat Prob. 8.22 for $F_1 = 0$, $F_2 = F_2 \sin \omega_f t$.

8.24 The system of Fig. 8.30 has $m = 0.05$ lb · s²/in, $k = 12$ lb/in, $F_1 = 10 \sin 15t$ lb, and $F_2 = 8 \sin 20t$ lb. Find the forced amplitude of each mass.

8.25 A component rotating at 1800 r/min has a 3 lb · in unbalance. The component is part of a machine having a total weight of 300 lb which is mounted on springs having a total spring constant of 400 lb/in, as shown in Fig. 8.31. An undamped vibration absorber is to be added for which $W_a = 40$ lb. What is the required value k_a of the absorber spring constant?

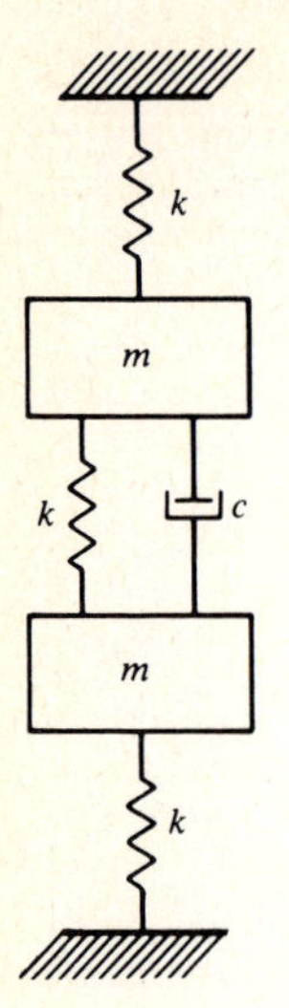

Figure 8.28

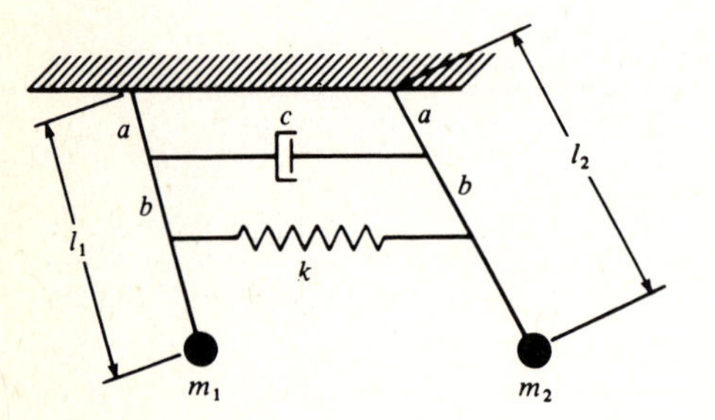

Figure 8.29

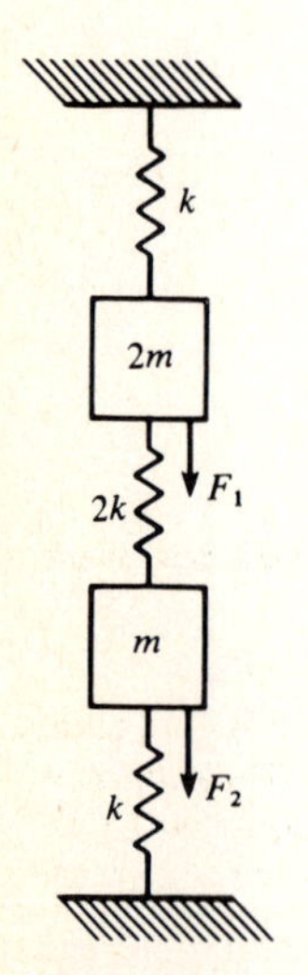

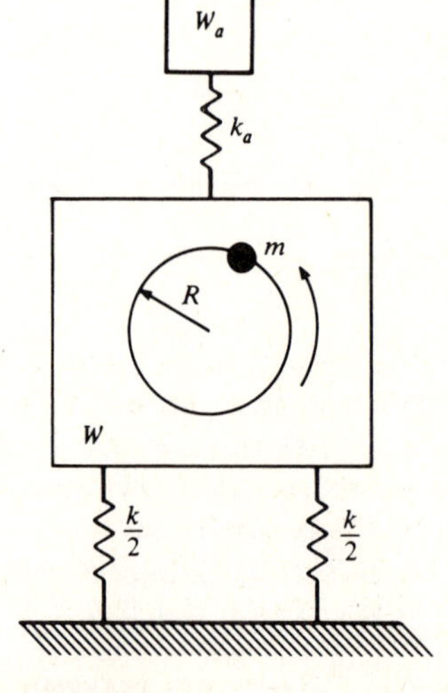

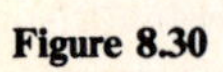

Figure 8.30

Figure 8.31

8.26 Solve Prob. 8.25 for a rotating speed of 1180 r/min, and find the forced amplitude of the absorber.

8.27 Figure 8.32 depicts a two-degree-of-freedom system which is excited by a moving support. Determine the differential equations of motion, and obtain expressions for the forced amplitudes.

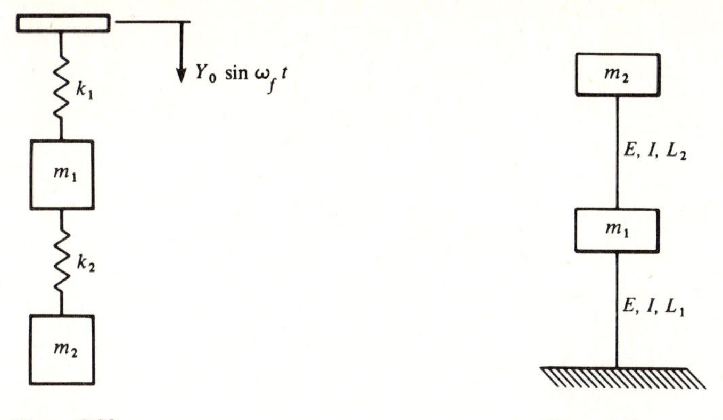

Figure 8.32

Figure 8.33

8.28 Two masses are attached to a slender circular rod which is held in a vertical plane as shown in Fig. 8.33. Derive the differential equations of motion of the system if it is subjected to harmonic motion of the support in the horizontal direction.

8.29 A commonly used model for studying the oscillations of a two-story building is shown in Fig. 8.34. For the case of equal masses and equal spring constants, (*a*) determine the natural frequencies of horizontal oscillations of the system, (*b*) plot the mode shapes, and (*c*) determine the forced response to harmonic ground motion.

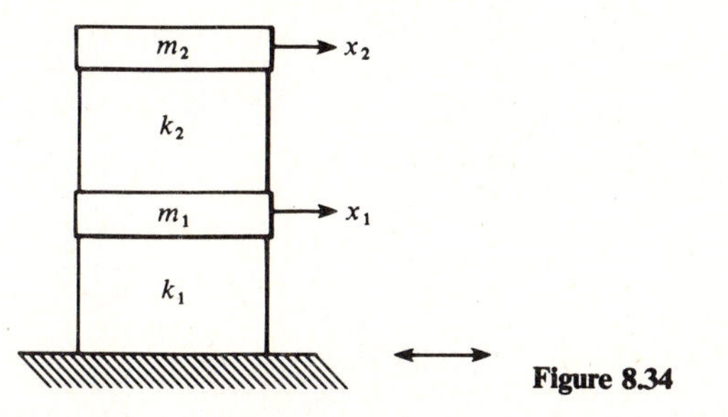

Figure 8.34

8.30 In the system of Fig. 8.34, $m_2 = 0.5\, m_1$, $k_2 = 0.5\, k_1$, and a force $F = F_o \sin \omega_f t$ acts in the horizontal direction on m_2. Determine the forced response of each mass.

8.31 Repeat Prob. 8.30 if the force acts on m_1.

8.32 An unbalanced rotor which has a total mass M is symmetrically mounted between identical bearings having spring constants k relative to vertical motion, as shown in Fig. 8.35. The unbalance mass m is located at radius R from the rotor axis and distance d from the axis of symmetry O. Denoting by I_o the moment of inertia of the shaft about the axis of symmetry, determine the equations of motion in the vertical plane. Let the rotational speed be N r/min.

8.33 Discuss Prob. 8.32 relative to the two-plane dynamic balancing procedure discussed in Chap. 6.

8.34 Determine the differential equations of motion and obtain expressions for the forced response of the system in Fig. 8.36 assuming small angular oscillations.

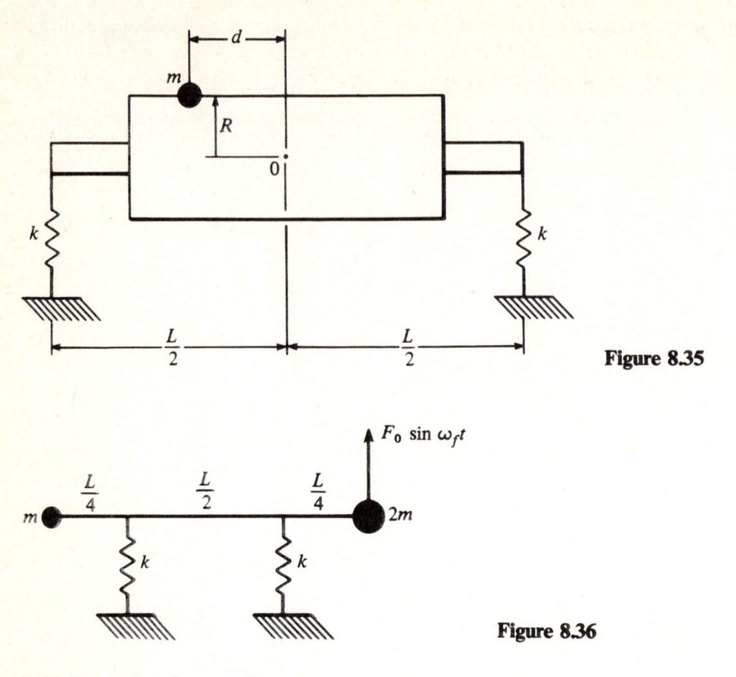

Figure 8.35

Figure 8.36

8.35 Prove that the equations

$$C_2^{(1)} = (a + bi)C_1^{(1)} \qquad C_2^{(2)} = (a - bi)C_1^{(2)}$$

hold for real a and b only if $C_1^{(1)}$ and $C_1^{(2)}$ are complex conjugates and $C_2^{(1)}$ and $C_2^{(2)}$ are complex conjugates.

8.36 A spring-mass system having mass m_1 and elastic constant k_1 is excited near resonance by an external force $F_o \sin \omega_f t$, where $\omega_f^2 = 0.98\, k_1/m_1$. The vibration is to be reduced by adding a *damped vibration absorber* as shown in Fig. 8.37. If $m_1 = m_2 = m$, $k_1 = k_2 = k$, and $c = \sqrt{2mk}$, determine the resulting reduction in the forced amplitude of m_1.

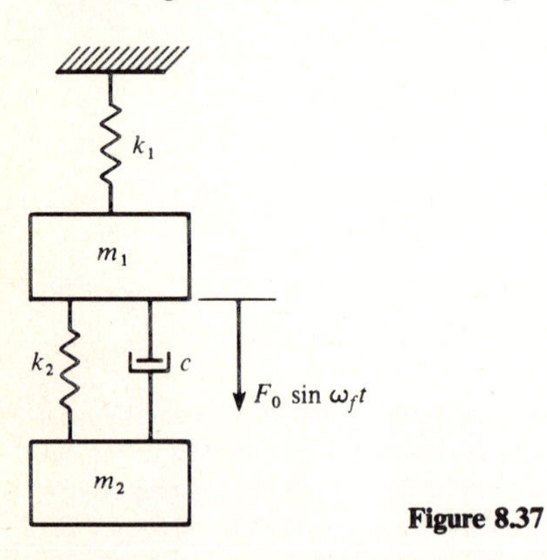

Figure 8.37

8.37 The damped system of Fig. 8.37 is subjected to a harmonic force having an amplitude of 40 N. The physical parameters are $m_1 = 8$ kg, $m_2 = 4$ kg, $k_1 = 1750$ N/m, and $c = 120$ N · s/m. Determine the steady-state amplitude of the oscillation of each mass if the forcing frequency is the same as the fundamental frequency of the undamped system.

8.38 Repeat Prob. 8.37 for $\omega_f = \omega_2$.

8.39 The undamped vibration absorber of Prob. 8.25 is to be replaced by a viscous vibration absorber. All physical constants are as given, except that the rotational speed may vary.

(*a*) Determine the optimum damping constant c for this system if the absorber spring k_a is replaced by a dashpot.

(*b*) Find the corresponding peak amplitude of the main mass and the frequency at which this amplitude occurs.

(*c*) What is the peak amplitude of the absorber mass?

CHAPTER

NINE

ANALOG-COMPUTER TECHNIQUES IN VIBRATION ANALYSIS

According to Webster, an analogy is "an explaining of something by comparing it point by point with something else." An analogy for vibrating mechanical systems is found in electric circuits. Many circuits which exhibit oscillations in current and voltage are characterized by differential equations of the same type as those governing vibrations of mechanical systems. The analysis of such circuits is therefore similar to that of a vibrating system, and results obtained by studying the response of a given circuit can be applied to a corresponding mechanical system.

Let us consider the circuit shown in Fig. 9.1, composed of an inductance L, resistance R, and capacitance C connected in series with an alternating voltage source $E = -(E_o/\omega_f)\cos\omega_f t$. We denote the current by i and express the fact that the algebraic sum of the applied voltage and the drops in potential around the circuit is zero:

$$-\frac{E_o}{\omega_f}\cos\omega_f t - L\frac{di}{dt} - Ri - \frac{1}{C}\int i\,dt = 0 \tag{9.1}$$

Differentiating Eq. (9.1) with respect to time and rearranging result in

$$L\frac{d^2 i}{dt^2} + R\frac{di}{dt} + \frac{i}{C} = E_o \sin\omega_f t \tag{9.2}$$

as the governing differential equation for current in the circuit. Recalling the equation governing the displacement of a damped spring-mass system excited by

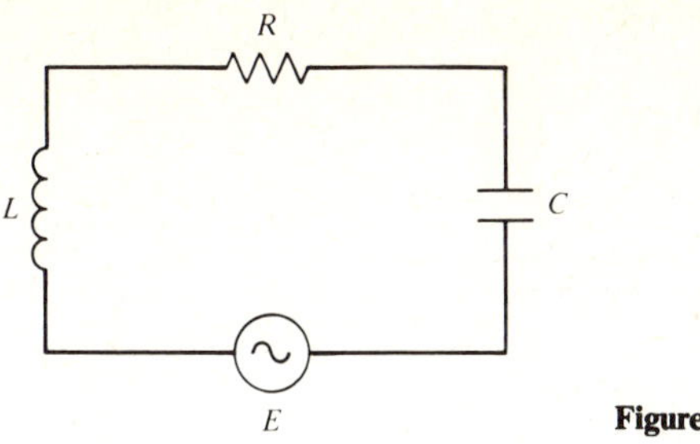

Figure 9.1

a harmonic force,

$$m\frac{d^2x}{dt^2} + c\frac{dx}{dt} + kx = F_o \sin \omega_f t \tag{9.3}$$

we observe a striking similarity to Eq. (9.2). We have a perfect analogy between current in the circuit of Fig. 9.1 and displacement of the mass in the mechanical system governed by Eq. (9.3). Many other circuits could be constructed which are analogous to some mechanical system. In practice, however, individual circuits are not designed each time a solution by analogy is desired. Rather, a device known as an *analog computer* is used to construct analogous circuits. The characteristics and operation of analog computers and their application to vibrating systems will be detailed in this chapter.

9.1 CHARACTERISTICS OF OPERATIONAL AMPLIFIERS

We would find it most inconvenient and impractical if we were forced to begin from scratch each time an analog circuit was required. Fortunately, the *electronic analog computer* (EAC) contains, in compact and convenient form, all the components normally required for obtaining analog solutions to many mechanical problems. The various components of an analog computer are building blocks which may be interconnected to form the desired analog circuit. The dependent variables in an analog-computer circuit are the voltages at various points, and the independent variable is time. The voltages can be monitored by voltmeter observations, with oscilloscopes, or by plotting with electromechanical recorders.

The basic component of an analog computer is the *operational amplifier*, which can be used to obtain the operations of addition, subtraction, multiplication by a constant, and integration. Operational amplifiers used in analog computers are direct-current (dc) amplifiers which have a high *negative gain* and are designed for stability and accuracy. The *gain* of an amplifier is defined as the ratio of output voltage to input voltage and will be denoted by G. In addition, *negative gain* means that a reversal of polarity occurs between input and output voltages. For the operational amplifier shown schematically in

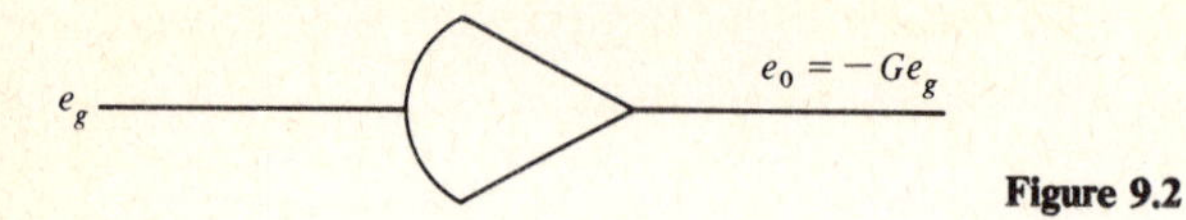

Figure 9.2

Fig. 9.2, the voltage to the input stage is e_g, and the output voltage is given by

$$e_o = -Ge_g \tag{9.4}$$

The subscript g is a holdover from days past when the input was to the grid of a vacuum-tube amplifier. Modern operational amplifiers are solid-state devices. For most commercial analog computers, the gain G is 10^5 or higher. Used alone as in Fig. 9.2, an operational amplifier is of little value for computational purposes. To achieve practical operations, additional components must be connected to the amplifier.

9.2 MATHEMATICAL FUNCTIONS OF OPERATIONAL AMPLIFIERS

Modifications can be made to an operational amplifier circuit to provide the capability of performing mathematical operations on an input voltage. In the following, we discuss the operational amplifier circuit needed for solving linear differential equations with constant coefficients.

Voltage Inverter

To construct a *voltage inverter*, we add a resistor R_1 to the input side of an operational amplifier and connect an identical feedback resistor between output and input, as in Fig. 9.3*a*. If an input voltage e_1 is applied as in Fig. 9.3*b*, currents i_1, i_g, and i_f will exist in the circuit as shown. Applying Kirchhoff's law gives

$$i_1 + i_f = i_g \tag{9.5}$$

which can be written in terms of voltages as

$$\frac{e_1 - e_g}{R_1} + \frac{e_o - e_g}{R_1} = i_g \tag{9.6}$$

where e_1, e_g, and e_o are the input, grid, and output voltages, respectively. Another characteristic of the operational amplifiers used in analog computers is that they have a very high internal input impedance. This results in very small input current so that $i_g \simeq 0$ and Eq. (9.6) becomes

$$\frac{e_1 - e_g}{R_1} + \frac{e_o - e_g}{R_1} = 0 \tag{9.7}$$

Substituting for e_g from Eq. (9.4) gives

$$e_1 + \frac{2e_o}{G} + e_o = 0 \tag{9.8}$$

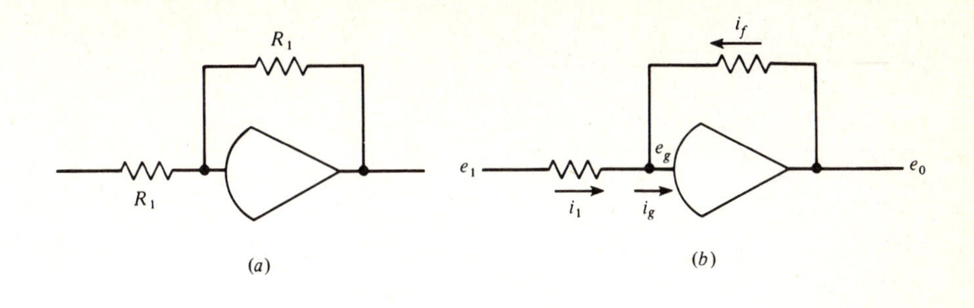

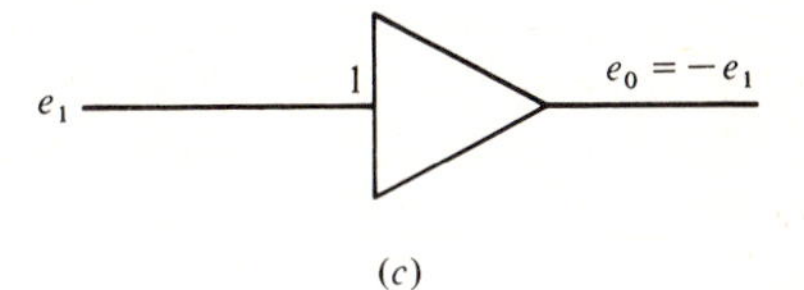

Figure 9.3

Most analog computers are designed for amplifier operations within the range of $\pm$ 10 V dc (older models may be $\pm$ 100 V dc). Recalling that $G \geq 10^5$, we see that the term $2e_o/G$ is small and may be neglected to obtain

$$e_o = -e_1 \tag{9.9}$$

as the circuit equation for Fig. 9.3*a*. The output voltage has the same magnitude as the input voltage but is of opposite polarity. Such a circuit is called a *voltage inverter* or simply an *inverter*, and its computer symbol is as shown in Fig. 9.3*c*. Note that the *net gain* of this amplifier circuit is unity, and this is indicated on the computer symbol.

Multiplication by a Constant

Consider next the circuit shown in Fig. 9.4, in which the feedback resistance R_f may have any value. The circuit equation for this amplifier is obtained by replacing R_1 with R_f in the second term of Eq. (9.7). This gives

$$\frac{e_1 - e_g}{R_1} + \frac{e_o - e_g}{R_f} = 0 \tag{9.10}$$

Again, e_g is small in comparison to the other voltages, and we may neglect it to obtain

$$e_o = -\frac{R_f}{R_1} e_1 \tag{9.11}$$

as the circuit equation for Fig. 9.4. We note that the input voltage is inverted in polarity *and* its magnitude is multiplied by the constant R_f/R_1 to produce the output voltage e_o. The computer symbol for this circuit is the same as that in Fig. 9.3*c*, except that the net gain would be indicated as the numerical value of

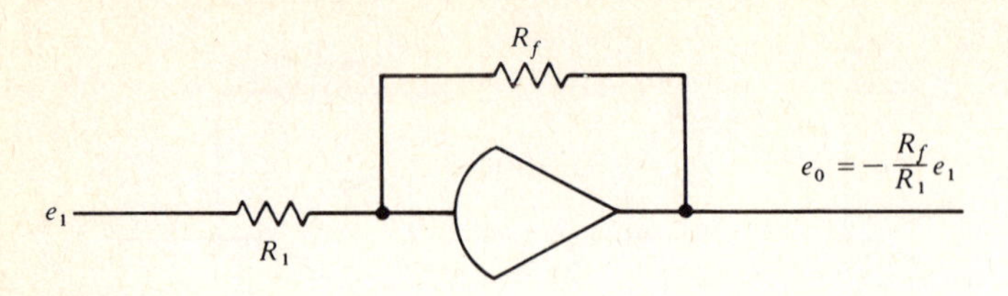

Figure 9.4

R_f/R_1. In theory, the net gain of such an amplifier circuit can be made equal to any desired value by the proper combination of resistances. In practice, analog-computer amplifiers are provided with multiple input terminals which are internally connected to fixed resistors to give net gains of 1, or 10. Other gains are obtained through the use of coefficient potentiometers, which will be discussed in Sec. 9.3.

Summing Amplifier

When an operational amplifier is connected as shown in Fig. 9.5*a*, the amplifier circuit is known as a *summing amplifier*. Two external input voltages e_1 and e_2 are applied through external resistances R_1 and R_2, respectively. For this circuit, Kirchhoff's law gives

$$i_1 + i_2 + i_f = i_g \simeq 0 \tag{9.12}$$

or

$$\frac{e_1 - e_g}{R_1} + \frac{e_2 - e_g}{R_2} + \frac{e_o - e_g}{R_f} = 0 \tag{9.13}$$

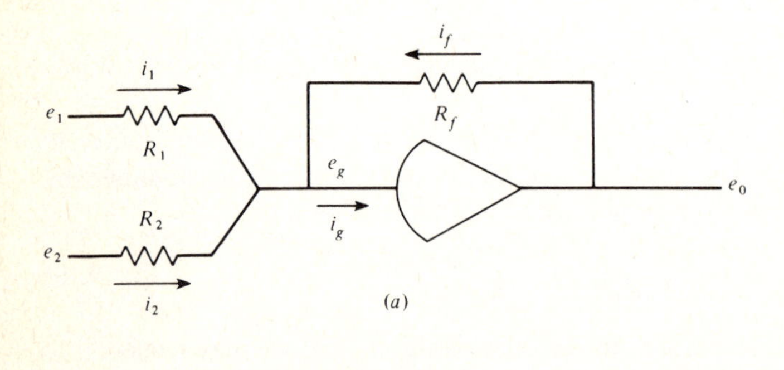

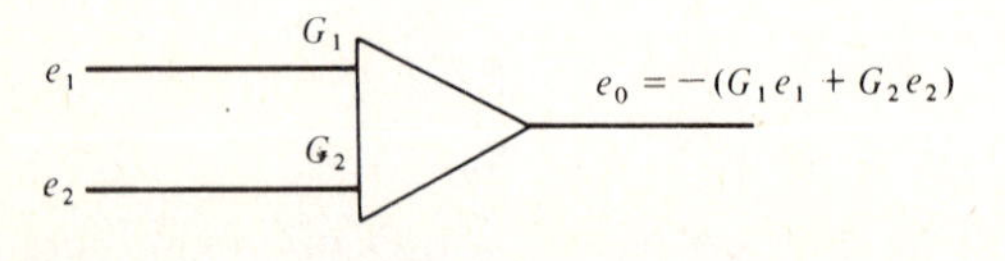

(*b*)

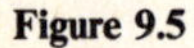
Figure 9.5

Using $e_g = -e_o/G \simeq 0$ results in

$$e_o = -\left(\frac{R_f}{R_1}e_1 + \frac{R_f}{R_2}e_2\right) \tag{9.14}$$

as the circuit equation for a summing amplifier. Each input voltage is multiplied by a constant, the two resulting voltages are added numerically, and the polarity of the sum is reversed. If e_1 and e_2 have the same polarity, the amplifier performs addition; if e_1 and e_2 are of opposite polarity, the operation is subtraction. It is important to note that the net gain of the amplifier may be different with respect to each input voltage. In analog computation, the gain of an amplifier refers to the gain for a particular input. The computer symbol for a summing amplifier is shown in Fig. 9.5*b*.

A summing amplifier is not limited to two input voltages. In general, if there are n input voltages through n input resistances, the output of a summing amplifier is

$$e_o = -\left(\frac{R_f}{R_1}e_1 + \frac{R_f}{R_2}e_2 + \cdots + \frac{R_f}{R_n}e_n\right) \tag{9.15}$$

Most analog-computer amplifiers provide terminals for up to five inputs. Additional capability is obtained by using multiple amplifiers in series and parallel combinations.

Example 9.1 For the circuit shown in Fig. 9.6*a*, draw the corresponding computer symbol and determine the output voltage.

SOLUTION The net gain is

$$G_1 = \frac{R_f}{R_1} = \frac{10^6}{10^5} = 10 \text{ V/V}$$

and the computer symbol is shown in Fig. 9.6*b*. From Eq. (9.11), the output

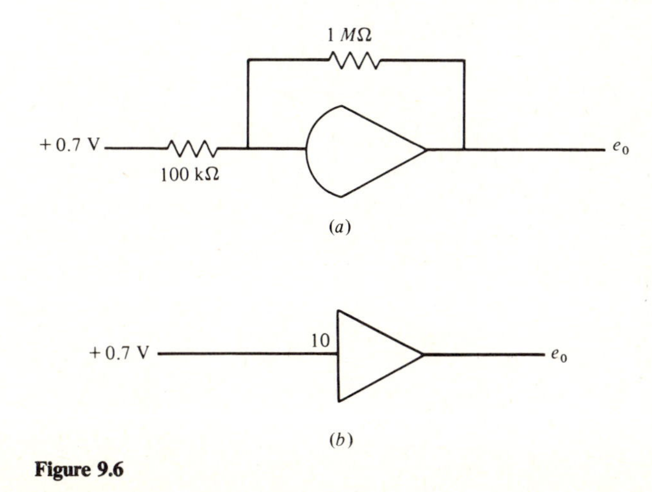

Figure 9.6

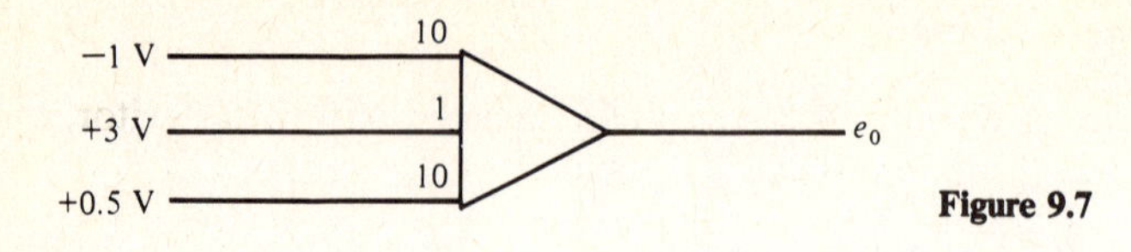

Figure 9.7

voltage is

$$e_o = -\frac{R_f}{R_1} e_1 = -10 \times 0.7 = -7 \text{ V}$$

Example 9.2 Determine the output voltage for the amplifier circuit of Fig. 9.7.

SOLUTION This is an example of a summing amplifier with three inputs. In this case we have $G_1 = 10$, $G_2 = 1$, and $G_3 = 10$ so that Eq. (9.15) gives

$$e_o = -[10(-1) + 1(3) + 10(0.5)] = 2 \text{ V}$$

Example 9.3 Determine the output voltage e_3 of the amplifier circuit shown in Fig. 9.8.

SOLUTION We first calculate e_1 and e_2 using Eq. (9.14):

$$e_1 = -[10(-1.5) + 1(6)] = 9 \text{ V}$$

$$e_2 = -[1(-3) + 1(6.5)] = -3.5 \text{ V}$$

Output voltages e_1 and e_2 are the inputs for amplifier 3, so a third application of Eq. (9.14) gives

$$e_3 = -[1(9) + 1(-3.5)] = -5.5 \text{ V}$$

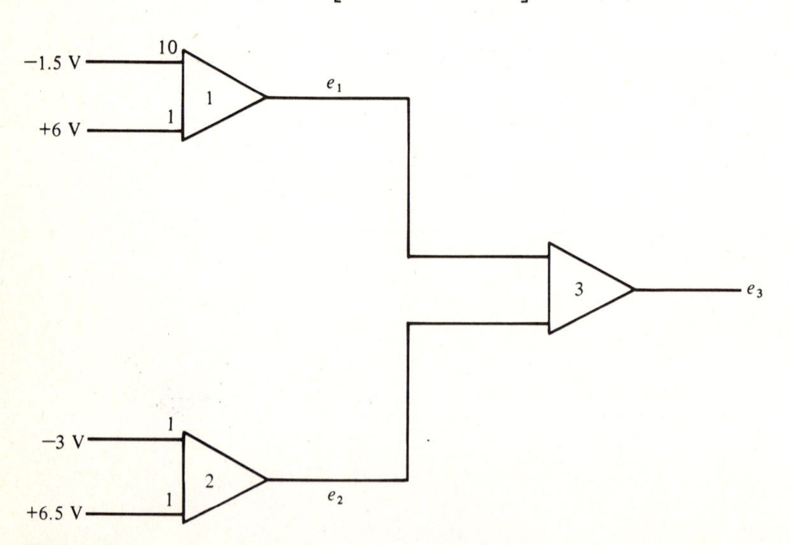

Figure 9.8

Integrating Amplifier

If the feedback resistor of an amplifier circuit is replaced with a capacitor, the mathematical function of the amplifier is altered significantly. Consider an operational amplifier which has a feedback capacitance C and an input voltage e_1 through resistance R_1, as in Fig. 9.9*a*. Applying Kirchhoff's law to the circuit gives

$$i_1 + i_f = i_g \simeq 0 \tag{9.16}$$

where we have again utilized the fact that the current to the input stage of the amplifier is negligible. Feedback current through the capacitor is proportional to the time rate of change of potential across the capacitor, so Eq. (9.16) can be written

$$\frac{e_1 - e_g}{R_1} + C\frac{d}{dt}(e_o - e_g) = 0 \tag{9.17}$$

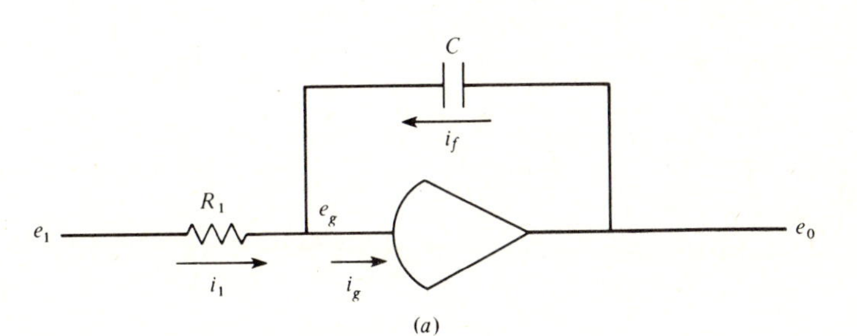

(*a*)

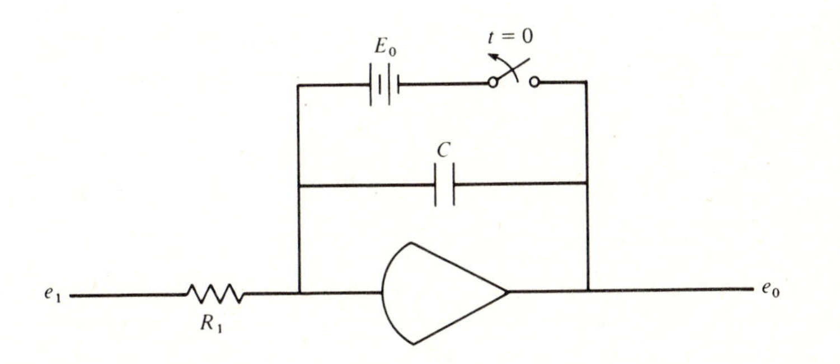

(*b*)

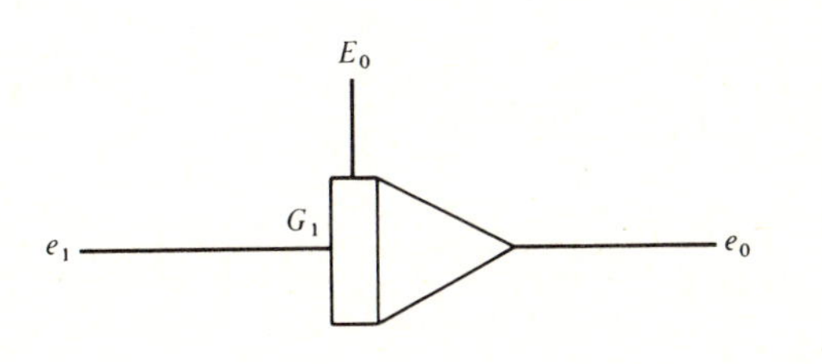

(*c*)

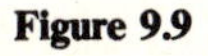
Figure 9.9

Since e_g is also negligible, this becomes

$$\frac{e_1}{R_1} + C\frac{de_o}{dt} = 0 \tag{9.18}$$

or

$$\frac{de_o}{dt} = -\frac{e_1}{R_1C} \tag{9.19}$$

Integrating Eq. (9.19) with respect to time gives the output voltage as

$$e_o = -\int_0^t \frac{e_1}{R_1C}\,dt + E_o \tag{9.20}$$

where E_o is a constant of integration which represents the value of the output voltage at $t = 0$. Equation (9.20) is the circuit equation for an *integrating amplifier*. The input voltage is multiplied by the gain $1/R_1C$ and integrated with respect to time, and the result is inverted in polarity. Integration begins at the instant an input voltage is applied and proceeds in real time until the input is removed.

To obtain a specified initial condition E_o, it is necessary to modify the circuit as shown in Fig. 9.9*b*. A voltage source equal to E_o is connected across the capacitor and in series with a switch which opens at time zero. In this manner, an initial charge corresponding to $e_o = E_o$ is placed on the capacitor. At $t = 0$, an external input is applied to the amplifier, and the switch opens, removing voltage source E_o from the circuit. The charge on the capacitor ensures that the effect of the initial condition is present. All analog computers utilize a main control switch which automatically takes care of switching the initial-condition voltage source out of the circuit at time zero. Generally, the main control is a three-position switch labeled *reset/hold/compute*. In the *reset* position, all external inputs to the amplifiers are removed and normally closed relays are used to apply initial-condition voltages. When the main control switch is set to the *compute* position, all external inputs are applied and the relays automatically open, thus removing the initial-condition voltage sources. In the *hold* position, all circuit voltages are "frozen" at the values attained when the switch was placed at hold. Regardless of the particular construction details and nomenclature used, most analog computers automatically accomplish removal of initial-condition voltage sources. One additional point must be made with regard to initial conditions, however. In the majority of commercial analog computers, *initial-condition voltages are inverted in polarity*. Since the integrating amplifier preserves the inverting character of the summing amplifier, inversion of initial-condition voltages is simply a matter of consistency. In practice this means that if the specified initial value of e_o is positive, applied voltage E_o must be negative, and vice versa. This convention will be used in the examples which follow. The computer symbol for the integrating amplifier will be as shown in Fig. 9.9*c*, which explicitly shows the applied initial voltage E_o as well as the net gain for the input.

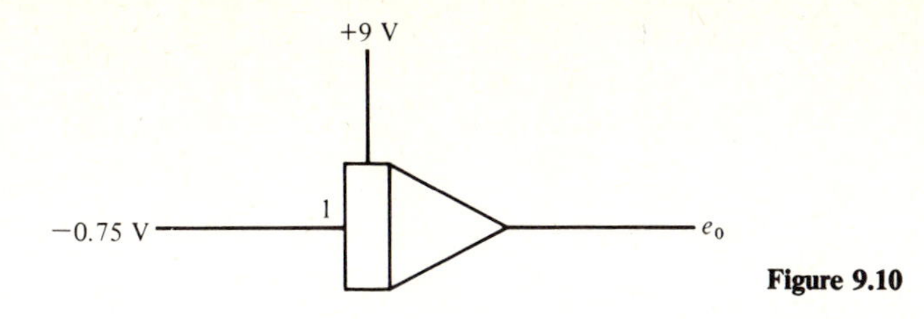

Figure 9.10

Example 9.4 For the integrating amplifier shown in Fig. 9.10, determine the output voltage as a function of time and the value of the output voltage when $t = 2$ s.

SOLUTION From the diagram we have $G_1 = 1/R_1C = 1$, $e_1 = -0.75$ V, and $E_o = -9$ (that is, $-E_o = +9$), so that Eq. (9.20) gives

$$e_o = -\int_0^t 1(-0.75)\,dt - 9$$

or

$$e_o = 0.75t - 9$$

At $t = 2$ s,

$$e_o = 0.75(2) - 9 = -7.5 \text{ V}$$

Summing Integrator

The integrating amplifier is not limited to a single input voltage. If the feedback resistor of the summing amplifier of Fig. 9.5*a* is replaced with a capacitor (and an initial-condition voltage is added), the amplifier becomes a *summing integrator*. An analysis similar to that used for a single input will show that the circuit equation is

$$e_o = -\int_0^t \left(\frac{e_1}{R_1C} + \frac{e_2}{R_2C}\right) dt + E_o \tag{9.21}$$

Input voltages are modified by their respective gains, added together algebraically, and integrated with respect to time, and the polarity of the resultant voltage function is inverted. The number of inputs to a summing integrator is limited only by the number of input terminals available on the particular amplifier. For n input voltages, the general circuit equation for a summing integrator is

$$e_o = -\int_0^t \left(\frac{e_1}{R_1C} + \frac{e_2}{R_2C} + \cdots + \frac{e_n}{R_nC}\right) dt + E_o \tag{9.22}$$

Example 9.5 Given the summing integrator shown in Fig. 9.11, find the output voltage at $t = 3$ s.

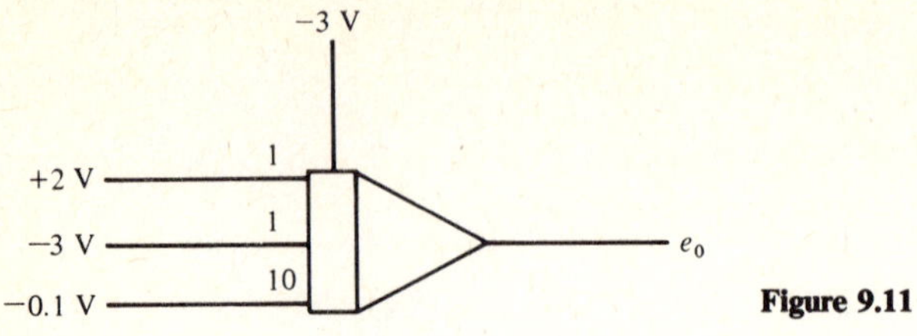

Figure 9.11

SOLUTION From the diagram we have

$$e_1 = 2 \qquad G_1 = 1$$

$$e_2 = -3 \qquad G_2 = 1$$

$$e_3 = -0.1 \qquad G_3 = 10$$

$$E_o = 3$$

Substitution into Eq. (9.22) with $n = 3$ gives

$$e_o = -\int_0^t [1(2) + 1(-3) + 10(-0.1)]\, dt + 3$$

$$= -\int_0^t -2\, dt + 3 = 2t + 3$$

At $t = 3$ s,

$$e_o = 2(3) + 3 = 9 \text{ V}$$

9.3 COEFFICIENT POTENTIOMETERS

As previously mentioned, the amplifiers used in most analog computers are internally circuited to provide gains of unity and 10. In addition, precision voltage supplies of ± 10 V dc (or ± 100 V dc) are provided for use as input voltages. To provide necessary flexibility, analog computers incorporate a number of variable resistors, referred to as *coefficient potentiometers*, which may be used to modify amplifier gains and voltage supplies to any value between 0 and 10. As shown in the circuit diagram of Fig. 9.12*a*, a potentiometer is composed of a resistor connected between a voltage source and ground, and a movable arm which "picks off" the voltage at any position along the length of the resistor. Thus, the arm voltage is some fraction of the applied voltage. A coefficient potentiometer provides a means for multiplying a voltage by any constant value between 0 and 1 inclusive. The coefficient potentiometers used in analog computers may be set accurately to three decimal places. They are represented symbolically as shown in Fig. 9.12*b*, where the value of K is the potentiometer setting, with $0 \leq K \leq 1$.

As an example of how the gain of an amplifier may be modified through the use of a coefficient potentiometer, consider the circuit depicted in Fig. 9.13 in which a potentiometer with a setting K has been placed in the input line of an

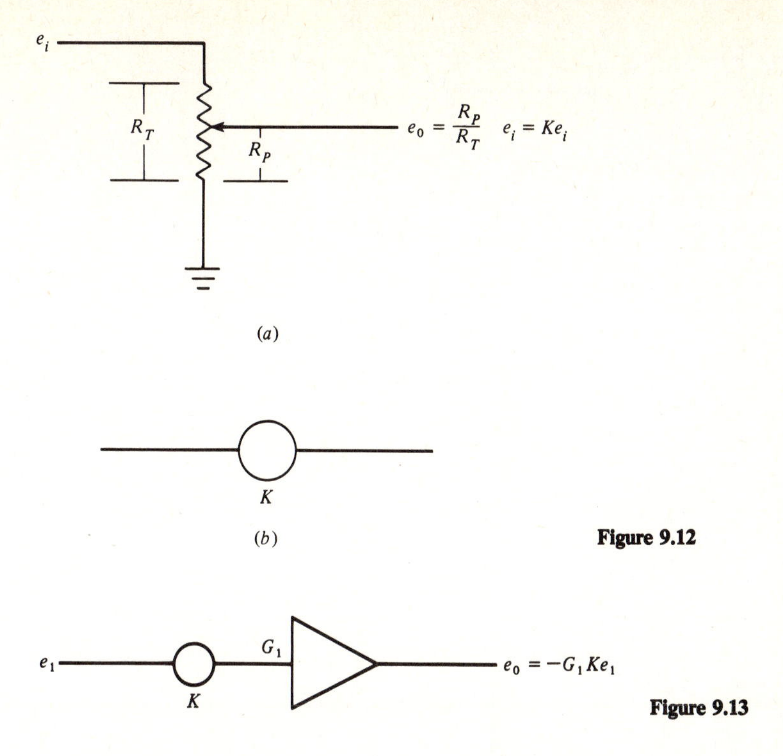

Figure 9.12

Figure 9.13

amplifier. The input is connected to the amplifier such that the gain would normally be 10. The effect of the potentiometer is to reduce the voltage e_1 such that the actual input voltage to the amplifier is

$$e_1' = Ke_1 \qquad 0 \le K \le 1 \tag{9.23}$$

The output voltage is then

$$e_o = -G_1 e_1' = -G_1 K e_1 \tag{9.24}$$

so that the effective gain of the amplifier is

$$G_1' = G_1 K \qquad 0 \le G_1' \le G_1 \tag{9.25}$$

Example 9.6 Determine the output voltage of amplifier 3 of the system in Fig. 9.14*a*.

SOLUTION Labeling the various voltages as shown in Fig. 9.14*b*, we have

$$e_{11}' = 10(0.125) = 1.25 \text{ V}$$
$$e_{22}' = -10(0.361) = -3.61 \text{ V}$$
$$e_{12}' = 10(0.500) = 5 \text{ V}$$

The output of amplifier 1 is

$$e_{o1} = -[10(1.25) + 1(-3.61)] = -8.89 \text{ V}$$

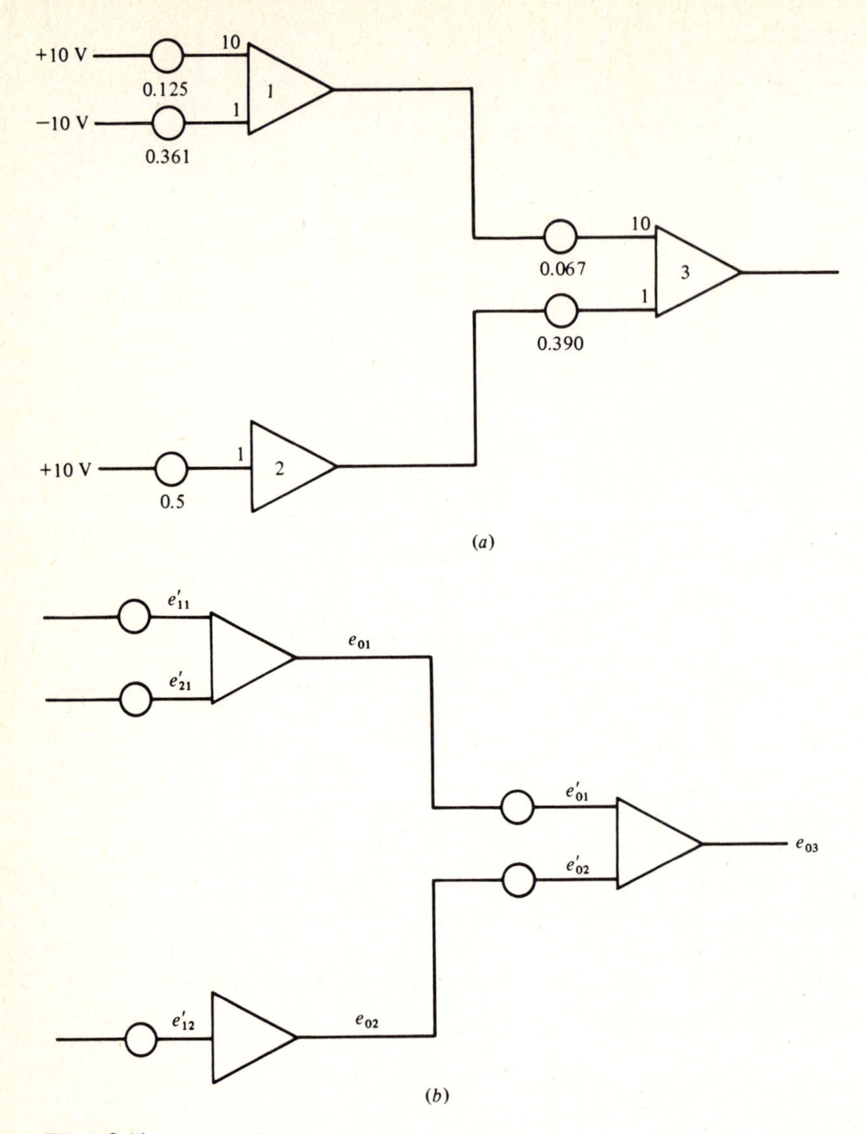

Figure 9.14

and the output of amplifier 2 is

$$e_{o2} = -10(0.500) = -5 \text{ V}$$

The input voltage to amplifier 3 becomes

$$e'_{o1} = -8.89(0.067) = -0.596 \text{ V}$$

and
$$e'_{o2} = -5(0.390) = -1.95 \text{ V}$$

giving an output of

$$e_{o3} = -[10(-0.596) + 1(-1.95)] = 7.91 \text{ V}$$

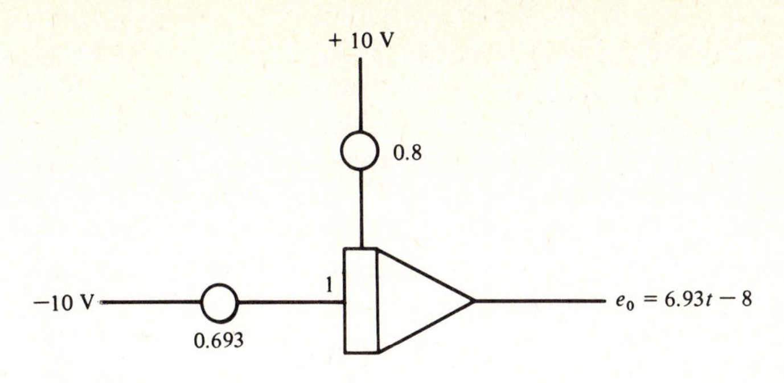

Figure 9.15

Example 9.7 Draw the computer diagram for an integrating amplifier which has as its output the function $e_o = 6.93t - 8$.

SOLUTION To obtain the proper amplifier characteristics, we must determine the required input voltage, the required gain, and the initial condition. Setting $t = 0$ gives the initial condition as

$$E_o = 6.93 \times 0 - 8 = -8$$

Substituting all known values into Eq. (9.20), we have

$$6.93t - 8 = -\int_0^t G_1' e_1 \, dt - 8$$

where G_1' has been used to denote the effective gain of the integrator. Differentiating with respect to time gives

$$6.93 = -G_1' e_1$$

or

$$G_1' e_1 = -6.93$$

Assuming that the voltage sources available are ± 10 V dc, we must select $e_1 = -10$ V, from which we deduce that the effective gain must be 0.693. This is best accomplished by using a nominal amplifier gain of unity and a potentiometer set at 0.693. The computer diagram is shown in Fig. 9.15. Note the use of the second potentiometer to obtain the value of E_o.

9.4 CIRCUIT DIAGRAMS FOR PROBLEM SOLVING

With only the few very basic components discussed so far, the analog computer provides a powerful tool for solving many differential equations. Of particular importance to the study of vibrations, the analog computer is capable of solving the differential equation

$$\ddot{x} + a_1\dot{x} + a_2 x = f(t) \tag{9.26}$$

for constant values of a_1 and a_2 and for certain functions $f(t)$.

First we shall consider the homogeneous equation

$$\ddot{x} + a_1\dot{x} + a_2 x = 0 \tag{9.27}$$

subject to the initial conditions $x(0) = x_o$ and $\dot{x}(0) = \dot{x}_o$. To obtain an analog solution, a computer circuit must be constructed such that the differential equation governing the voltages in the circuit is identical to Eq. (9.27) and the given initial conditions are satisfied. As a recommended general approach for constructing analog diagrams, assume that a voltage corresponding to the solution function $x = x(t)$ is obtained at a certain point in the circuit. Mathematically, the solution function x is the integral with respect to time of $\dot{x} = dx/dt$. Given the analog-computer components that are available, it follows that the solution will be the output of an integrating amplifier, as shown in Fig. 9.16*a*. Note that the input to the integrator is *negative* $\dot{x}$, to account for the

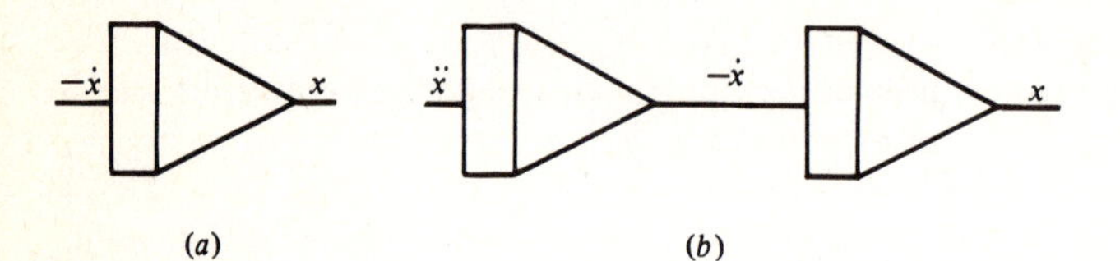

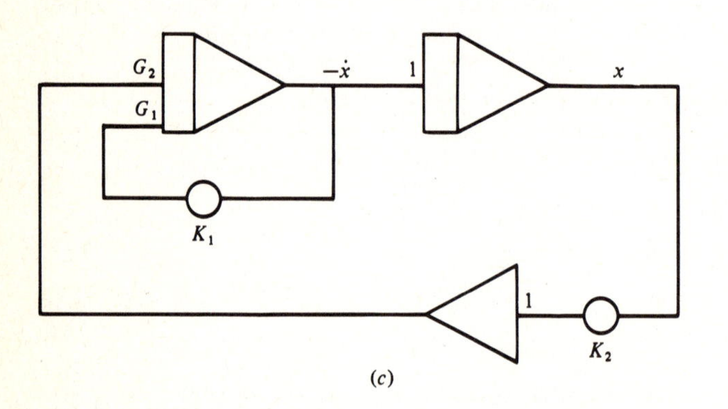

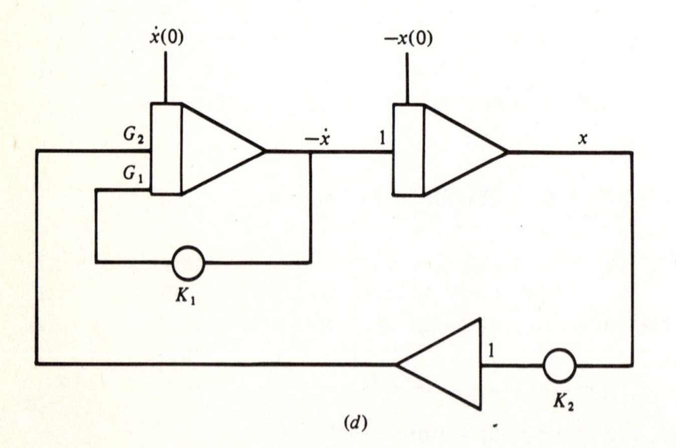

Figure 9.16

inverting character of the amplifier. Continuing this reasoning, we are led to the addition of a second integrator (Fig. 9.16*b*) having $\ddot{x}$ as its input and negative $\dot{x}$ as its output. This process could continue indefinitely and to no avail. This is unnecessary, however, since the solution to a second-order differential equation requires only two integrations, and this is precisely what our diagram shows so far. If we can now obtain a voltage or voltages equivalent to $\ddot{x}$, the computer circuit will be complete except for satisfying the initial conditions. To obtain $\ddot{x}$ and ensure that the circuit corresponds to the given differential equation, we rewrite Eq. (9.27) as

$$\ddot{x} = -a_1\dot{x} - a_2 x \tag{9.28}$$

which shows that $\ddot{x}$ is a function of the assumed outputs of the integrators. The diagram shown in Fig. 9.16*c* is now drawn to satisfy Eq. (9.28). For this equation to be satisfied, the amplifier gains and potentiometer settings must be such that

$$G_1K_1 = a_1 \tag{9.29}$$

and

$$G_2K_2 = a_2 \tag{9.30}$$

To complete the analog circuit diagram requires only the addition of the initial-condition voltages as shown in Fig. 9.16*d*. The resulting circuit applies to any equation having the form of Eq. (9.27) and may be used for each such equation by satisfying Eqs. (9.29) and (9.30) and the initial conditions.

To obtain the appropriate computer circuit for Eq. (9.26) for $f(t) \neq 0$ is a straightforward extension. The solution still requires only two integrations, so the basic circuit is unchanged. For $f(t) \neq 0$, Eq. (9.28) becomes

$$\ddot{x} = -a_1\dot{x} - a_2 x + f(t) \tag{9.31}$$

so that an additional voltage input corresponding to $f(t)$ is required. The computer diagram for the inhomogeneous equation is shown in Fig. 9.17. Depending upon the complexity of $f(t)$, it may not be possible to obtain a voltage representation for $f(t)$ with the computer components discussed thus far. Examples using some common functions will be presented later.

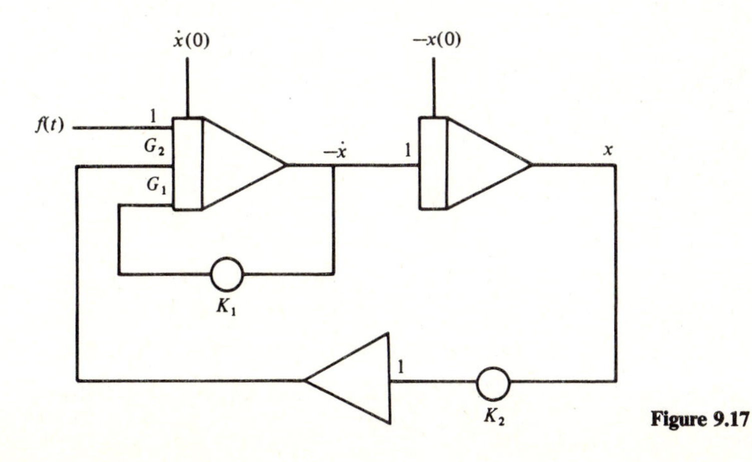

Figure 9.17

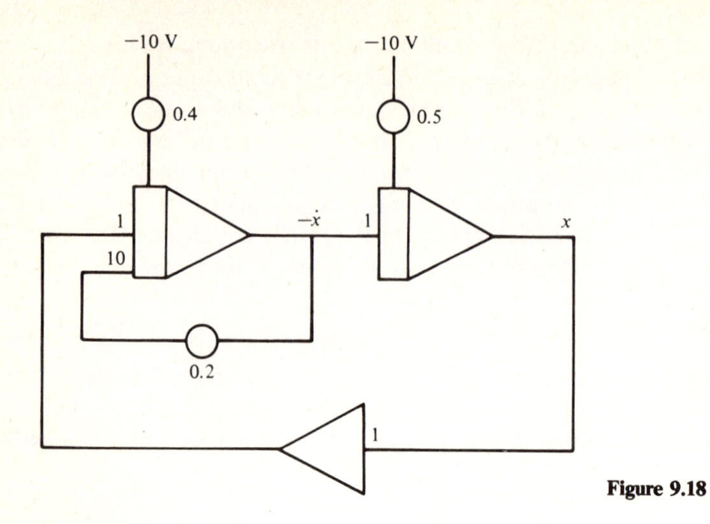

Figure 9.18

Example 9.8 Draw the analog-computer diagram for the differential equation

$$\ddot{x} + 2\dot{x} + x = 0$$

subject to $x(0) = 5$ and $\dot{x}(0) = -4$. Assume that the nominal amplifier gains available are unity and 10, and that the voltage supplies are ± 10 V.

SOLUTION Adapting the general diagram of Fig. 9.16*d* to the problem at hand, we have

$$G_1K_1 = 2 \qquad G_2K_2 = 1$$

$$x_0 = 5 \qquad \dot{x}_0 = -4$$

The equation will be satisfied if $G_1 = 10$ and $K_1 = 0.2$ while $G_2 = 1$ and $K_2 = 1$. The completed computer diagram is depicted in Fig. 9.18.

Example 9.9 Draw the analog-computer diagram for the differential equation

$$\ddot{x} + x = 0$$

subject to $x(0) = 0$ and $\dot{x}(0) = 5$.

SOLUTION The differential equation does not contain the first derivative, so $a_1 = 0$. In terms of the computer diagram of Fig. 9.16*d*, this means that $G_1K_1 = 0$ or, more simply, that one of the inputs to the first integrator is unnecessary. The computer diagram is as shown in Fig. 9.19*a*.

Let us further examine the significance of this example by noting that the general solution to $\ddot{x} + x = 0$ is

$$x(t) = A \sin t + B \cos t$$

which, after we apply the initial conditions $x(0) = 0$ and $\dot{x}(0) = 5$, becomes

$$x(t) = 5 \sin t$$

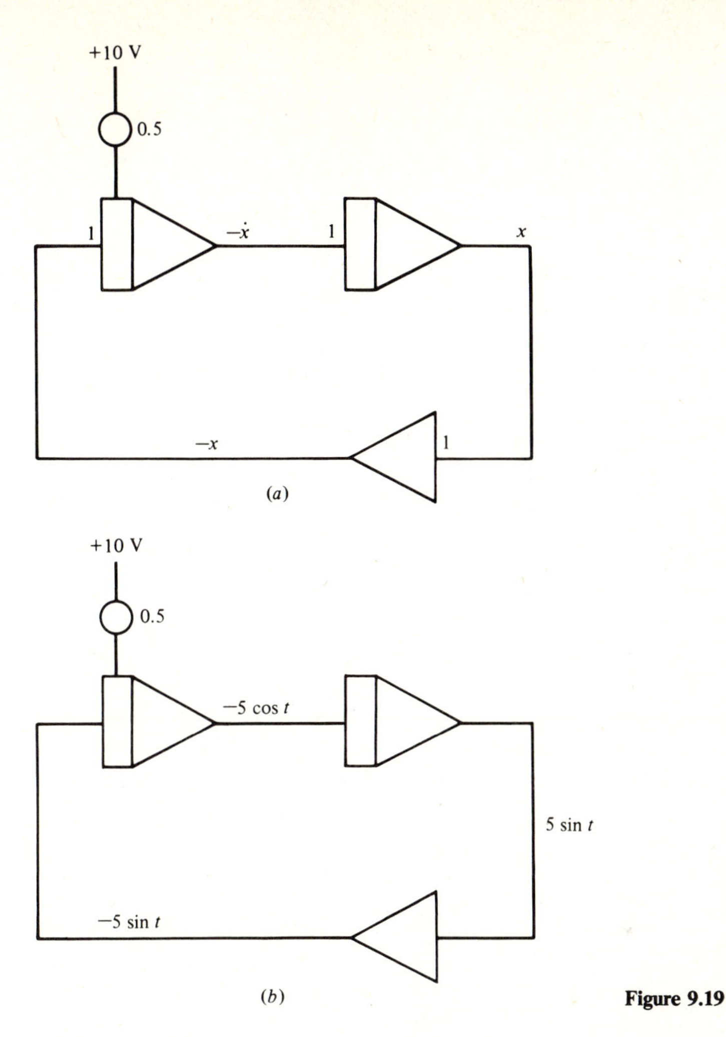

Figure 9.19

The output voltage of the second integrator is a sine function, and

$$\dot{x}(t) = 5 \cos t$$

shows that the output of the first integrator will be a cosine function. The functional form of the voltage at each point in the circuit is as indicated in Fig. 9.19*b*. This is an extremely important point as it shows the capability of the analog computer to generate certain functions. This capability will be used to obtain $f(t)$ for the solution of inhomogeneous differential equations.

The results of Example 9.9 can be extended to a more general case by considering the differential equation

$$\ddot{x} + \omega^2 x = 0 \tag{9.32}$$

which has the solution

$$x(t) = A \sin \omega t + B \cos \omega t \tag{9.33}$$

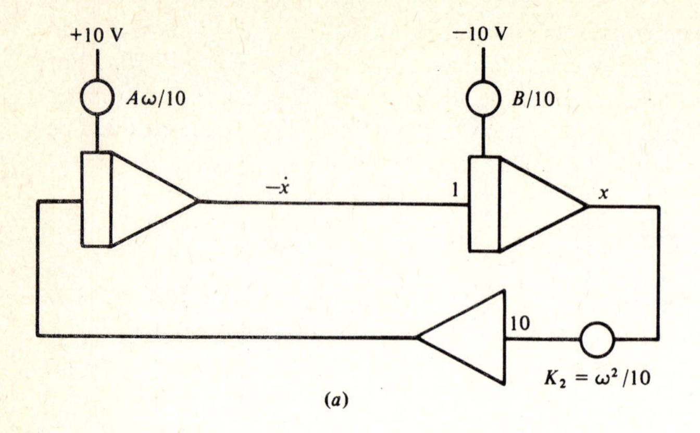

(*a*)

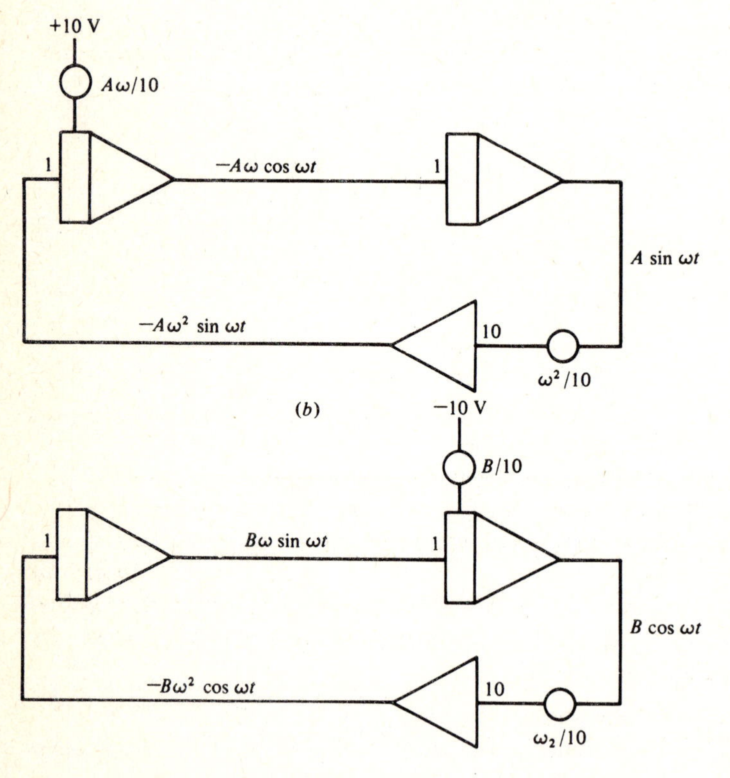

(*c*)

Figure 9.20

In terms of the arbitrary constants, the initial conditions are $x(0) = B$ and $\dot{x}(0) = A\omega$. The corresponding computer diagram is shown in Fig. 9.20*a*. For the moment we must assume that $\omega^2 \leq 10$ in order that the potentiometer setting K_2 not be greater than unity. We shall now show that this circuit can be used to generate sine *or* cosine functions for any frequency.

Case 1: Sine-Function Generator

If the initial conditions corresponding to Eq. (9.32) are specified as $x(0) = 0$ and $\dot{x}(0) = A\omega$, where A is any nonzero constant, the solution becomes

$$x(t) = A \sin \omega t \tag{9.34}$$

from which

$$\dot{x}(t) = A\omega \cos \omega t \tag{9.35}$$

and

$$\ddot{x}(t) = -A\omega^2 \sin \omega t = -\omega^2 x \tag{9.36}$$

The computer diagram for this case is labeled in Fig. 9.20*b*, from which it can be seen that the output of the second integrator is the sine function.

Case 2: Cosine-Function Generator

If the initial conditions are specified as $x(0) = B$ and $\dot{x}(0) = 0$, where B is any nonzero constant, the solution and its derivatives become

$$x(t) = B \cos \omega t \tag{9.37}$$

$$\dot{x}(t) = -B\omega \sin \omega t \tag{9.38}$$

$$\ddot{x}(t) = -B\omega^2 \cos \omega t = -\omega^2 x \tag{9.39}$$

The corresponding computer diagram is shown in Fig. 9.20*c*, from which it may be seen that the cosine function is obtained.

The sine- and cosine-function generators are very necessary to the study of vibration problems by analog-computer techniques since the external forces which act on mechanical systems are often harmonic functions of time. Function generation with analog-computer circuits is by no means limited to harmonic functions, however. Essentially any function which is known to be the solution of an ordinary differential equation may be generated, as will be illustrated by an additional example.

Example 9.10 For the first-order differential equation

$$\dot{x} + x = 0$$

with the initial condition $x(0) = 1$, draw the analog-computer diagram and determine the functional form of $x(t)$.

SOLUTION For a first-order differential equation, only one integration is required to obtain the solution. Therefore the computer circuit will contain only a single integrator connected such that $\dot{x} = -x$ and $x(0) = 1$. The computer diagram is as shown in Fig. 9.21. To determine the functional

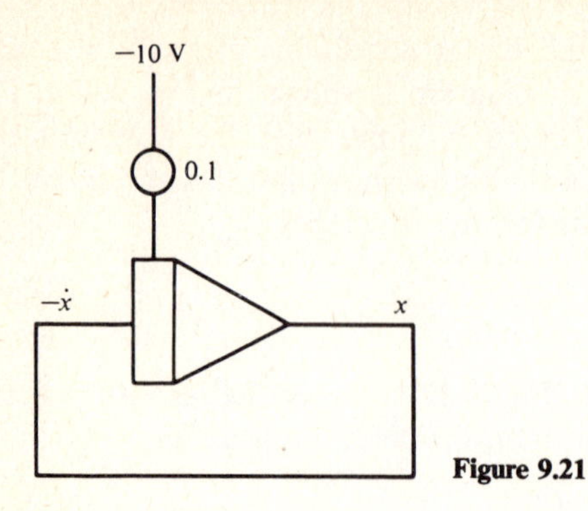

Figure 9.21

form of $x(t)$, we shall solve the differential equations using the basic method of Chap. 2. Assuming that the solution is

$$x(t) = Ce^{st}$$

we obtain

$$\dot{x}(t) = Cse^{st}$$

Substituting into the differential equation gives

$$Cse^{st} + Ce^{st} = 0$$

or

$$(s + 1)Ce^{st} = 0$$

from which we obtain $s = -1$. The solution is then

$$x(t) = Ce^{-t}$$

and the initial condition $x(0) = 1$ will be satisfied if $C = 1$. The solution is the negative exponential function $x(t) = e^{-t}$, and we have thus obtained a computer circuit for generating this function. Note that by changing the value of the initial condition, any constant multiple of the exponential function can be obtained.

9.5 COMPUTER SCALING

The operational amplifiers used in most analog computers are limited to an operating voltage range of ± 10 V and perform integration with *real time* as the independent variable. These analog-computer features often become severely restrictive when physical problems are to be solved. Since the dependent variables involved in a physical system are represented by voltages in the analog computer, *magnitude scaling* is often required to ensure that the voltage limits of the computer are not exceeded. Additionally, it may be necessary to change the time base through *time scaling* to expand or contract actual computing time. Mathematically the scaling operations are simply changes of the dependent or independent variables to suit the computer limitations.

Magnitude Scaling

The values of physical problem variables most often numerically exceed the voltage limitations of analog-computer amplifiers. Thus it becomes necessary to relate the problem variables to their voltage representations through the use of *magnitude scale factors*. In addition to maintaining voltages within appropriate operating ranges, such scale factors should be chosen so that computer voltages are as near the maximum allowable value as possible. This is because the computer-generated solution is more susceptible to error caused by electrical noise at lower voltage levels. As an example, consider a vibrating spring-mass system in which the initial displacement is 25 mm. If this problem is to be solved on an analog computer which has a range of ± 10 V, a one-to-one relationship would produce an overload condition of 25 V. Rather, a scale factor of 2.5 mm/V must be used to relate the computer variable (voltage) to the physical displacement variable.

In scaling physical problems for analog-computer solution, it is neither necessary nor desirable to use the same scale factor for each dependent variable. For vibration problems in particular, it is appropriate to use three different scale factors for displacement, velocity, and acceleration. The scale factors are determined by relating the maximum *anticipated* values of the problem variables to the maximum allowable voltages for the computer. Since it is not always possible to estimate accurately the maximum values of the variables, it may be necessary to determine scale factors through a trial-and-error adjustment procedure. The techniques involved in magnitude scaling are illustrated by the following example. Throughout this chapter, we assume an operating range of ± 10 V.

Example 9.11 Draw the analog-computer diagram for a circuit which will generate a voltage corresponding to the function $x(t) = 10 \sin 3t$ in.

SOLUTION From the discussion of Sec. 9.4, the given function $x(t)$ is the solution to the differential equation

$$\ddot{x} + 9x = 0$$

with initial conditions $x(0) = 0$ and $\dot{x}(0) = 30$. Since the analog-computer solution of a second-order differential equation requires two integrations, the dependent variables in the problem are

$$x(t) = 10 \sin 3t$$

$$\dot{x}(t) = 30 \cos 3t$$

$$\ddot{x}(t) = -90 \sin 3t$$

and we observe that both the first- and second-derivative terms will exceed the maximum allowable computer voltage. Magnitude scaling is therefore required. We introduce the new variables

$$x_c = S_0 x \qquad \dot{x}_c = S_1 \dot{x} \qquad \ddot{x}_c = S_2 \ddot{x}$$

where the subscript c denotes computer variables, and S_0, S_1, and S_2 are the scale factors for the zeroth, first, and second derivatives, respectively. We seek scale factors such that the values of the computer variables fall within the operating range. In this case, the variables are oscillatory so we actually will use scale factors for which the oscillations utilize as much of the operating voltage range as possible. Substituting the known functional forms, we have

$$x_c = S_0(10 \sin 3t)$$
$$\dot{x}_c = S_1(30 \cos 3t)$$
$$\ddot{x}_c = -S_2(90 \sin 3t)$$

If we choose the scale factors

$$S_0 = 1 \text{ V/in} \qquad S_1 = \tfrac{1}{3}\text{V/ (in/s)} \qquad S_2 = \tfrac{1}{9}V/\text{ (in/s}^2)$$

the computer variables become

$$x_c = 10 \sin 3t$$
$$\dot{x}_c = 10 \cos 3t$$
$$\ddot{x}_c = -10 \sin 3t$$

with initial conditions $x_c(0) = 0$ and $\dot{x}_c(0) = 10$.

At this point, we must caution the reader by pointing out that $\dot{x}_c$ and $\ddot{x}_c$ are *not* the first and second derivatives of x_c. This can be seen from the equations above. The correct mathematical relations are obtained by use of the proper amplifier gains in satisfying the differential equation $\ddot{x} + 9x = 0$ on the analog computer. To illustrate, consider Fig. 9.22*a*, which shows the analog-computer diagram for the *unscaled* problem. In this diagram, $x(t) = 10 \sin 3t$ is represented by the voltage output of amplifier 2. The output voltages of amplifiers 1 and 3 represent the first and second derivatives, respectively, and these amplifiers will be overloaded. The computer diagram for the scaled problem is shown in Fig. 9.22*b*. Note that the diagram still includes two integrators and an inverter, but the gains must be adjusted so that both the scaling factors and the original differential equation are satisfied. Since

$$x = \frac{x_c}{S_0} \qquad \text{and} \qquad \dot{x} = \frac{\dot{x}_c}{S_1}$$

the mathematical function of amplifier 2 is described by

$$x = -\int_0^t \dot{x}\, dt + x(0)$$

or

$$x_c = -\int_0^t \frac{K_2 G_2 \dot{x}_c}{S_1}\, dt + x_c(0)$$

A comparison of the previous two equations yields $K_2 G_2 = 3$, which will be

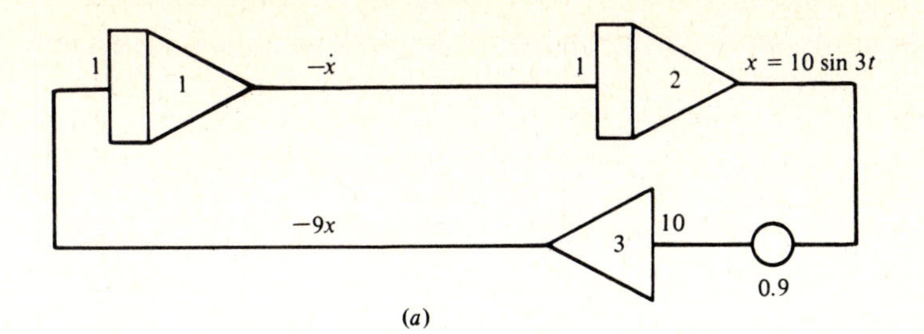

(a)

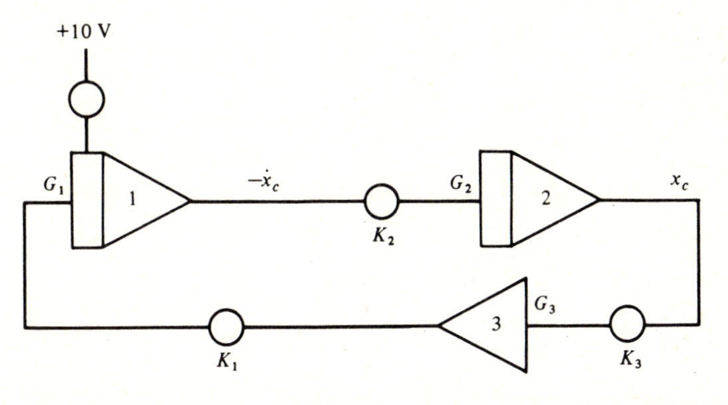

(b)

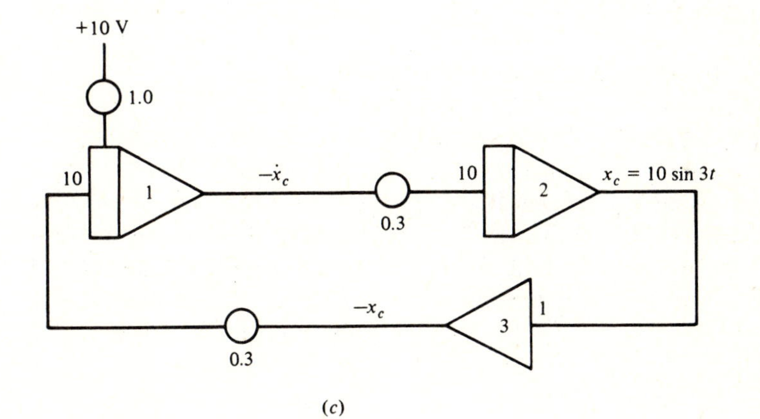

(c)

Figure 9.22

satisfied if $K_2 = 0.3$ and $G_2 = 10$. Similarly, for amplifier 1 we use

$$\ddot{x} = \frac{\ddot{x}_c}{S_2}$$

$$\dot{x} = -\int_0^t \ddot{x}\, dt + \dot{x}(0)$$

$$\frac{\dot{x}_c}{S_1} = -\int_0^t \frac{K_1 G_1 \ddot{x}_c}{S_2}\, dt + \dot{x}_c(0)$$

with $S_1 = \frac{1}{3}$ and $S_2 = \frac{1}{9}$ to obtain $K_1G_1 = 3$. So $K_1 = 0.3$ and $G_1 = 10$ will suffice here also. Finally, to determine the required net gain of amplifier 3, we write the differential equation

$$\ddot{x} + 9x = 0$$

in the computer variables

$$\frac{\ddot{x}_c}{S_2} + 9\frac{x_c}{S_1} = 0$$

to obtain

$$\ddot{x}_c = -\frac{9S_2}{S_1}x_c = -x_c$$

The last equation will be satisfied by the computer diagram if $K_3G_3 = 1$. The complete scaled diagram is shown in Fig. 9.22*c*. Comparison of Fig. 9.22*a* and *c* shows that the product of the net gains around the loop is 9 for either diagram. This is indicative of the fact that each diagram represents the differential equation $\ddot{x} + 9x = 0$. In the scaled diagram, the gain is evenly distributed so as to avoid overloading an individual amplifier.

The previous example was such that the maximum values attained by the problem variables could be explicitly determined beforehand. This luxury makes magnitude scaling an easy task, but it seldom occurs. Additional methods of magnitude scaling for analog solutions of vibration systems will be discussed in Example 9.13.

Time Scaling

Analog solutions for vibrating systems involving high-frequency oscillations often exhibit variables that fluctuate widely over a short period of time. In such cases, computer voltages can vary so rapidly that the response capabilities of the computing components or recording equipment are exceeded. This is particularly true when electromechanical plotters are used to record solution results. The converse is also true. Some systems respond so slowly that error is introduced by amplifier drift or capacitor leakage in integrating circuits. In either case, *time scaling* is required to lengthen or shorten the time span required for problem solution. Time scaling is accomplished by changing the independent variable from real time t to computer time T using

$$T = \alpha t \tag{9.40}$$

where α is the *time-scaling factor*. If $\alpha > 1$, the computer time required for problem solution will be greater than the actual problem time. If $\alpha < 1$, the computer time will be less than the corresponding problem time.

To illustrate time scaling, let us consider the homogeneous differential equation

$$\frac{d^2x}{dt^2} + a_1\frac{dx}{dt} + a_2x = 0 \tag{9.41}$$

and introduce the change of variable defined by Eq. (9.40). By use of the chain rule, the derivatives are

$$\frac{dx}{dt} = \frac{dx}{dT}\frac{dT}{dt} = \alpha\frac{dx}{dT} \tag{9.42}$$

and

$$\frac{d^2x}{dt^2} = \frac{d^2x}{dT^2}\left(\frac{dT}{dt}\right)^2 = \alpha^2\frac{d^2x}{dT^2} \tag{9.43}$$

The *time-scaled equation* is then

$$\alpha^2\frac{d^2x}{dT^2} + \alpha a_1\frac{dx}{dT} + a_2x = 0 \tag{9.44}$$

or

$$\frac{d^2x}{dT^2} + \frac{a_1}{\alpha}\frac{dx}{dT} + \frac{a_2}{\alpha_2}x = 0 \tag{9.45}$$

Thus a time-scaled analog-computer solution for Eq. (9.41) can be obtained by programming the computer to simulate Eq. (9.45). The only additional consideration necessary is that of properly accounting for the initial conditions for the time-scaled equation. Since the scaling affects only the time required to attain a given value of $x(t)$ and not the value itself, the initial condition of the primary variable is not affected; that is,

$$x(t = 0) = x(T = 0) \tag{9.46}$$

On the other hand, Eq. (9.42) shows a direct effect of time scaling on the first derivative. Thus the initial condition for the first derivative in the time-scaled equation becomes

$$\left(\frac{dx}{dT}\right)_{T=0} = \frac{1}{\alpha}\left(\frac{dx}{dt}\right)_{t=0} \tag{9.47}$$

Example 9.12 Given the differential equation

$$\frac{d^2x}{dt^2} + 3\frac{dx}{dt} + x = 0$$

and the initial conditions $x(0) = 2$ and $\dot{x}(0) = 3$, obtain the time-scaled equation and the corresponding analog-computer diagram required to obtain 20 s of problem time in 10 s of computing time.

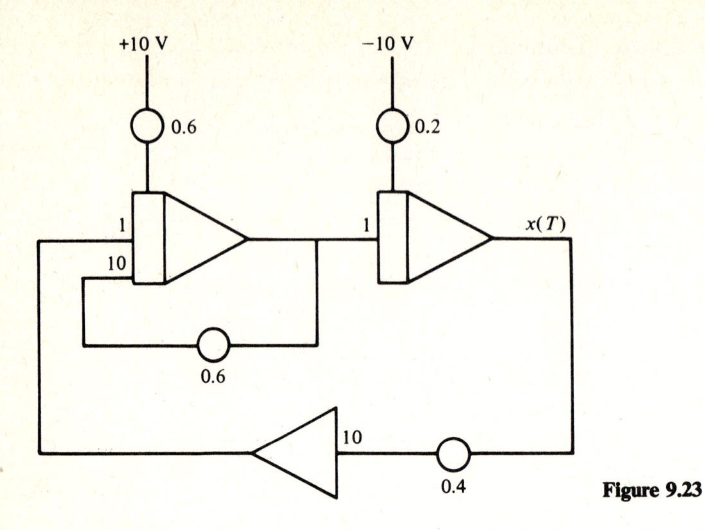

Figure 9.23

SOLUTION First we note that the required time scaling "speeds up" the computation by a factor of 2. Computing time must then be related to problem time by $T = \frac{1}{2}t$, from which the time-scaling factor is $\alpha = \frac{1}{2}$. By use of Eqs. (9.42) and (9.43), the derivatives become

$$\frac{dx}{dt} = \alpha\frac{dx}{dT} = \frac{1}{2}\frac{dx}{dT}$$

$$\frac{d^2x}{dt^2} = \alpha^2\frac{d^2x}{dT^2} = \frac{1}{4}\frac{d^2x}{dT^2}$$

so that the time-scaled equation is

$$\frac{d^2x}{dT^2} + 6\frac{dx}{dT} + 4x = 0$$

subject to the initial conditions

$$x(T = 0) = 2 \qquad \left(\frac{dx}{dT}\right)_{T=0} = 6$$

The analog-computer diagram for the time-scaled equation is shown in Fig. 9.23.

The foregoing discussions of magnitude and time scaling were treated separately for purposes of clarity. When it is necessary to use both magnitude scaling and time scaling to set up a problem for analog-computer solution, it is usually best to do both operations simultaneously. The primary reason for this is that both procedures affect amplifier gains, and it is easier to ensure that computer voltages are maintained within acceptable limits if scaling is accomplished through an integrated procedure.

Let us consider the inhomogeneous, second-order differential equation with constant coefficients

$$\frac{d^2x}{dt^2} + a_1\frac{dx}{dt} + a_2x = f(t) \tag{9.48}$$

where we assume that $f(t)$ is obtainable from a separate computer circuit or an external function generator. Equation (9.48) is to be magnitude scaled using

$$x_c = S_0 x \qquad \frac{dx_c}{dt} = S_1 \frac{dx}{dt} \qquad \frac{d^2x_c}{dt^2} = S_2 \frac{d^2x}{dt^2} \tag{9.49}$$

and time scaled according to

$$T = \alpha t \tag{9.50}$$

Substituting the new dependent and independent variables into Eq. (9.48) gives

$$\frac{\alpha^2}{S_2}\frac{d^2x_c}{dT^2} + \frac{\alpha a_1}{S_1}\frac{dx_c}{dT} + \frac{a_2 x_c}{S_0} = f\left(\frac{T}{\alpha}\right) \tag{9.51}$$

which can be rewritten as

$$\frac{d^2x_c}{dT^2} + \frac{a_1 S_2}{\alpha S_1}\frac{dx_c}{dT} + \frac{a_2 S_2}{\alpha^2 S_0} x_c = \frac{S_2}{\alpha^2} f\left(\frac{T}{\alpha}\right) \tag{9.52}$$

From our previous study of this type of differential equation, we know that the time response of x_c and its derivatives will be similar in form to the function $f(t)$. As a starting point, then, we may choose a value of α such that the time-scaled function $f(T/\alpha)$ varies reasonably with time, that is, neither too fast nor too slow. Next we can obtain a value of S_2 such that the term $S_2 f(T/\alpha)/\alpha^2$ has an acceptable magnitude, given the computer voltage limits. The remaining scale factors S_0 and S_1 are determined by estimating the maximum values to be attained by x and dx/dt. No specific general rules are available, and proper scaling often involves making trial computer runs and adjusting scale factors based on the results. The following example will illustrate some of the techniques used.

Example 9.13 Perform the necessary scaling and draw the computer diagram for an analog solution to the differential equation

$$\frac{d^2x}{dt^2} + 6\frac{dx}{dt} + 350x = 200 \sin 20t$$

with the initial conditions $x = 2$ in and $dx/dt = 20$ in/s at $t = 0$.

SOLUTION After the introduction of magnitude- and time-scaling factors, the differential equation becomes, in terms of computer variables,

$$\frac{d^2x_c}{dT^2} + \frac{6S_2}{\alpha S_1}\frac{dx_c}{dT} + \frac{350S_2}{\alpha^2 S_0} x_c = \frac{200S_2}{\alpha^2} \sin \frac{20}{\alpha} T$$

From Sec. 9.4, we know that the sine function on the right-hand side can easily be generated by an analog circuit, provided the circular frequency is less than $\sqrt{10}$. Let us arbitrarily select $\alpha = 10$ to obtain the time-scaled forcing function

$$f(T) = 2S_2 \sin 2T$$

From this result, we can see that choosing $S_2 = 5$ will provide a forcing function with magnitude equal to the computer limit of 10 V, which is the ideal situation. With $S_2 = 5$ and $\alpha = 10$, the differential equation becomes

$$\frac{d^2x_c}{dT^2} + \frac{3}{S_1}\frac{dx}{dT} + \frac{17.5}{S_0}x_c = 10 \sin 2T$$

To determine the scale factors S_0 and S_1, we estimate the maximum values of $x(t)$ and dx/dt as follows: Since the original differential equation is that of a damped spring-mass system excited by a harmonic force, the maximum possible displacement and velocity will occur when damping is zero. The undamped system is governed by

$$\frac{d^2x}{dt^2} + 350x = 200 \sin 20t$$

which, by Eq. (4.10), has maximum displacement

$$x_{\max} = \left|\frac{F_0/m}{k/m - \omega_f^2}\right| = \frac{200}{400 - 350} = 4 \text{ in}$$

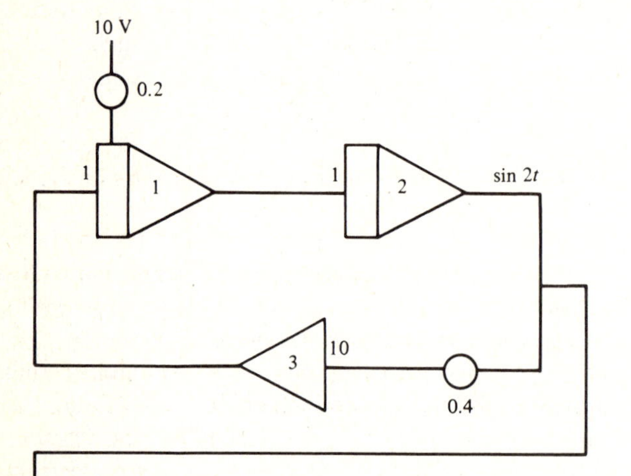

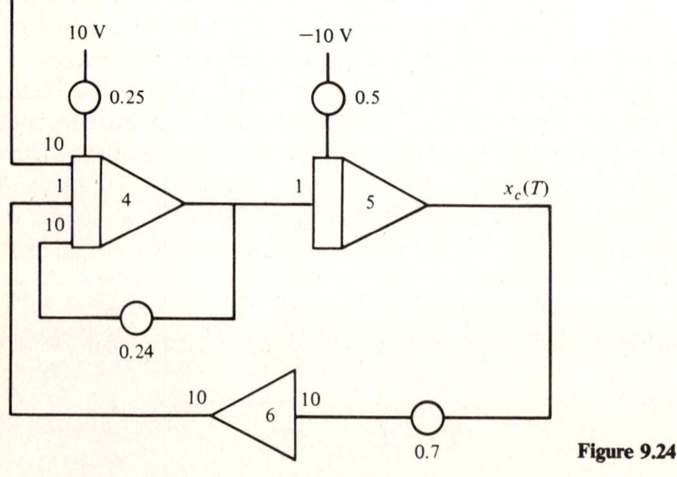

Figure 9.24

and maximum velocity

$$\left(\frac{dx}{dt}\right)_{max} = \omega_f x_{max} = 80 \text{ in/s}$$

For accuracy, we want the computer variables to have maximum values near the computer limit of 10 V. In equation form,

$$(x_c)_{max} = S_0 x_{max} \simeq 10$$

$$\left(\frac{dx_c}{dT}\right)_{max} = \frac{S_1}{\alpha}\left(\frac{dx}{dt}\right)_{max} \simeq 10$$

from which we choose $S_0 = 2.5$ V/in and $S_1 = 1.25$ V/(in/s). By substituting these values, the final form of the scaled equation is found to be

$$\frac{d^2x_c}{dT^2} + 2.4\frac{dx_c}{dT} + 7x_c = 10 \sin 2T$$

The initial conditions for the computer variables are

$$(x_c)_{T=0} = (S_0 x)_{t=0} = 2.5 \times 2 = 5 \text{ V}$$

$$\left(\frac{dx_c}{dT}\right)_{T=0} = \frac{S_1}{\alpha}\left(\frac{dx}{dt}\right)_{t=0} = \frac{1.25}{10}20 = 2.5 \text{ V}$$

The analog-computer diagram is shown in Fig. 9.24.

9.6 APPLICATIONS TO SYSTEMS WITH TWO DEGREES OF FREEDOM

The major advantages of the application of the analog computer to the study of vibrating systems are the ease and speed with which one can obtain the effects on system response of changes in physical parameters. By simply rewiring a computer circuit or adjusting amplifier gains, the vibration response of almost any combination of mass, damping, and spring elements can be found. Using the analog computer eliminates the necessity of formal mathematical solution of the equations of motion and/or construction of physical systems for experimental study. The latter is very important in reducing both the time and the cost required for the construction of prototypes for alternative system designs.

Since the previous portions of this chapter dealt with analog-computer solutions of differential equations of the type which govern the vibration of systems having one degree of freedom, the remainder of this section is devoted to analog simulation of systems with two degrees of freedom. In Chap. 8 we found that such systems are governed by coupled differential equations which must be solved simultaneously to obtain the solutions. We also found that the mathematical solution procedures can be quite laborious. To illustrate the power of the analog computer in eliminating tedious calculations, let us consider the system shown in Fig. 9.25. The differential equations of motion can be written as

$$\begin{aligned} m_1\ddot{x}_1 &= -k_1x_1 - c_1\dot{x}_1 + k_2(x_2 - x_1) + c_2(\dot{x}_2 - \dot{x}_1) \\ m_2\ddot{x}_2 &= -k_2(x_2 - x_1) - c_2(\dot{x}_2 - \dot{x}_1) + F(t) \end{aligned} \tag{9.53}$$

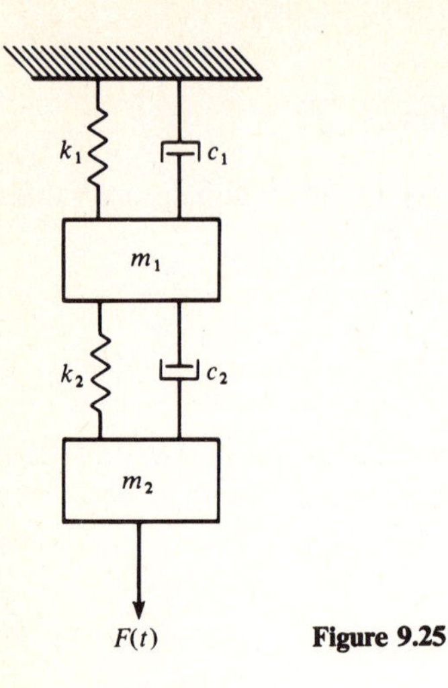

Figure 9.25

For analog-computer solution, Eqs. (9.53) are most easily programmed when written as

$$\ddot{x}_1 = -\frac{k_1}{m_1}x_1 - \frac{c_1}{m_1}\dot{x}_1 + \frac{k_2}{m_1}(x_2 - x_1) + \frac{c_2}{m_1}(\dot{x}_2 - \dot{x}_1)$$
$$\ddot{x}_2 = -\frac{k_2}{m_2}(x_2 - x_1) - \frac{c_2}{m_2}(\dot{x}_2 - \dot{x}_1) + \frac{F(t)}{m_2} \tag{9.54}$$

Let us temporarily ignore the fact that Eqs. (9.54) are coupled and proceed as follows: Assuming that $\ddot{x}_1$ can be generated as an analog voltage, we could perform two integrations on this voltage to obtain x_1 as in Fig. 9.26*a*. A similar assumption regarding $\ddot{x}_2$ produces the partial diagram of Fig. 9.26*b*. Next we note from Eqs. (9.54) that $\ddot{x}_1$ and $\ddot{x}_2$ are composed of various constant multiples of x_1, x_2, $\dot{x}_1$, and $\dot{x}_2$, plus the external forcing function $F(t)/m_2$. If we assume further that $F(t)$ can be generated externally, the coupled differential equations are simulated on the analog computer by "coupling" the partial computer diagrams of Fig. 9.26*a* and *b* so that the voltages corresponding to $\ddot{x}_1$ and $\ddot{x}_2$ are produced according to Eqs. (9.54). The resulting computer diagram is shown in Fig. 9.26*c*. The complete analog circuit would be obtained by adding the appropriate initial-condition voltages.

At first glance, the computer diagram may seem complicated. However, the reader is urged to consider that construction of the analog circuit represented by the diagram is simply a matter of wiring the amplifiers together and adjusting a few potentiometers. A review of the mathematical procedure required to solve this two-degrees-of-freedom problem should further convince the reader of the simplicity of the analog approach. In this context, we observe that the analog solution is the complete solution, since it automatically includes both free and forced vibration response.

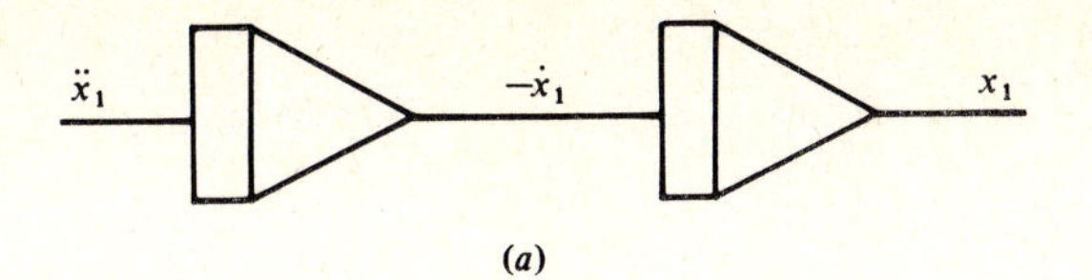

(a)

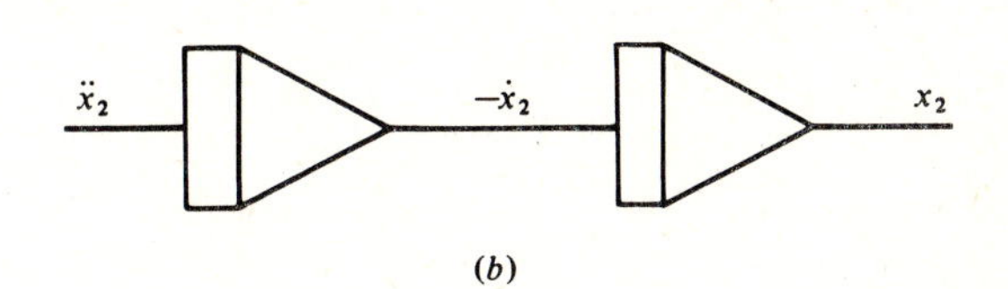

(b)

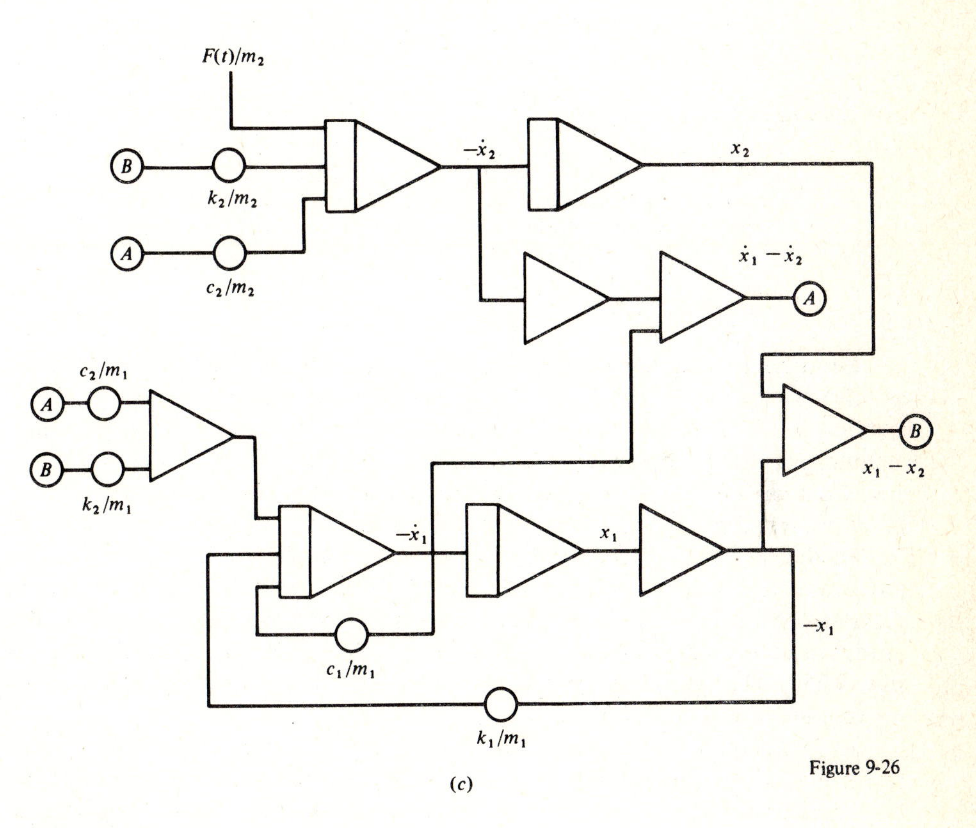

(c)

Figure 9.26

Example 9.14 The free motion of a mechanical system is governed by

$$0.1\ddot{x}_1 + \dot{x}_1 + 40x_1 - 20x_2 = 0$$

$$0.1\ddot{x}_2 + 20x_2 - 20x_1 - \dot{x}_1 = 0$$

with initial conditions $x_1(0) = 1$ in, $x_2(0) = 2$ in, $\dot{x}_1(0) = \dot{x}_2(0) = 0$. Draw an analog-computer diagram for solving this two-degrees-of-freedom problem.

SOLUTION For convenience, we rewrite the equations of motion as

$$\ddot{x}_1 = -10\dot{x}_1 - 200x_1 - 200(x_1 - x_2)$$

$$\ddot{x}_2 = -200(x_2 - x_1) + 10\dot{x}_1$$

The appearance of the relatively large coefficient (200) indicates that some form of scaling is required. Considering the small magnitudes of the initial conditions and the fact that this is a damped system, we are led to time scaling the problem first. A computer run will then be used to indicate whether magnitude scaling is required. Arbitrarily selecting $\alpha = 20$ as the time-scaling factor, we rewrite the equations of motion as

$$\ddot{y}_1 = -0.5\dot{y}_1 - 10y_1 - 10(y_1 - y_2)$$

$$\ddot{y}_2 = -10(y_2 - y_1) + 0.5\dot{y}_1$$

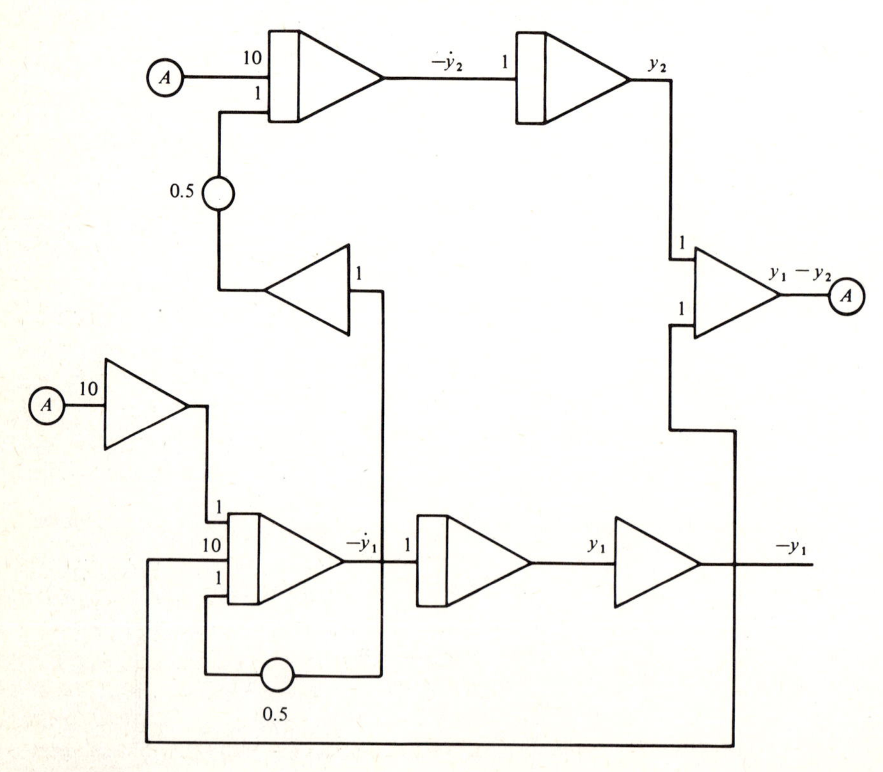

Figure 9.27

where $y_1 = y_1(T)$ and $y_2 = y_2(T)$ are the computer variables. The initial conditions are now $y_1(0) = 1$ V, $y_2(0) = 2$ V, and $\dot{y}_1(0) = \dot{y}_2(0) = 0$. The corresponding computer diagram is shown in Fig. 9.27.

PROBLEMS*

Note: In each of the following problems, assume the amplifier operating range is ± 10 V.

In Probs. 9.1 to 9.5, determine the indicated amplifier gains, input and output voltages, and potentiometer settings.

9.1 See Fig. 9.28.

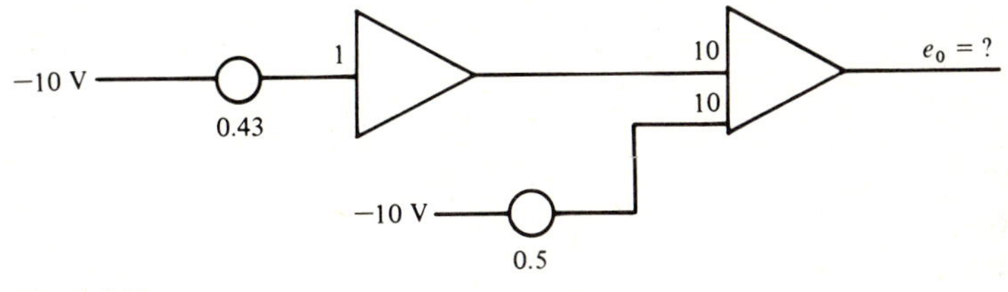

Figure 9.28

9.2 See Fig. 9.29.

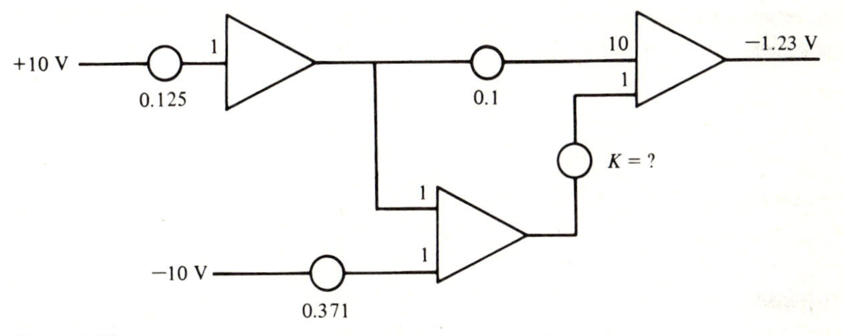

Figure 9.29

9.3 See Fig. 9.30.

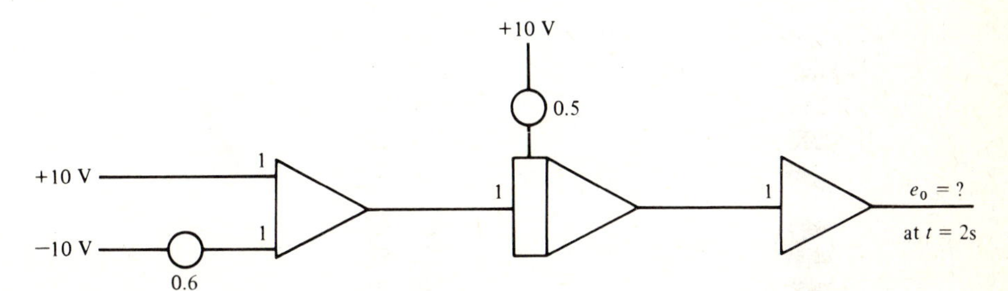

Figure 9.30

* Additional problems for analog-computer simulation may be assigned from Chaps. 2 to 5 and 8.

9.4 See Fig. 9.31.

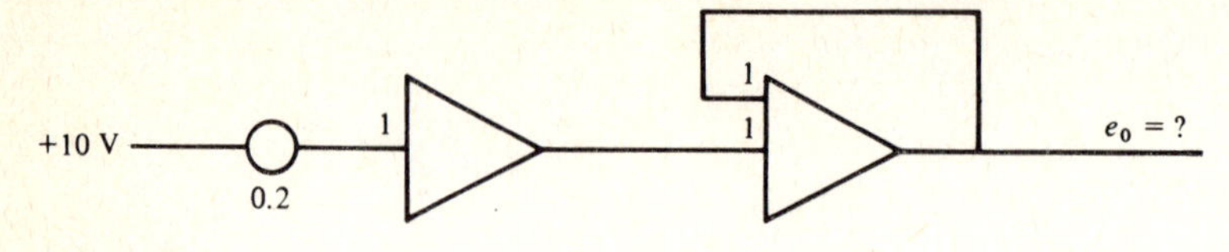

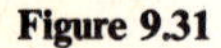
Figure 9.31

9.5 See Fig. 9.32.

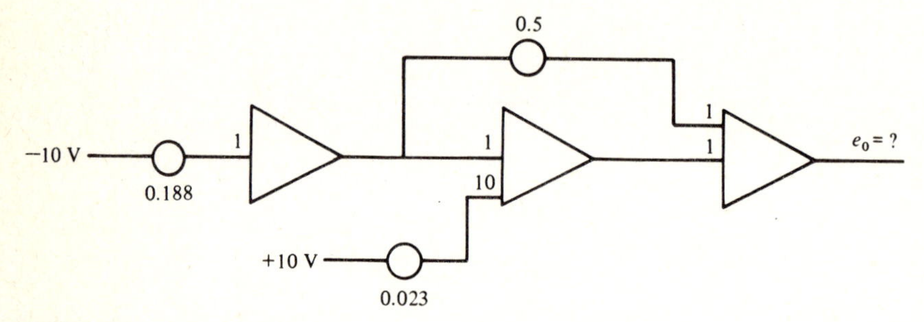

Figure 9.32

For Probs. 9.6 to 9.12 draw a computer diagram for an analog circuit which will generate a voltage corresponding to the given function.

9.6 $f(t) = 4 + 3t$

9.7 $f(t) = t^2 - 2$

9.8 $f(t) = 4 \sin 3t$

9.9 $f(t) = 2 \cos 1.62t$

9.10 $f(t) = 2\, e^{-t}$

9.11 $f(t) = 3 \sin 1.5t + 2 \cos 1.5t$

9.12 $f(t) = (e^t - e^{-t})/2 = \sinh t$

In Probs. 9.13 to 9.16, determine the differential equation and initial conditions corresponding to the given analog diagram.

9.13 See Fig. 9.33.

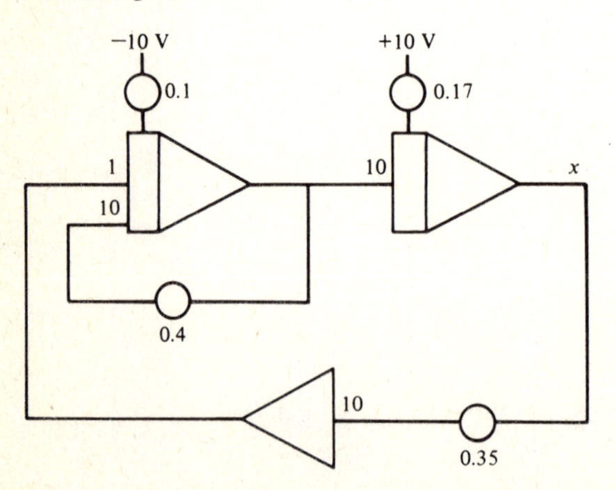

9.14 See Fig. 9.34.

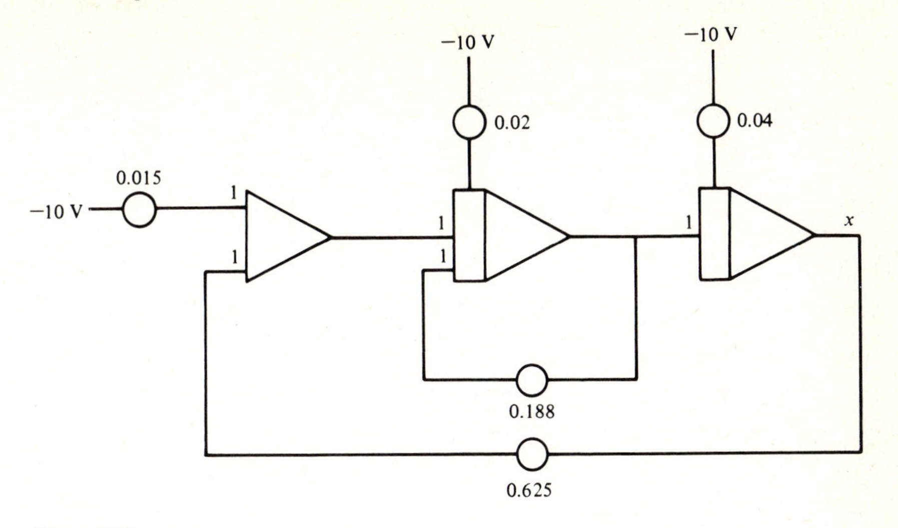

Figure 9.34

9.15 See Fig. 9.35.

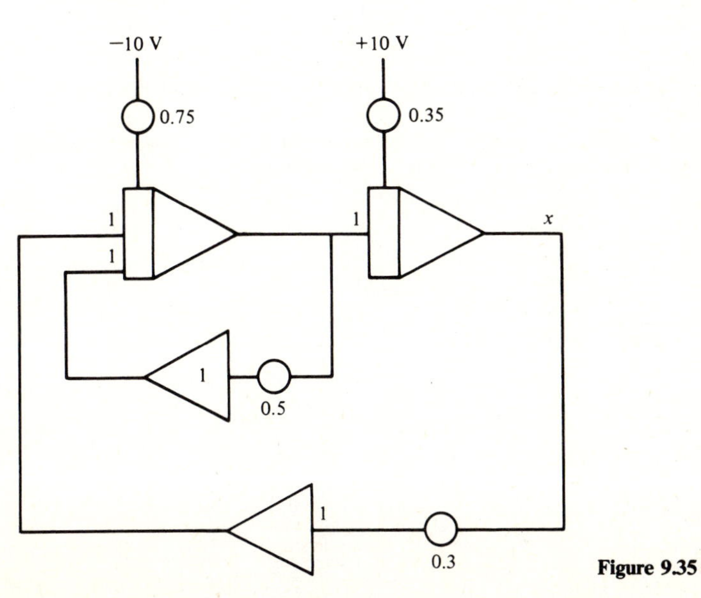

Figure 9.35

9.16 See Fig. 9.36.

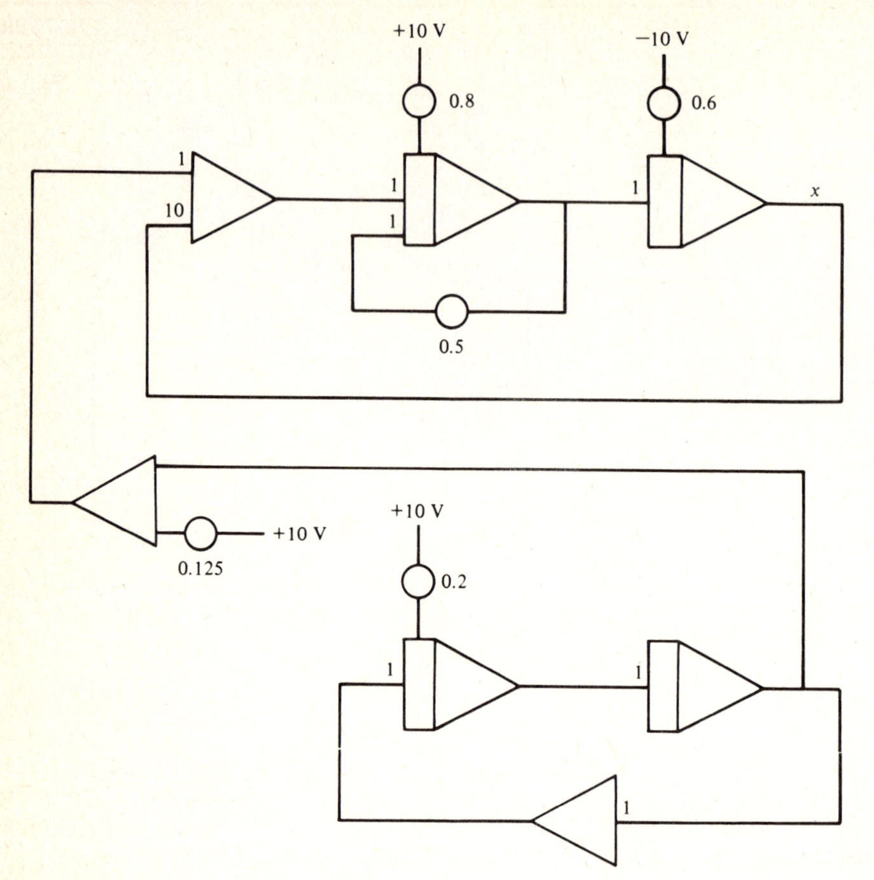

Figure 9.36

9.17 Draw an analog-computer diagram for the differential equation $\ddot{x} + 2\dot{x} + 3x = t$ with initial conditions $x(0) = 4$ and $\dot{x}(0) = 1.5$.

9.18 For the system shown in Fig. 9.37, set up an analog-computer diagram which explicitly produces a voltage corresponding to the force transmitted to the foundation.

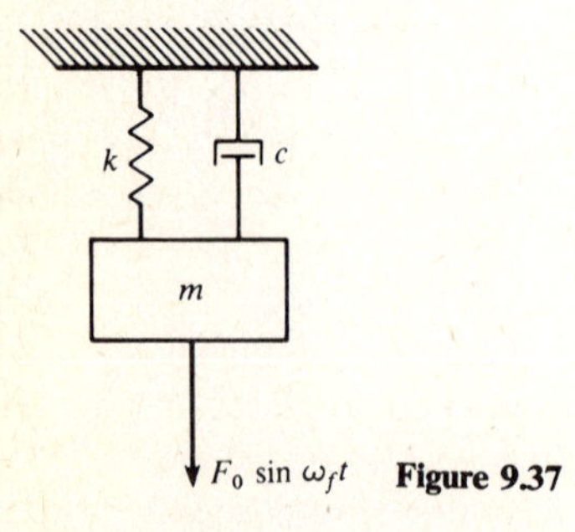

Figure 9.37

CHAPTER

TEN

DIGITAL-COMPUTER TECHNIQUES FOR SYSTEMS WITH MULTIPLE DEGREES OF FREEDOM

Systems having only one or two degrees of freedom are idealized mathematical models of mechanical systems, and these models are quite useful for obtaining an understanding of dynamic behavior. Many physical systems are too complex for such simple models, and more detailed analysis is required. In fact, all mechanical systems are composed of elements which have distributed mass and elasticity. However, in many systems the distribution of properties is highly nonuniform, with some elements being massive and quite stiff while other elements have low mass but are relatively flexible. Such systems can be modeled as combinations of discrete mass and elastic elements which possess many degrees of freedom; these models are often called *lumped-mass* systems.

The two-degrees-of-freedom system discussed in Chap. 8 is but a special case of the more general system having many degrees of freedom. All the vibration theory and the mathematical techniques discussed in Chap. 8 are applicable to systems having a larger number of degrees of freedom. A linear system having n degrees of freedom exhibits motion governed by a set of n simultaneous second-order differential equations. Except under certain very special circumstances, the equations of motion are coupled. For systems having many degrees of freedom, the sheer volume of calculations required in obtaining a numerical solution makes the use of digital computers almost imperative. Since matrices are extremely convenient for digital computation purposes, the emphasis of this chapter is on matrix formulation of the equations of motion.

10.1 EQUATIONS OF MOTION—DIRECT APPROACH

The development of the equations of motion via direct application of Newton's second law will be illustrated using the three-degrees-of-freedom system shown in Fig. 10.1. We assume that each mass is constrained to execute vertical motion only and that an external force F_i acts on each mass. As has been shown for similar systems with one and two degrees of freedom, the gravity forces acting on each mass and the forces due to initial spring deformations are self equilibrating. Thus, we shall consider only dynamic forces in the following analysis.

The positions of the masses are denoted by x_1, x_2, and x_3, measured positive downward from the respective equilibrium positions. The equations of motion are obtained from the free-body diagrams shown in Fig. 10.2 which are drawn by assuming that $x_3 > x_2 > x_1$ and $\dot{x}_3 > \dot{x}_2 > \dot{x}_1$, and that each is positive. These assumptions are made without loss of generality, as sign changes will be accounted for automatically.

Applying Newton's second law to each free-body diagram gives

$$m_1\ddot{x}_1 = F_1 - k_1x_1 - c_1\dot{x}_1 + k_2(x_2 - x_1) + c_2(\dot{x}_2 - \dot{x}_1)$$

$$m_2\ddot{x}_2 = F_2 - k_2(x_2 - x_1) - c_2(\dot{x}_2 - \dot{x}_1) + k_3(x_3 - x_2) + c_3(\dot{x}_3 - \dot{x}_2)$$

$$m_3\ddot{x}_3 = F_3 - k_3(x_3 - x_2) - c_3(\dot{x}_3 - \dot{x}_2) - k_4x_3 - c_4\dot{x}_3 \tag{10.1}$$

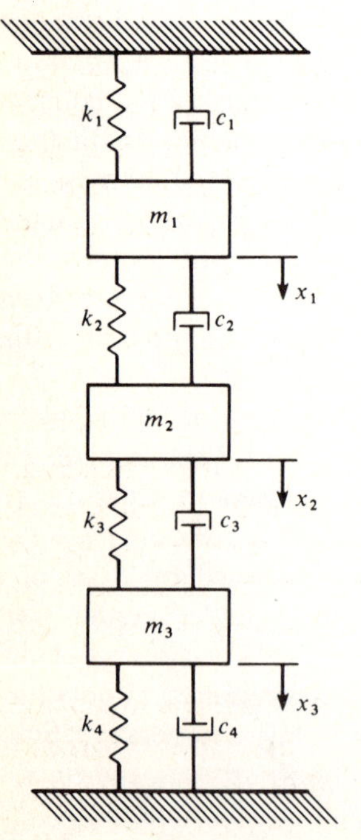

Figure 10.1

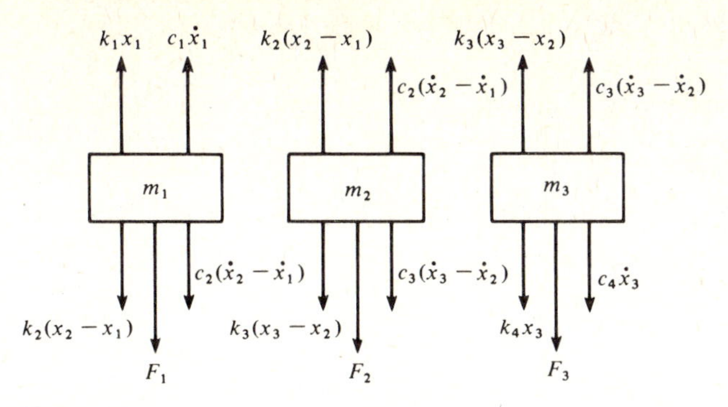

Figure 10.2

By expanding and rearranging, Eqs. (10.1) can be written as

$$m_1\ddot{x}_1 + (c_1 + c_2)\dot{x}_1 + (k_1 + k_2)x_1 - k_2x_2 - c_2\dot{x}_2 = F_1$$

$$m_2\ddot{x}_2 + (c_2 + c_3)\dot{x}_2 + (k_2 + k_3)x_2 - k_2x_1 - c_2\dot{x}_1 - k_3x_3 - c_3\dot{x}_3 = F_2$$

$$m_3\ddot{x}_3 + (c_3 + c_4)\dot{x}_3 + (k_3 + k_4)x_3 - k_3x_2 - c_3\dot{x}_2 = F_3 \tag{10.2}$$

Equations (10.2) show that system motion is governed by three second-order, linear differential equations with constant coefficients. As with two degrees of freedom, we note that the equations are coupled and must be solved simultaneously since each contains more than one dependent variable. To obtain the solution $x_1(t)$, $x_2(t)$, and $x_3(t)$ requires that we find both homogeneous and particular solutions for each displacement such that all three equations are simultaneously satisfied for all values of time. Methods of solution will be discussed in subsequent sections of this chapter.

10.2 GENERALIZED COORDINATES

To describe the motion of a mechanical system, a set of coordinates is required. In some cases (Fig. 10.1, for example) the choice of suitable coordinates is obvious. Often, however, multiple sets of coordinates exist, each of which could be used to describe the system configuration. A proper choice among several possible coordinate sets can often simplify the mathematical analysis of the motion of a system.

As an elementary example, consider circular motion of a particle in a plane (Fig. 10.3). If we choose a set of rectangular coordinate axes with the origin located at the center of the path of motion, the position of the particle is specified by the coordinate point (x, y). The values of two coordinates are required, and this implies that there are two degrees of freedom. Applying the

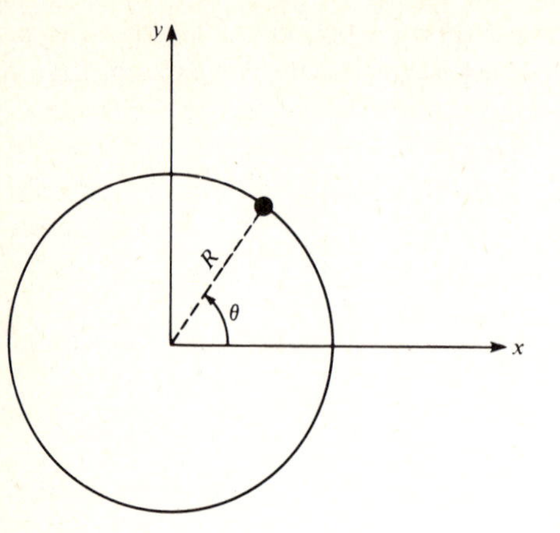

Figure 10.3

constraint equation

$$x^2 + y^2 = R^2 \tag{10.3}$$

shows that the coordinates are not independent, however. Specification of either x or y is sufficient to locate the particle. Thus, there is a single degree of freedom.

On the other hand, if we choose to describe the motion with polar coordinates (r, θ), we have $r = R =$ constant, and the polar angle θ is the only required coordinate. Note that in this case, the coordinate is an angular value. The two systems are related by the usual relations $x = R \cos \theta$ and $y = R \sin \theta$.

The preceding example leads us to the concept of *generalized coordinates*, which may be defined as any set of parameters which define the configuration of a system and are independent of each other. The requirement of independence implies that the generalized coordinates are also independent of any constraint conditions. Again referring to Fig. 10.3, we note that the position vector of the particle is

$$r = x\mathbf{i} + y\mathbf{j} + z\mathbf{k} \tag{10.4}$$

where x, y, and z are the rectangular coordinates. In actuality, then, there are three coordinates involved. The plane-motion constraint is represented by

$$z = 0 \tag{10.5}$$

while the circular-motion constraint is

$$y = \sqrt{R^2 - x^2} \tag{10.6}$$

Combination of Eqs. (10.4) through (10.6) shows that there exist a single independent coordinate x and a single degree of freedom. This is a general result; the number of degrees of freedom is the same as the number of generalized coordinates.

The generalized coordinates of a physical system are not unique. Often several such sets may exist, and the choice of one set over another is a matter of convenience. Consider a uniform slender bar mounted on linear springs at each end, as shown in Fig. 10.4*a*. The bar is assumed to be constrained such that it can execute small oscillations about its equilibrium position in the plane represented by the paper. A possible position of motion is shown in Fig. 10.4*b*, which also shows four coordinates of interest. The coordinates are the positions of each end of the bar x_1 and x_2, the position of the center of mass x_G, and the angular orientation θ of the bar from the horizontal. If these four coordinates form a set of generalized coordinates, the system has four degrees of freedom. That this is not so is shown by considering geometric constraints. For example, if we know x_1 and θ, then

$$x_2 = x_1 + L_o \sin\theta \tag{10.7}$$

and

$$x_G = x_1 + \frac{L_o}{2} \sin\theta \tag{10.8}$$

Thus x_1 and θ form a set of generalized coordinates, and the system has two degrees of freedom. In fact, any combination of two of the four coordinates is a set of generalized coordinates. Thus, the system of Fig. 10.4*a* has six possible sets of generalized coordinates.

At this point, we shall adopt the standard notation q_i, $i = 1, 2, \ldots, n$, where n is the number of degrees of freedom, to represent generalized coordinates. For uniformity of notation and later use, we shall also introduce the concept of *generalized forces* as follows. If during a small change δq_i in generalized coordinate q_i, the amount of work done is U_i, an associated generalized force Q_i is defined as

$$Q_i = \frac{U_i}{\delta q_i} \tag{10.9}$$

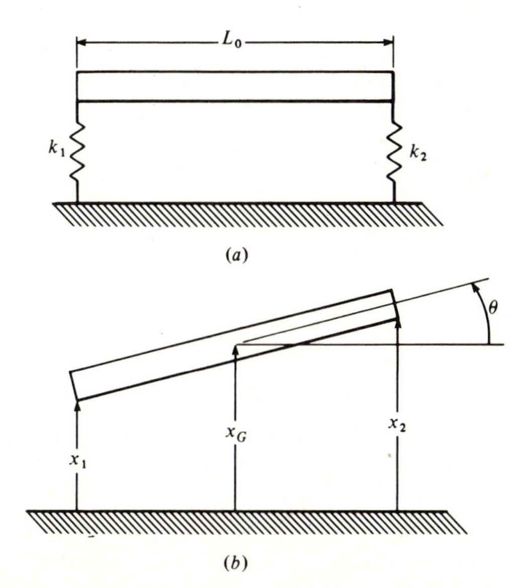

Figure 10.4

Note that the product $Q_i \delta q_i$ has the units of work. This means that Q_i is not necessarily a force in the usual sense; Q_i can be any quantity provided it produces work during a change in the corresponding generalized coordinate.

10.3 LAGRANGE'S EQUATIONS

In Sec. 3.6, we developed the equation of motion for a single-degree-of-freedom system using the work-energy equation for the system. The concepts of generalized coordinates and generalized forces allow a somewhat similar approach to be used to obtain equations of motion for more complex systems. The technique is known as *Lagrange's method*, and it utilizes both the principle of virtual displacements and d'Alembert's principle. For simplicity, we shall develop Lagrange's equations for the motion of a single particle and extend the result to systems without rigorous proof.

Consider the motion of a particle in space, defined by

$$F_x = m\ddot{x} \qquad F_y = m\ddot{y} \qquad F_z = m\ddot{z} \tag{10.10}$$

According to d'Alembert's principle, Eqs. (10.10) may be considered as static-equilibrium equations of the form

$$F_x - m\ddot{x} = 0 \qquad F_y - m\ddot{y} = 0 \qquad F_z - m\ddot{z} = 0 \tag{10.11}$$

The method of virtual displacements may now be applied to Eqs. (10.11). This principle of statics states that if a body in equilibrium is displaced a small distance, the increment of work done by the forces is zero since the resultant force system is zero. Thus if the particle undergoes arbitrary displacements δx, δy, and δz, we can write

$$(F_x - m\ddot{x})\delta x + (F_y - m\ddot{y})\,\delta y + (F_z - m\ddot{z})\,\delta z = 0 \tag{10.12}$$

The displacements δx, δy, and δz are small displacements having arbitrary values consistent with physical restraints and are independent of each other.

We now assume that the motion of the particle is to be described in terms of generalized coordinates q_1, q_2, q_3 which are related to the rectangular coordinates through known relations

$$\begin{aligned} x &= f_1(q_1, q_2, q_3) \\ y &= f_2(q_1, q_2, q_3) \\ z &= f_3(q_1, q_2, q_3) \end{aligned} \tag{10.13}$$

In terms of generalized coordinates, the displacements are

$$\begin{aligned} \delta x &= \frac{\partial x}{\partial q_1}\delta q_1 + \frac{\partial x}{\partial q_2}\delta q_2 + \frac{\partial x}{\partial q_3}\delta q_3 \\ \delta y &= \frac{\partial y}{\partial q_1}\delta q_1 + \frac{\partial y}{\partial q_2}\delta q_2 + \frac{\partial y}{\partial q_3}\delta q_3 \\ \delta z &= \frac{\partial z}{\partial q_1}\delta q_1 + \frac{\partial z}{\partial q_2}\delta q_2 + \frac{\partial z}{\partial q_3}\delta q_3 \end{aligned} \tag{10.14}$$

Since by definition the generalized coordinates are independent, let us examine the case in which $\delta q_2 = \delta q_3 = 0$. This assumption is made to simplify the algebra, and it involves no loss of generality. Setting $\delta q_2 = \delta q_3 = 0$ in Eq. (10.14) and substituting for δx, δy, and δz in Eq. (10.12) give

$$(F_x - m\ddot{x})\frac{\partial x}{\partial q_1}\delta q_1 + (F_y - m\ddot{y})\frac{\partial y}{\partial q_1}\delta q_1 + (F_z - m\ddot{z})\frac{\partial z}{\partial q_1}\delta q_1 = 0 \quad (10.15)$$

or

$$\left(F_x\frac{\partial x}{\partial q_1} + F_y\frac{\partial y}{\partial q_1} + F_z\frac{\partial z}{\partial q_1}\right)\delta q_1 = \left(m\ddot{x}\frac{\partial x}{\partial q_1} + m\ddot{y}\frac{\partial y}{\partial q_1} + m\ddot{z}\frac{\partial z}{\partial q_1}\right)\delta q_1 \quad (10.16)$$

To complete the conversion to generalized coordinates, we must write the time derivatives in this system. From Eq. (10.13) we have

$$\dot{x} = \frac{dx}{dt} = \frac{\partial x}{\partial q_1}\dot{q}_1 + \frac{\partial x}{\partial q_2}\dot{q}_2 + \frac{\partial x}{\partial q_3}\dot{q}_3 \quad (10.17)$$

and similar expressions for $\dot{y}$ and $\dot{z}$. The determination of expressions for the second derivatives is more complex, but the task is simplified somewhat by noting that, by the chain rule for differentiation of a product,

$$\frac{d}{dt}\left(\dot{x}\frac{\partial x}{\partial q_1}\right) = \ddot{x}\frac{\partial x}{\partial q_1} + \dot{x}\frac{d}{dt}\left(\frac{\partial x}{\partial q_1}\right) \quad (10.18)$$

which can be written in a form more suitable for our purposes as

$$\ddot{x}\frac{\partial x}{\partial q_1} = \frac{d}{dt}\left(\dot{x}\frac{\partial x}{\partial q_1}\right) - \dot{x}\frac{d}{dt}\left(\frac{\partial x}{\partial q_1}\right) \quad (10.19)$$

Differentiating Eq. (10.17) with respect to $\dot{q}_1$ shows that

$$\frac{\partial \dot{x}}{\partial \dot{q}_1} = \frac{\partial x}{\partial q_1} \quad (10.20)$$

which can be substituted in the first term on the right of Eq. (10.19) while reversing the order of differentiation in the second term to obtain

$$\ddot{x}\frac{\partial x}{\partial q_1} = \frac{d}{dt}\left(\dot{x}\frac{\partial \dot{x}}{\partial \dot{q}_1}\right) - \dot{x}\frac{\partial \dot{x}}{\partial q_1} \quad (10.21)$$

which is equivalent to

$$\ddot{x}\frac{\partial x}{\partial q_1} = \frac{d}{dt}\left[\frac{\partial}{\partial \dot{q}_1}\left(\frac{\dot{x}^2}{2}\right)\right] - \frac{\partial}{\partial q_1}\left(\frac{\dot{x}^2}{2}\right) \quad (10.22)$$

Utilizing a completely analogous procedure, one can show that

$$\ddot{y}\frac{\partial y}{\partial q_1} = \frac{d}{dt}\left[\frac{\partial}{\partial \dot{q}_1}\left(\frac{\dot{y}^2}{2}\right)\right] - \frac{\partial}{\partial q_1}\left(\frac{\dot{y}^2}{2}\right) \quad (10.23)$$

and

$$\ddot{z}\frac{\partial z}{\partial q_1} = \frac{d}{dt}\left[\frac{\partial}{\partial \dot{q}_1}\left(\frac{\dot{z}^2}{2}\right)\right] - \frac{\partial}{\partial q_1}\left(\frac{\dot{z}^2}{2}\right) \quad (10.24)$$

Substituting and collecting terms transform Eq. (10.16) to

$$\left(F_x \frac{\partial x}{\partial q_1} + F_y \frac{\partial y}{\partial q_1} + F_z \frac{\partial z}{\partial q_1}\right)\delta q_1$$
$$= m\left\{\frac{d}{dt}\left[\frac{\partial}{\partial \dot{q}_1}\left(\frac{\dot{x}^2}{2} + \frac{\dot{y}^2}{2} + \frac{\dot{z}^2}{2}\right)\right] - \frac{\partial}{\partial q_1}\left(\frac{\dot{x}^2}{2} + \frac{\dot{y}^2}{2} + \frac{\dot{z}^2}{2}\right)\right\}\delta q_1 \tag{10.25}$$

At this point in the development we can begin to relate physical significance to the equation. First note that the quantity on the left side of Eq. (10.25) has the units of work. This term represents the work done by all external forces as the system undergoes virtual displacement δq_1. Recalling Eq. (10.9), we observe that the generalized force associated with generalized coordinate q_1 is

$$Q_1 = F_x \frac{\partial x}{\partial q_1} + F_y \frac{\partial y}{\partial q_1} + F_z \frac{\partial z}{\partial q_1} \tag{10.26}$$

Second, the kinetic energy of the particle

$$T = \tfrac{1}{2}m(\dot{x}^2 + \dot{y}^2 + \dot{z}^2) \tag{10.27}$$

is contained explicitly on the right side of Eq. (10.25). Applying these observations, we have

$$Q_1\delta q_1 = \left[\frac{d}{dt}\left(\frac{\partial T}{\partial \dot{q}_1}\right) - \frac{\partial T}{\partial q_1}\right]\delta q_1 \tag{10.28}$$

from which

$$\frac{d}{dt}\left(\frac{\partial T}{\partial \dot{q}_1}\right) - \frac{\partial T}{\partial q_1} = Q_1 \tag{10.29}$$

This is Lagrange's equation for motion of the particle with respect to generalized coordinate q_1. For the particle having three degrees of freedom, two additional equations of similar form will be obtained for q_2 and q_3. For a system of particles or rigid bodies having n degrees of freedom, the corresponding Lagrange's equations are

$$\frac{d}{dt}\left(\frac{\partial T}{\partial \dot{q}_i}\right) - \frac{\partial T}{\partial q_i} = Q_i \qquad i = 1, 2, \ldots, n \tag{10.30}$$

where T is the expression representing the total kinetic energy of the system.

An additional form of Lagrange's equation can be obtained by separating the generalized force Q_i into conservative and nonconservative parts. Since conservative forces can be derived from a potential function V, we can write

$$Q_i = -\frac{\partial V}{\partial q_i} + \overline{Q}_i \tag{10.31}$$

where $\overline{Q}_i$ is the nonconservative portion of the generalized force. Substituting into Eq. (10.30) gives

$$\frac{d}{dt}\left(\frac{\partial T}{\partial \dot{q}_1}\right) - \frac{\partial T}{\partial q_i} = -\frac{\partial V}{\partial q_i} + \overline{Q}_i \qquad i = 1, 2, \ldots, n \tag{10.32}$$

Since potential energy V is not a function of the generalized velocities $\dot{q}_i$, this is most often written as

$$\frac{d}{dt}\left(\frac{\partial L}{\partial \dot{q}_1}\right) - \frac{\partial L}{\partial q_i} = \overline{Q}_i \qquad i = 1, 2, \ldots, n \tag{10.33}$$

where $L = T - V$ is known as the *Lagrangian function* of the system.

Example 10.1 Using Lagrange's method and the generalized coordinates $q_1 = x_1$ and $q_2 = x_2$, derive the equations of motion for the system shown in Fig. 10.4.

SOLUTION The kinetic energy of the system is

$$T = \tfrac{1}{2} m\dot{x}_G^2 + \tfrac{1}{2} I_G \dot{\theta}^2$$

while the potential energy is given by

$$V = \tfrac{1}{2} k_1 (x_1 - L_1)^2 + \tfrac{1}{2} k_2 (x_2 - L_2)^2 + mgx_G$$

where L_1 and L_2 are the undeformed lengths of springs 1 and 2, respectively. For small oscillations we can write

$$x_G = \frac{x_1 + x_2}{2} \qquad \text{and} \qquad \theta = \frac{x_2 - x_1}{L_o}$$

Then

$$L = T - V = \frac{1}{2} m \left(\frac{\dot{x}_1 + \dot{x}_2}{2}\right)^2 + \frac{1}{2} I_G \left(\frac{\dot{x}_2 - \dot{x}_1}{L_o}\right)^2$$

$$- \frac{1}{2} k_1 (x_1 - L_1)^2 - \frac{1}{2} k_2 (x_2 - L_2)^2 - mg \frac{x_1 + x_2}{2}$$

Since there are no nonconservative forces, Lagrange's equations are

$$\frac{d}{dt}\left(\frac{\partial L}{\partial \dot{x}_i}\right) - \frac{\partial L}{\partial x_i} = 0 \qquad i = 1, 2$$

For $i = 1$, performing the indicated differentiation gives

$$\frac{1}{2} m(\ddot{x}_1 + \ddot{x}_2) - \frac{I_G}{L_o}(\ddot{x}_2 - \ddot{x}_1) + k_1(x_1 - L_1) + \frac{mg}{2} = 0$$

while for $i = 2$,

$$\frac{1}{2} m(\ddot{x}_1 + \ddot{x}_2) + \frac{I_G}{L_o}(\ddot{x}_2 - \ddot{x}_1) + k_2(x_2 - L_2) + \frac{mg}{2} = 0$$

Note that the weight mg of the bar, as well as the free length of each spring, appears in the equations of motion. This is because coordinates x_1 and x_2 are not measured from the equilibrium position of the bar.

10.4 MATRIX FORMULATION

The algebraic manipulations involved in solving simultaneous ordinary differential equations are relatively straightforward but cumbersome and time-consuming if the number of degrees of freedom is large. Fortunately, many systems are governed by equations which are readily amenable to solution by matrix methods.

To illustrate, let us return to the system of Fig. 10.1, which is governed by Eqs. (10.2). First, we define the 3×1 column matrix x_i as

$$\{x_i\} = \begin{Bmatrix} x_1(t) \\ x_2(t) \\ x_3(t) \end{Bmatrix} \tag{10.34}$$

which will be referred to as the *displacement* or *solution* matrix. Similarly, the *velocity* and *acceleration* matrices may be defined as

$$\{\dot{x}_i\} = \begin{Bmatrix} \dot{x}_1(t) \\ \dot{x}_2(t) \\ \dot{x}_3(t) \end{Bmatrix} \tag{10.35}$$

and

$$\{\ddot{x}_i\} = \begin{Bmatrix} \ddot{x}_1(t) \\ \ddot{x}_2(t) \\ \ddot{x}_3(t) \end{Bmatrix} \tag{10.36}$$

respectively.

To replace Eqs. (10.2) with an equivalent matrix equation, we must find matrices $[m]$, $[c]$, and $[k]$ such that

$$[m]\{\ddot{x}_i\} + [c]\{\dot{x}_i\} + [k]\{x_i\} = \{F_i\} \tag{10.37}$$

where

$$\{F_i\} = \begin{Bmatrix} F_1(t) \\ F_2(t) \\ F_3(t) \end{Bmatrix} \tag{10.38}$$

is the 3×1 matrix of external forcing functions. Since the right-hand side of Eq. (10.38) is a 3×1 matrix, it follows that each term on the left must also be 3×1. This fact and the known form of the governing equation make it easy to show that $[m]$ is a 3×3 diagonal matrix given by

$$\lceil m_{ij} \rfloor = \begin{bmatrix} m_{11} & 0 & 0 \\ 0 & m_{22} & 0 \\ 0 & 0 & m_{33} \end{bmatrix} = \begin{bmatrix} m_1 & 0 & 0 \\ 0 & m_2 & 0 \\ 0 & 0 & m_3 \end{bmatrix} \tag{10.39}$$

where $\lceil\ \rfloor$ denotes a matrix having all off-diagonal terms equal to zero.

Next we must determine the elements of $[c]$ such that the matrix multiplication

$$[c_{ij}]\{\dot{x}_i\} = \begin{bmatrix} c_{11} & c_{12} & c_{13} \\ c_{21} & c_{22} & c_{23} \\ c_{31} & c_{32} & c_{33} \end{bmatrix} \begin{Bmatrix} \dot{x}_1 \\ \dot{x}_2 \\ \dot{x}_3 \end{Bmatrix} \tag{10.40}$$

gives results equivalent to the velocity terms in the governing equations. Specifically, we must have

$$\begin{aligned} c_{11}\dot{x}_1 + c_{12}\dot{x}_2 + c_{13}\dot{x}_3 &= (c_1 + c_2)\dot{x}_1 - c_2\dot{x}_2 \\ c_{21}\dot{x}_1 + c_{11}\dot{x}_2 + c_{23}\dot{x}_3 &= -c_2\dot{x}_1 + (c_2 + c_3)\dot{x}_2 - c_3\dot{x}_3 \\ c_{31}\dot{x}_1 + c_{32}\dot{x}_2 + c_{33}\dot{x}_3 &= -c_3\dot{x}_2 + (c_3 + c_4)\dot{x}_3 \end{aligned} \tag{10.41}$$

Equating coefficients of $\dot{x}_i$ in each of Eqs. (10.41) gives

$$[c_{ij}] = \begin{bmatrix} c_1 + c_2 & -c_2 & 0 \\ -c_2 & c_2 + c_3 & -c_3 \\ 0 & -c_3 & c_3 + c_4 \end{bmatrix} \tag{10.42}$$

A completely analogous procedure (left as an exercise for the student) will give

$$[k] = [k_{ij}] = \begin{bmatrix} k_1 + k_2 & -k_2 & 0 \\ -k_2 & k_2 + k_3 & -k_3 \\ 0 & -k_3 & k_3 + k_4 \end{bmatrix} \tag{10.43}$$

as the stiffness matrix. In general, any element k_{ij} of the stiffness matrix is defined as the force required at position i to produce a unit displacement at position j and such that the displacements at all positions except j are zero.

The matrices $[m]$, $[c]$, and $[k]$ are known as the *mass* (or *inertia*), *damping*, and *stiffness* matrices, respectively. We note that each of these matrices is symmetric about its diagonal. This is expressed in terms of matrix algebra as

$$[m] = [m]^T \qquad [c] = [c]^T \qquad [k] = [k]^T \tag{10.44}$$

where the superscript T indicates the *transpose* of the matrix. The transpose of a matrix is formed by interchanging the rows and columns of the matrix.

If the motion of the system is described in a set of generalized coordinates, the matrix form will be

$$[m]\{\ddot{q}\} + [c]\{\dot{q}\} + [k]\{q\} = \{Q\} \tag{10.45}$$

where $\{q\}$, $\{\dot{q}\}$, and $\{\ddot{q}\}$ are column matrices of generalized coordinates, velocities, and accelerations, respectively, and $\{Q\}$ is the column matrix of generalized forces. It must be noted at this point that the mass matrix is not necessarily diagonal for a particular set of generalized coordinates.

Example 10.2 Determine the mass, damping, and stiffness matrices for the system of Fig. 10.5.

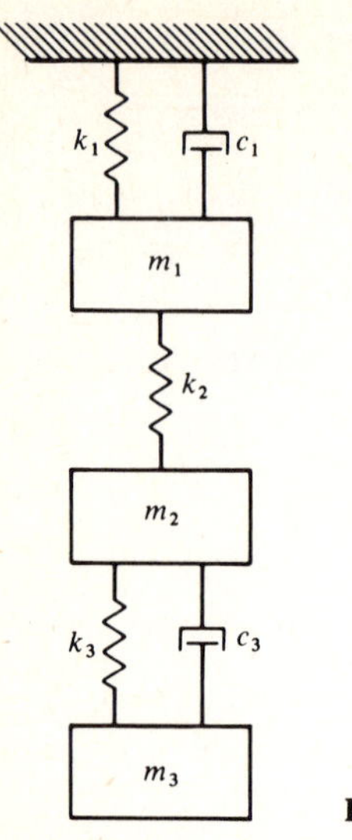

Figure 10.5

SOLUTION Noting that the system is the same as that in Fig. 10.1 with $c_2 = c_4 = k_4 = 0$, we have

$$[m] = \begin{bmatrix} m_1 & 0 & 0 \\ 0 & m_2 & 0 \\ 0 & 0 & m_3 \end{bmatrix} \qquad [c] = \begin{bmatrix} c_1 & 0 & 0 \\ 0 & c_3 & -c_3 \\ 0 & -c_3 & c_3 \end{bmatrix}$$

$$[k] = \begin{bmatrix} k_1 + k_2 & -k_2 & 0 \\ -k_2 & k_2 + k_3 & -k_3 \\ 0 & -k_3 & k_3 \end{bmatrix}$$

Example 10.3 Determine the stiffness matrix for the system shown in Fig. 10.6*a* if the mass of the beam is negligible. The beam has uniform flexural rigidity *EI*.

SOLUTION To determine each element of the stiffness matrix, we apply forces F_1, F_2, and F_3 as shown in Fig. 10.6*b*. By using beam-deflection tables and the principle of superposition, the deflections can be written as

$$\frac{L^3}{12EI}(9F_1 + 11F_2 + 7F_3) = x_1$$

$$\frac{L^3}{12EI}(11F_1 + 16F_2 + 11F_3) = x_2$$

$$\frac{L^3}{12EI}(7F_1 + 11F_2 + 9F_3) = x_3$$

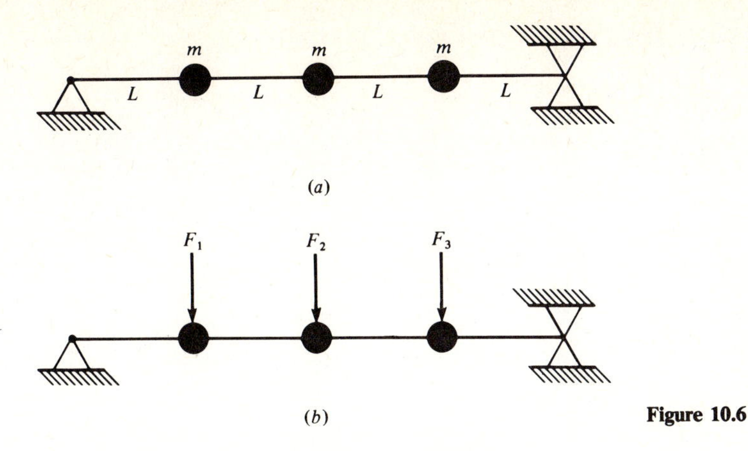

Figure 10.6

In matrix form the deflection equations are

$$\frac{L^3}{12EI}\begin{bmatrix} 9 & 11 & 7 \\ 11 & 16 & 11 \\ 7 & 11 & 9 \end{bmatrix}\begin{Bmatrix} F_1 \\ F_2 \\ F_3 \end{Bmatrix} = \begin{Bmatrix} x_1 \\ x_2 \\ x_3 \end{Bmatrix}$$

We first set $x_1 = 1$ and $x_2 = x_3 = 0$ and solve for F_1, F_2, and F_3. Referring to the definition of the elements of the stiffness matrix, we note that the resulting force values are equal to k_{11}, k_{21}, and k_{31}, respectively. For this calculation the matrix equation becomes

$$\frac{L^3}{12EI}\begin{bmatrix} 9 & 11 & 7 \\ 11 & 16 & 11 \\ 7 & 11 & 9 \end{bmatrix}\begin{Bmatrix} F_1 \\ F_2 \\ F_3 \end{Bmatrix} = \begin{Bmatrix} 1 \\ 0 \\ 0 \end{Bmatrix}$$

Solving by Cramer's rule gives

$$F_1 = k_{11} = \frac{69EI}{7L^3} \qquad F_2 = k_{21} = -\frac{66EI}{7L^3} \qquad F_3 = k_{31} = \frac{27EI}{7L^3}$$

Next we set $x_2 = 1$ and $x_1 = x_3 = 0$ and repeat the procedure to obtain

$$k_{12} = -\frac{66EI}{7L^3} \qquad k_{22} = \frac{96EI}{7L^3} \qquad k_{32} = -\frac{66EI}{7L^3}$$

Finally, we set $x_3 = 1$ and $x_1 = x_2 = 0$ to obtain

$$k_{13} = \frac{27EI}{7L^3} \qquad k_{23} = -\frac{66EI}{7L^3} \qquad k_{33} = \frac{69EI}{7L^3}$$

The stiffness matrix can now be written as

$$\frac{EI}{7L^3}\begin{bmatrix} 69 & -66 & 27 \\ -66 & 96 & -66 \\ 27 & -66 & 69 \end{bmatrix}$$

and we observe that the matrix is symmetric.

10.5 NATURAL FREQUENCIES OF UNDAMPED FREE VIBRATIONS

To begin our study of the methods of solving problems involving multiple degrees of freedom, we shall restrict the analysis to systems containing no damping elements and free of external forces. Under these restrictions and in terms of a set of generalized coordinates, the equations of motion are represented in matrix form by

$$[m]\{\ddot{q}\} + [k]\{q\} = \{0\} \tag{10.46}$$

which is equivalent to the system of simultaneous homogeneous differential equations

$$\sum_{j=1}^{n} m_{ij}\ddot{q}_j + \sum_{j=1}^{n} k_{ij}q_j = 0 \qquad i = 1, 2, \ldots, n \tag{10.47}$$

where n is the number of degrees of freedom. At this point we make no assumptions regarding the mass and stiffness matrices except that each is symmetric.

From our general discussion of differential equations in Chap. 2, Eqs. (10.47) would appear to be of the form for which exponential solutions exist. Further, of such equations studied to this point, those containing no damping terms had solutions in the form of sine and cosine functions. To test this logic, we shall assume solutions in the form

$$q_j(t) = A_j \sin(\omega t + \phi) \tag{10.48}$$

where A_j is a constant, and proceed to prove or disprove this assumption. Differentiating twice, we obtain

$$\ddot{q}_j(t) = -A_j\omega^2 \sin(\omega t + \phi) \tag{10.49}$$

Comparing Eqs. (10.48) and (10.49), in matrix notation we have

$$\{\ddot{q}\} = -\omega^2\{q\} \tag{10.50}$$

which converts Eq. (10.46) to

$$-\omega^2[m]\{q\} + [k]\{q\} = 0 \tag{10.51}$$

Equation (10.51) is equivalent to

$$\sum_{j=1}^{n} (k_{ij} - \omega^2 m_{ij})A_j = 0 \qquad i = 1, 2, \ldots, n \tag{10.52}$$

This represents a system of n simultaneous algebraic equations in n unknowns A_j, $j = 1, 2, \ldots, n$. Such a system has a nontrivial solution if and only if the determinant of the coefficient matrix is identically zero. The trivial solution $A_j = 0$ simply corresponds to static equilibrium. Thus we must determine the value (or values) of ω for which

$$|k_{ij} - \omega^2 m_{ij}| = 0 \tag{10.53}$$

A problem of this type is known as a *characteristic-value* (or *eigenvalue*) problem, and Eq. (10.53) is referred to as the *characteristic equation.* This terminology implies that the value of ω satisfying the equation is a characteristic of the particular system in question.

The characteristic equation is a polynomial of degree n in ω^2 and can be shown to have n roots (assumed to be distinct)

$$\omega_1^2 < \omega_2^2 < \cdots < \omega_n^2$$

where we have followed the customary practice of numbering the roots in ascending order. The positive square roots of the characteristic values or eigenvalues are the *natural frequencies* of the system and represent the circular frequencies at which the system can oscillate. In actuality, the motion of a system is most often a combination of many or all of its natural frequencies. The smallest natural frequency ω_1 is designated as the *fundamental frequency* of the system.

We have now determined that a system having n degrees of freedom has n natural frequencies ω_r, $r = 1, 2, \ldots, n$, corresponding to the form of solution of Eq. (10.48). For each of the natural frequencies ω_r, there exists a set of amplitudes A_j, $j = 1, 2, \ldots, n$, satisfying Eq. (10.52). Then for each generalized coordinate q_j, there are actually n independent solutions, each of which can now be written as

$$q_j = A_j^{(r)}(\omega_r t + \phi_r) \tag{10.54}$$

where the superscript r is used to indicate the amplitude associated with a particular natural frequency ω_r. These are known as *principal-mode* solutions, since each represents a possible mode of oscillation of the system. Equation (10.51) can now be written as

$$\omega_r^2[m]\{A^{(r)}\} = [k]\{A^{(r)}\} \qquad r = 1, 2, \ldots, n \tag{10.55}$$

where the vectors $\{A^{(r)}\}$ are known as the *characteristic vectors*, or *eigenvectors*. The eigenvectors are also called the *modal vectors*, since they define the mode shapes of the principal modes. The principal-mode solutions satisfy Eq. (10.46), and since this equation is linear, the principle of superposition applies. We can then write the general solutions of Eq. (10.46) in the form

$$q_j = \sum_{r=1}^{n} A_j^{(r)} \sin(\omega_r t + \phi_r) \qquad j = 1, 2, \ldots, n \tag{10.56}$$

or in matrix form as

$$\{q\} = \sum_{r=1}^{n} \{A^{(r)}\} \sin(\omega_r t + \phi_r) \tag{10.57}$$

We must now address the problem of determining the amplitudes. If we substitute a particular natural frequency, say ω_1, into Eqs. (10.53), we obtain a system of n simultaneous, homogeneous, algebraic equations in the unknowns

$A_j^{(1)}, j = 1, 2, \ldots, n$. Unfortunately, only $n - 1$ of these equations are independent; the remaining equation is useful only as a check. The best we can do at this point is solve for $n - 1$ amplitudes in terms of the remaining amplitude. This is accomplished by setting one amplitude, say $A_1^{(1)}$, equal to unity and then solving $n - 1$ equations for $A_j^{(1)}, j = 2, \ldots, n$. Note that the resulting values are not absolute but are ratios of the $n - 1$ amplitudes to $A_1^{(1)}$.

If this procedure is carried out for each natural frequency, we obtain n sets of amplitude ratios, and we still have n unknowns $A_1^{(r)}, r = 1, 2, \ldots, n$. These remaining unknowns are effectively constants of integration, and together with the phase angles $\phi_r, r = 1, 2, \ldots, n$ are the $2n$ arbitrary constants which depend on initial conditions. The details of the procedure are illustrated by the following numerical example.

Example 10.4 For the system shown in Fig. 10.5, let $c_1 = c_3 = 0$, $k_1 = k_3 = k$, $k_2 = 2k$, $m_1 = m_2 = m$, and $m_3 = 2m$. Determine (*a*) the natural frequencies of the system and (*b*) the general solution.

SOLUTION Utilizing the results of Example 10.2, we have

$$[m] = \begin{bmatrix} m & 0 & 0 \\ 0 & m & 0 \\ 0 & 0 & 2m \end{bmatrix} \qquad [k] = \begin{bmatrix} 3k & -2k & 0 \\ -2k & 3k & -k \\ 0 & -k & k \end{bmatrix}$$

The characteristic determinant [Eq. (10.53)] is

$$\begin{vmatrix} 3k - \omega^2 m & -2k & 0 \\ -2k & 3k - \omega^2 m & -k \\ 0 & -k & k - 2\omega^2 m \end{vmatrix} = 0$$

After expanding and simplifying, the characteristic equation is found to be

$$\omega^6 - 6.5\frac{k}{m}\omega^4 + 7.5\left(\frac{k}{m}\right)^2\omega^2 - \left(\frac{k}{m}\right)^3 = 0$$

From the form of the characteristic equation, it appears suitable to assume that

$$\omega^2 = C\frac{k}{m}$$

and attempt to determine C. Substituting for ω^2 in the characteristic equation results in

$$(C^3 - 6.5C^2 + 7.5C - 1)\left(\frac{k}{m}\right)^3 = 0$$

or

$$C^3 - 6.5C^2 + 7.5C - 1 = 0$$

since $k/m \neq 0$. This equation has the three real roots $C_1 = 0.1532$, $C_2 = 1.2912$, and $C_3 = 5.0556$. The natural frequencies of the system are then

$$\omega_1 = \sqrt{C_1 \frac{k}{m}} = 0.3914\sqrt{\frac{k}{m}} \text{ rad/s}$$

$$\omega_2 = \sqrt{C_2 \frac{k}{m}} = 1.1363\sqrt{\frac{k}{m}} \text{ rad/s}$$

$$\omega_3 = \sqrt{C_3 \frac{k}{m}} = 2.2485\sqrt{\frac{k}{m}} \text{ rad/s}$$

(For a fuller appreciation of the mathematical complexity involved in solving the characteristic equation for even this simple system with three degrees of freedom, the reader is invited to determine the roots in this case independently. The author used a trigonometric solution procedure which can be described as tedious at best. Digital-computer techniques to be discussed shortly can be used to greatly reduce the complexity.)

To determine the amplitude ratios, we must substitute each natural frequency in turn into Eq. (10.52). For ω_1 we obtain

$$(3k - \omega_1^2 m)A_1^{(1)} - 2kA_2^{(1)} = 0$$

$$-2kA_1^{(1)} + (3k - \omega_1^2 m)A_2^{(1)} - kA_3^{(1)} = 0$$

$$-kA_2^{(1)} + (k - 2\omega_1^2 m)A_3^{(1)} = 0$$

or

$$2.847kA_1^{(1)} - 2kA_2^{(1)} = 0$$

$$-2kA_1^{(1)} + 2.847kA_2^{(1)} - kA_3^{(1)} = 0$$

$$-kA_2^{(1)} + 0.694kA_3^{(1)} = 0$$

Setting $A_1^{(1)} = 1$ and solving any two of the three equations gives

$$A_2^{(1)} = 1.4235 \qquad A_3^{(1)} = 2.0511$$

Following the same procedure using ω_2 and ω_3 gives

$$A_1^{(2)} = 1 \qquad A_2^{(2)} = 0.8544 \qquad A_3^{(2)} = -0.5399$$

and

$$A_1^{(3)} = 1 \qquad A_2^{(3)} = -1.0279 \qquad A_3^{(3)} = 0.1128$$

The general solution for motion of the system is then

$$x_1(t) = A_1^{(1)} \sin(\omega_1 t + \phi_1) + A_1^{(2)} \sin(\omega_2 t + \phi_2) + A_1^{(3)} \sin(\omega_3 t + \phi_3)$$

$$x_2(t) = 1.4235A_1^{(1)} \sin(\omega_1 t + \phi_1) + 0.8544A_1^{(2)} \sin(\omega_2 t + \phi_2) - 1.0279A_1^{(3)} \sin(\omega_3 t + \phi_3)$$

$$x_3(t) = 2.0511A_1^{(1)} \sin(\omega_1 t + \phi_1) - 0.5399A_1^{(2)} \sin(\omega_2 t + \phi_2) + 0.1128A_1^{(3)} \sin(\omega_3 t + \phi_3)$$

The values of $A_1^{(1)}$, $A_1^{(2)}$, $A_1^{(3)}$, ϕ_1, ϕ_2, and ϕ_3 depend on the initial conditions of oscillation of the system.

A qualitative description of the principal-mode oscillations is obtained by considering the algebraic signs of the amplitude ratios. For $\omega_1 = 0.3914\sqrt{k/m}$, each amplitude has the same sign. This means that the direction of motion of the three masses is the same at all times. In other words, the masses oscillate in phase but with different amplitudes. In the second principal mode, masses 1 and 2 move together while mass 3 moves in the opposite direction. For the third principal mode, masses 1 and 3 are in phase while mass 2 is out of phase. The actual motion of the system will most likely be a combination of these modes, since only a particular set of initial conditions will produce a principal-mode oscillation.

10.6 ORTHOGONALITY OF THE PRINCIPAL MODES

A fundamental mathematical relation known as *orthogonality* exists between the principal-mode solutions of systems having multiple degrees of freedom. We shall illustrate this relation with a two-degrees-of-freedom system for simplicity and then generalize the result to more complex systems.

Let us consider any two modes of an undamped system with two degrees of freedom, for which Eqs. (10.55) can be written as

$$\omega_1^2(m_{11}A_1^{(1)} + m_{12}A_2^{(1)}) = k_{11}A_1^{(1)} + k_{12}A_2^{(1)}$$

$$\omega_1^2(m_{21}A_1^{(1)} + m_{22}A_2^{(1)}) = k_{21}A_1^{(1)} + k_{22}A_2^{(1)} \tag{10.58}$$

for the first mode, and

$$\omega_2^2(m_{11}A_1^{(2)} + m_{12}A_2^{(2)}) = k_{11}A_1^{(2)} + k_{12}A_2^{(2)}$$

$$\omega_2^2(m_{21}A_1^{(2)} + m_{22}A_2^{(2)}) = k_{21}A_1^{(2)} + k_{22}A_2^{(2)} \tag{10.59}$$

for the second mode. Multiplying the first of Eqs. (10.58) by $A_1^{(2)}$ and the second by $A_2^{(2)}$ and adding give

$$\omega_1^2(m_{11}A_1^{(1)}A_1^{(2)} + m_{12}A_2^{(1)}A_1^{(2)} + m_{21}A_1^{(1)}A_2^{(2)} + m_{22}A_2^{(1)}A_2^{(2)})$$
$$= k_{11}A_1^{(1)}A_1^{(2)} + k_{12}A_2^{(1)}A_1^{(2)} + k_{21}A_1^{(1)}A_2^{(2)} + k_{22}A_2^{(1)}A_2^{(2)} \tag{10.60}$$

Similarly, multiplication of the first of Eqs. (10.59) by $A_1^{(1)}$ and the second by $A_2^{(1)}$ and adding result in

$$\omega_2^2(m_{11}A_1^{(2)}A_1^{(1)} + m_{12}A_2^{(2)}A_1^{(1)} + m_{21}A_1^{(2)}A_2^{(1)} + m_{22}A_2^{(2)}A_2^{(1)})$$
$$= k_{11}A_1^{(2)}A_1^{(1)} + k_{12}A_2^{(2)}A_1^{(1)} + k_{21}A_1^{(2)}A_2^{(1)} + k_{22}A_2^{(2)}A_2^{(1)} \tag{10.61}$$

Subtracting Eq. (10.61) from (10.60) and using the fact that $m_{ij} = m_{ji}$ and

$k_{ij} = k_{ji}$ (since the mass and stiffness matrices are symmetric[1]) gives

$$\left(m_{11}A_1^{(1)}A_1^{(2)} + m_{12}A_1^{(2)}A_2^{(1)} + m_{21}A_2^{(2)}A_1^{(1)} + m_{22}A_2^{(1)}A_2^{(2)}\right)\left(\omega_1^2 - \omega_2^2\right) = 0 \tag{10.62}$$

Since $\omega_1 \neq \omega_2$, we have $\omega_1^2 - \omega_2^2 \neq 0$, and Eq. (10.62) can be written as

$$m_{11}A_1^{(1)}A_1^{(2)} + m_{12}A_1^{(2)}A_2^{(1)} + m_{21}A_2^{(2)}A_1^{(1)} + m_{22}A_2^{(1)}A_2^{(2)} = 0 \tag{10.63}$$

which is the *orthogonality condition* for the two-degrees-of-freedom system.

The orthogonality condition for a system having n degrees of freedom can be similarly obtained and is given by

$$\sum_{i=1}^{n}\sum_{j=1}^{n} m_{ij}A_i^{(r)}A_j^{(s)} = 0 \qquad r \neq s \tag{10.64}$$

The orthogonality condition is a relation between the amplitudes of any two different principal-mode vibrations. This is also often referred to as orthogonality of the *modal vectors* and is quite useful when one desires to uncouple the equations of motion. For the latter purpose it is convenient to write the orthogonality condition in the matrix form

$$\{A^{(s)}\}^T[m]\{A^{(r)}\} = 0 \qquad r \neq s \tag{10.65}$$

The reader is urged to obtain Eq. (10.63) for the two-degrees-of-freedom system by applying Eq. (10.65) to verify that the results are identical.

Example 10.5 Verify that the orthogonality conditions are satisfied by the solution obtained for Example 10.4.

SOLUTION Referring back to Example 10.4, we have

$$[m] = \begin{bmatrix} m & 0 & 0 \\ 0 & m & 0 \\ 0 & 0 & 2m \end{bmatrix}$$

and the modal vectors

$$\{A^{(1)}\} = \begin{Bmatrix} 1.0000 \\ 1.4235 \\ 2.0511 \end{Bmatrix} \qquad \{A^{(2)}\} = \begin{Bmatrix} 1.0000 \\ 0.8544 \\ -0.5399 \end{Bmatrix} \qquad \{A^{(3)}\} = \begin{Bmatrix} 1.0000 \\ -1.0279 \\ 0.1128 \end{Bmatrix}$$

Applying Eq. (10.65) with $r = 1$ and $s = 2$ gives

$$m(1)(1) + m(0.8544)(1.4235) + 2m(-0.5399)(2.0511) = 0.001m \cong 0$$

which is sufficiently accurate for values generated with a hand calculator. For $r = 1$ and $s = 3$, we obtain

$$m(1)(1) + m(-1.0279)(1.4235) + 2m(0.1128)(2.0511) = -0.0005m \cong 0$$

[1] For proof of this statement, see Leonard Meirovitch, *Elements of Vibration Analysis*, McGraw-Hill, New York, 1975, chap. 4.

The final orthogonality relation is for $r = 2$ and $s = 3$ and is given by

$$m(1)(1) + m(-1.0279)(0.8544) + 2m(0.1128)(-0.5399) = 0.00004m \cong 0$$

This example shows that the orthogonality relations serve the extremely important function of providing a check on the accuracy of the principal-mode solutions.

10.7 RAYLEIGH'S METHOD

When an undamped lumped-mass system vibrates in its fundamental mode, each mass executes simple harmonic motion about its equilibrium position. Since the frequencies of oscillation are the same for each mass, it follows that maximum displacements are obtained simultaneously. At that instant, every mass is motionless, and all the vibrational energy is stored as elastic potential energy. Also, in simple harmonic motion, all masses pass through their respective equilibrium positions simultaneously, and the total energy at this instant is in the form of kinetic energy. These observations form the basis of *Rayleigh's method*, which is a technique for obtaining an estimate of the fundamental frequency of a conservative mechanical system.

Consider a system having n degrees of freedom which oscillates in its fundamental mode according to

$$q_j(t) = A_j^{(1)} \sin(\omega_1 t + \phi_1) \qquad j = 1, 2, \ldots, n \tag{10.66}$$

The kinetic and potential energies can be expressed as

$$T = \sum_{i=1}^{n} \sum_{j=1}^{n} a_{ij} \dot{q}_i \dot{q}_j$$

$$V = \sum_{i=1}^{n} \sum_{j=1}^{n} b_{ij} q_i q_j \tag{10.67}$$

where a_{ij} and b_{ij} are constants which depend on the choice of coordinates. Using Eq. (10.66), we can rewrite Eqs. (10.67) as

$$T = \left(\sum_{i=1}^{n} \sum_{j=1}^{n} a_{ij} A_i^{(1)} A_j^{(1)} \right) \omega_1^2 \cos^2(\omega_1 t + \phi_1)$$

$$V = \left(\sum_{i=1}^{n} \sum_{j=1}^{n} b_{ij} A_i^{(1)} A_j^{(1)} \right) \sin^2(\omega_1 t + \phi_1) \tag{10.68}$$

Since the system is conservative, $T + V =$ constant, and

$$\frac{d}{dt}(T + V) = 0$$

or

$$-\left(\sum_{i=1}^{n} \sum_{j=1}^{n} a_{ij} A_i^{(1)} A_j^{(1)} \right) 2\omega_1^3 \cos(\omega_1 t + \phi_1) \sin(\omega_1 t + \phi_1)$$

$$+ \left(\sum_{i=1}^{n} \sum_{j=1}^{n} b_{ij} A_i^{(1)} A_j^{(1)} \right) 2\omega_1 \cos(\omega_1 t + \phi_1) \sin(\omega_1 t + \phi_1) = 0 \tag{10.69}$$

Rearranging Eq. (10.69) gives

$$\omega_1^2 = \frac{\sum_{i=1}^{n}\sum_{j=1}^{n} b_{ij}A_i^{(1)}A_j^{(1)}}{\sum_{i=1}^{n}\sum_{j=1}^{n} a_{ij}A_1^{(1)}A_j^{(1)}} \tag{10.70}$$

where the right-hand side is known as *Rayleigh's quotient*. Equation (10.70) expresses the square of the fundamental frequency as the ratio of two quadratic functions of the principal-mode amplitudes. Our previous discussions of the free-vibration problem have shown that the eigenvectors $\{A^{(r)}\}$ for a particular mode can be determined precisely only if the frequency of the mode is known. In this sense, then, no simplification results from this approach. However, if an estimate of the fundamental frequency is desired, Eq. (10.70) will give a good approximation if a reasonable estimate of the mode shape for the fundamental mode can be obtained. By assuming the amplitudes comprising the eigenvector $\{A^{(1)}\}$ for the fundamental mode and calculating the kinetic and potential energies as outlined above, we then obtain

$$\omega_1^2 \cong \omega_R^2 = \frac{\sum_{i=1}^{n}\sum_{j=1}^{n} b_{ij}A_i^{(1)}A_j^{(1)}}{\sum_{i=1}^{n}\sum_{j=1}^{n} a_{ij}A_i^{(1)}A_j^{(1)}} \tag{10.71}$$

where ω_R is called the *Rayleigh frequency*. It can be shown that the estimated frequency so obtained is never less than the exact value, and that the error of the estimated frequency is less than the error involved in assuming the mode shape.

The foregoing discussion of the Rayleigh method proceeds from a physical basis in terms of the kinetic- and potential-energy functions of the system. An alternative approach provides a more convenient means of calculating the Rayleigh frequency, but the physical basis is somewhat obscured. Rewriting Eq. (10.55) for a system with n degrees of freedom, we have

$$\omega_1^2[m]\{A^{(1)}\} = [k]\{A^{(1)}\} \tag{10.72}$$

for the fundamental mode. Premultiplying both sides by $\{A^{(1)}\}^T$ gives

$$\omega_1^2\{A^{(1)}\}^T[m]\{A^{(1)}\} = \{A^{(1)}\}^T[k]\{A^{(1)}\} \tag{10.73}$$

Examination of the matrix products in Eq. (10.73) shows that each is a scalar, so we can write

$$\omega_1^2 = \frac{\{A^{(1)}\}^T[k]\{A^{(1)}\}}{\{A^{(1)}\}^T[m]\{A^{(1)}\}} \tag{10.74}$$

as an alternative form of Eq. (10.70). Again, if an assumed mode shape corresponding to eigenvector $\{A^{(1)}\}$ is used, we obtain an estimate of the fundamental frequency

$$\omega_1^2 \cong \omega_R^2 = \frac{\{A^{(1)}\}^T[k]\{A^{(1)}\}}{\{A^{(1)}\}^T[m]\{A^{(1)}\}} \tag{10.75}$$

Equations (10.71) and (10.75) are equivalent forms of Rayleigh's quotient. The latter form is much simpler to apply, as it does not require explicit formulation of the kinetic- and potential-energy functions.

Although we have discussed the application of Rayleigh's method only to the fundamental mode, it can be applied to higher modes as well. This is seldom done, however, since the configurations of the higher mode shapes cannot usually be estimated with sufficient accuracy. Also, an estimate of the fundamental frequency may be all that is required, especially if one desires to know if a system will operate at a frequency below resonance. In many cases an assumed fundamental mode shape corresponding to the static-equilibrium configuration of the system will give sufficient accuracy, as illustrated in the following example.

Example 10.6 Using Rayleigh's method, estimate the fundamental frequency of the system shown in Fig. 10.7*a*.

SOLUTION To use the static-equilibrium configuration as an estimate of the fundamental mode shape we draw the equilibrium free-body diagrams shown in Fig. 10.7*b* and write the equilibrium equations

$$m_1 g + k_2(\Delta_2 - \Delta_1) - k_1\Delta_1 = 0$$

$$m_2 g + k_3(\Delta_3 - \Delta_2) - k_2(\Delta_2 - \Delta_1) = 0$$

$$m_3 g - k_3(\Delta_3 - \Delta_2) = 0$$

Substituting $m_1 = m_2 = m$, $m_3 = 2m$, $k_1 = k_2 = k$, and $k_2 = 2k$ and solving the resulting equations give

$$\Delta_1 = \frac{4mg}{k} \qquad \Delta_2 = \frac{11mg}{2k} \qquad \Delta_3 = \frac{15mg}{2k}$$

Using these results, we assume the fundamental mode to be described by

$$\{A^{(1)}\} = \begin{Bmatrix} 1.000 \\ 1.375 \\ 1.875 \end{Bmatrix}$$

where we have taken $A_1^{(1)} = 1$, $A_1^{(2)} = \Delta_2/\Delta_1$, and $A_1^{(3)} = \Delta_3/\Delta_1$. It is left as an exercise for the reader to verify that the mass and stiffness matrices are

$$[m] = \begin{bmatrix} m & 0 & 0 \\ 0 & m & 0 \\ 0 & 0 & 2m \end{bmatrix} \qquad \text{and} \qquad [k] = \begin{bmatrix} 3k & -2k & 0 \\ -2k & 3k & -k \\ 0 & -k & k \end{bmatrix}$$

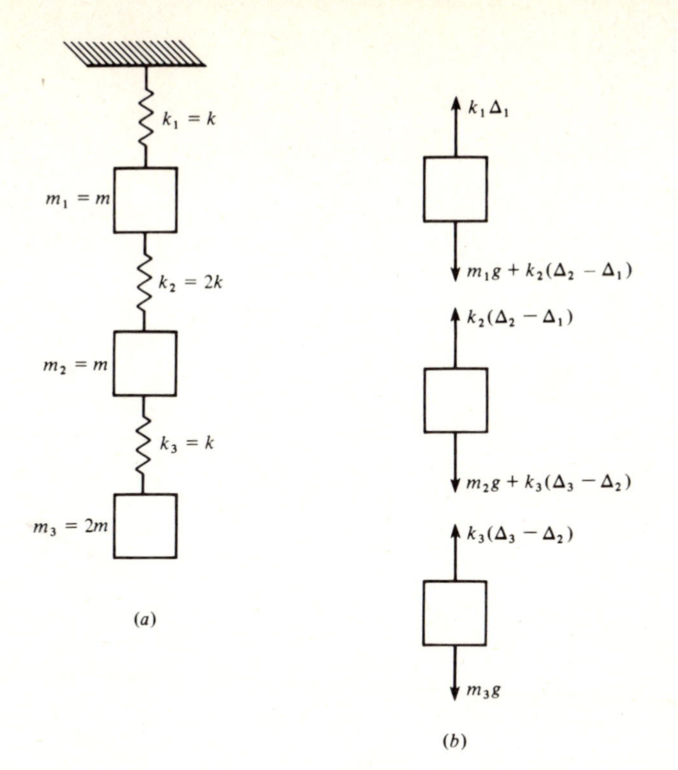

Figure 10.7

To apply Eq. (10.75), we calculate the numerator as

$$\{A^{(1)}\}^T[k]\{A^{(1)}\} = \{A^{(1)}\}^T\begin{bmatrix} 3k & -2k & 0 \\ -2k & 3k & -k \\ 0 & -k & k \end{bmatrix}\begin{Bmatrix} 1.000 \\ 1.375 \\ 1.875 \end{Bmatrix}$$

$$= \{1.000 \quad 1.375 \quad 1.875\}\begin{Bmatrix} 0.25k \\ 0.25k \\ 0.50k \end{Bmatrix} = 1.531k$$

and the denominator as

$$\{A^{(1)}\}^T[m]\{A^{(1)}\} = \{A^{(1)}\}^T\begin{bmatrix} m & 0 & 0 \\ 0 & m & 0 \\ 0 & 0 & 2m \end{bmatrix}\begin{Bmatrix} 1.000 \\ 1.375 \\ 1.875 \end{Bmatrix}$$

$$= \{1.000 \quad 1.375 \quad 1.875\}\begin{Bmatrix} m \\ 1.375m \\ 3.750m \end{Bmatrix} = 9.922m$$

The Rayleigh frequency is thus given by

$$\omega_R^2 = \frac{1.531k}{9.922m}$$

or

$$\omega_R = 0.393\sqrt{\frac{k}{m}} \quad \text{rad/s}$$

and the fundamental frequency is estimated as

$$\omega_1 \cong 0.393\sqrt{\frac{k}{m}} \quad \text{rad/s}$$

For purposes of comparison, the frequency equation for this system gives the true value $\omega_1 = 0.391\sqrt{k/m}$, which indicates an error of less than 1 percent.

10.8 MATRIX OPERATIONS IN FORTRAN

The increasing sophistication and speed of digital computers has greatly eased the burden of computation associated with the study of vibration systems having many degrees of freedom. In particular, the FORTRAN programming language is especially suitable for obtaining vibration solutions. The ease with which matrix calculations can be programmed and the existence of a large number of "canned" programs make the FORTRAN language an indispensable tool for vibration analysis. This section is devoted to a general discussion of the programming of matrix computations in FORTRAN as well as a brief discussion of some readily available subroutines.

Let us consider the general $n \times n$ matrix

$$U = \begin{bmatrix} U_{11} & U_{12} & \cdots & U_{1n} \\ \cdots & \cdots & \cdots & \cdots \\ U_{n1} & U_{n2} & \cdots & U_{nn} \end{bmatrix} = [U_{ij}] \tag{10.76}$$

which in computer applications is often called an *array*. In the FORTRAN language, any such matrix can be represented by a double-subscripted variable U(I, J), where the first subscript I indicates the row number and the second subscript J corresponds to the column number of the matrix element. Thus, the value of matrix element U_{23} is stored in the computer memory as the value of computer variable U(2, 3). The size of a matrix which can be represented in this manner is limited only by the memory capacity of the particular computer. To identify a FORTRAN variable as a subscripted variable and to reserve the correct amount of storage space requires a single program statement. For example, the FORTRAN statement

```
DIMENSION U(10,10)
```

identifies U as a double-subscripted variable and results in the assignment of

100 storage locations in computer memory. Then the computer variable U(I, J) can take on any of 100 different values specified by the subscripts.

Efficient use of subscripted variables requires proper input and output of numerical values. The most frequently occurring problems involve the order in which the elements of a matrix are read or printed by the computer. Although it is not our intent to cover every detail, a few examples will be used to illustrate various possibilities. Consider the sequence of statements

```
      DIMENSION U(10,10)
      READ (5,100)U
100   FORMAT (5F 10.4)
```

This particular READ statement appears to treat U as if it were an ordinary single variable. However the DIMENSION statement has previously identified U as a subscripted variable and reserved 100 storage locations. Thus, these three statements instruct the computer to read all 100 values, 5 per card, from a total of 20 data cards. More important, this form of input statement will result in storage of the 100 values in *column order*. That is, the first value read is assigned to U(1, 1), the second to U(2, 1), and so forth. The same is true on output. The statement

```
WRITE (6,200)U
```

of the same subscripted variable will result in the printing of the values in column order.

It is often more convenient to read or print matrix values in row order. This is easily accomplished through use of an implied DO loop within the appropriate input or output statement. For example,

```
READ (5,100)((U(I,J),J = 1,10),I = 1,10)
```

will cause the values read from data cards to be assigned in the row order U(1, 1), U(1, 2), U(1, 3) until all 100 values have been assigned.

Difficulty sometimes arises in distinguishing an $n \times 1$ (column) matrix from a $1 \times n$ (row) matrix, since each could be represented by a computer variable having a single subscript. This may seem to be a trivial concern, but it can cause real problems where matrix multiplications are required which involve both row and column matrices. Although double-subscripted variables could be used, it is more efficient to use single subscripts and simply consider the context carefully so as to avoid confusion.

To illustrate the preceding discussion and as a first application to vibrations, let us consider calculating an estimate of the fundamental frequency of a system having 10 degrees of freedom using Rayleigh's method. Referring to Eq. (10.75), we recall that this requires that we know the mass matrix $[m]$, the stiffness matrix $[k]$, and an estimate of the eigenvector $\{A^{(1)}\}$. Also, the transpose of the eigenvector is involved in the solution, so we must handle both a column matrix and a row matrix. The FORTRAN program to perform the required calculations is as follows:

```
       REAL M, K
       DIMENSION M(10,10),K(10,10),A(10)
       DATA SUM1,SUM2,B/3*0.0/
       READ (5,100)M,K,A
       DO 10 I = 1,10
       DO 15 J = 1,10
15     B = B + K(I,J)*A(J)
       SUM1 = SUM1 + B*A(I)
10     B = 0.0
       DO 20 I = 1,10
       DO 25 J = 1,10
25     B = B + M(I,J)*A(J)
       SUM2 = SUM2 + B*A(I)
20     B = 0.0
       WR = SQRT(SUM1/SUM2)
       WRITE (6,200)WR
100    FORMAT (10F8.4)
200    FORMAT (1H ,'OMEGA1 = ',E13.7)
       STOP
       END
```

In this program, note particularly the use of a DATA statement to initialize the value of certain variables as zero. This statement can also be used to advantage when it is required to predefine all elements of a matrix as zero. Also, observe that to use computer variables M and K to represent the mass and stiffness matrices, it is necessary to define these as real variables, since they are otherwise taken as integer variables according to standard FORTRAN conventions.

We now return to the general problem of obtaining the free response of a lumped-mass system having n degrees of freedom and discuss some computer techniques applicable to the solution steps required. A review of Secs. 10.4 and 10.5 shows that the solution sequence can be summarized as follows.

1. Derive the equations of motion using any convenient set of generalized coordinates. This can be accomplished by direct application of Newton's

second law and its rotational counterpart or by the energy approach using Lagrange's equations. Regardless of the method used, the equations of motion will be obtained in the matrix form

$$[m]\{\ddot{q}\} + [k]\{q\} = \{0\}$$

whereby the mass and stiffness matrices become known.

2. Assuming solutions of the form

$$\{q\} = \{A\} \sin(\omega t + \phi)$$

reduce the differential equations of motion to the homogeneous system of algebraic amplitude equations

$$-\omega^2[m]\{A\} + [k]\{A\} = \{0\}$$

3. To obtain the nontrivial solutions, write the frequency equation in determinant form as

$$|k_{ij} - \omega^2 m_{ij}| = 0$$

Solution of the frequency equation yields n natural frequencies ω_r ($r = 1, 2, \ldots, n$) corresponding to the n principal-mode solutions.

4. Substitute each of the natural frequencies, one at a time, into any $n - 1$ of the amplitude equations and solve simultaneously to obtain n amplitude vectors or eigenvectors $\{A^{(r)}\}$. Recall that one amplitude value for each frequency is arbitrarily set equal to 1 so that each amplitude vector contains one unknown constant.

5. Applying the principle of superposition, write the general solution as

$$\{q\} = \sum_{r=1}^{n} \{A^{(r)}\} \sin(\omega_r t + \phi_r)$$

Then obtain the complete solution for free oscillations by applying the $2n$ initial conditions to evaluate the $2n$ arbitrary constants, composed of n unknown amplitudes and n phase angles ϕ_r ($r = 1, 2, \ldots, n$). (We shall defer discussion of initial conditions until the next section, and concentrate here on the determination of the natural frequencies and amplitude vectors.)

In previous examples, we have explicitly followed this procedure to obtain numerical solutions. However, when the number of degrees of freedom exceeds four or five, determination of the natural frequencies by direct solution of the frequency equation becomes progressively more difficult. Since the problem of determining the n values of ω_r and the corresponding n principal modes $\{A^{(r)}\}$ is a well-known mathematical problem (the eigenvalue problem), many methods have been developed for its solution. Of particular interest is the fact that some methods for solving the eigenvalue problem simultaneously determine both the eigenvalues and the eigenvectors. Thus, the determination of natural frequencies and the associated amplitude vectors is combined into a single solution procedure.

To utilize some standard notation used for eigenvalue problems, we rewrite the amplitude equations as

$$[m]\{A\} - \lambda[k]\{A\} = \{0\} \qquad \lambda = \frac{1}{\omega^2} \tag{10.77}$$

Premultiplying Eq. (10.77) by $[k]^{-1}$, the *inverse of the stiffness matrix*,[2] gives

$$[k]^{-1}[m]\{A\} = \lambda[I]\{A\} \tag{10.78}$$

where $[I]$ is the identity matrix having all diagonal terms equal to one and all off-diagonal terms zero. By defining the *dynamical matrix* $[D]$ as

$$[D] = [k]^{-1}[m] \tag{10.79}$$

Eq. (10.78) can be written as

$$[D]\{A\} = \lambda\{A\} \tag{10.80}$$

This form allows us to see that the eigenvalue problem represents a set of simultaneous equations having a solution vector $\{A\}$ such that when $\{A\}$ is premultiplied by $[D]$, the result is a scalar multiple of $\{A\}$. We note that the dynamical matrix is, in general, *not* symmetric. An alternative form of Eq. (10.80) can be obtained by introducing the *modal matrix*

$$[A] = [\{A^{(1)}\} \quad \{A^{(2)}\} \quad \cdots \quad \{A^{(n)}\}] \tag{10.81}$$

which is seen to be the square matrix having the eigenvectors (principal-mode amplitude vectors) as its columns. We can then write the eigenvalue problem as

$$[D][A] = [A][\lambda] \tag{10.82}$$

where

$$[\lambda] = \begin{bmatrix} \lambda_1 & 0 & \cdots & 0 \\ 0 & \lambda_2 & \cdots & 0 \\ \cdots & \cdots & \cdots & \cdots \\ 0 & 0 & \cdots & \lambda_n \end{bmatrix} \tag{10.83}$$

is the diagonal matrix containing all eigenvalues.

There is a considerable amount of existing software in the form of FORTRAN subroutine programs which are quite useful in solving eigenvalue problems. Table 10.1 contains descriptions of several such subroutines that have been used often by the author and found to give excellent results for vibration problems having as many as 75 degrees of freedom. The remainder of this section will be devoted to a discussion of the applications of these subroutines.[3]

First let us consider the calculation of the dynamical matrix. According to Eq. (10.79), this requires that we obtain the inverse of the stiffness matrix and

[2] The inverse of the stiffness matrix is known as the *flexibility matrix* and is denoted by $[a] = [k]^{-1}$. Any element a_{ij} of the flexibility matrix is defined as the displacement of a point $x = x_i$ due to a unit force applied at $x = x_j$.

[3] This particular software package is no longer available from IBM. However, most moderate to large computing centers continue to maintain these or similar subroutines on a local basis. Readers are urged to check the matrix-analysis programs available through their computing centers. In addition, a series of FORTRAN programs and subroutines which perform similar calculations is included in App. B.

Table 10.1 FORTRAN subroutines for eigenvalue analysis

Subroutine	Description
SINV	Calculates the inverse of a positive-definite symmetric matrix.
MPRD	Calculates the matrix resulting from the multiplication of any two matrices which are comformable for multiplication
HSBG	Reduces a real square matrix to upper almost-triangular (Hessenberg) form while preserving the eigenvalues of the matrix
ATEIG	Calculates the eigenvalues of a real upper almost-triangular matrix using the QR iteration method
MTRA	Transposes a general matrix

*Reference: System/*360 *Scientific Subroutine Package, Programmer's Manual,* IBM Corporation, Technical Publications Department, White Plains, N.Y. (See fn. 3.)

then perform the matrix multiplication $[k]^{-1}[m]$. Since $[k]$ is a symmetric positive-definite square matrix, its inverse can be calculated using subroutine SINV. This subroutine takes the upper triangular part of a given matrix and the number of rows (columns) in the matrix as input and returns the upper triangular part of the inverse of the matrix as output. For proper input to this subroutine, the upper triangular part of $[k]$ must be stored in computer memory as a single-subscripted variable in column order. If the elements of the stiffness matrix are stored as the computer variables K(I, J), I = 1, 2, . . . , N, J = 1, 2, . . . , N, this can be accomplished by the sequence of FORTRAN statements

```
      L = 1
      DO 10 J = 1,N
      DO 10 I = 1,J
      B(L) = K(I,J)
10    L = L + 1
```

The result is a vector B containing the $n(n + 1)/2$ upper triangular elements of the stiffness matrix. The subroutine accepts this vector as its input and returns B as the upper triangular elements of the inverse. After the subroutines are called, the inverse matrix can be converted to double-subscripted form by

```
   L = 1
   DO 20 J = 1,N
   DO 20 I = 1,J
   KI(I,J) = B(L)
   KI(J,I) = KI(I,J)
20 L = L + 1
```

where the computer variables contain the elements of $[k]^{-1}$.

Having obtained the inverse of the stiffness matrix, we need only perform the matrix multiplication $[k]^{-1}[m]$ to obtain the dynamical matrix. This is a very simple operation in FORTRAN and is accomplished by

```
      DO 30 I = 1,N
      DO 30 J = 1,N
      DO 30 L = 1,N
   30 D(I,J) = D(I,J) + KI(I,L)*M(L,J)
```

provided all elements of D(I, J) have been initialized at zero as previously discussed. Subroutine MPRD (Table 10.1) is a more general matrix-multiplication program which is not restricted to square matrices. The latter routine is capable of accepting a symmetric matrix stored in vector mode, as discussed in the preceding paragraph. This feature can be used to reduce storage requirements for large programs.

We must now address the problem of calculating the eigenvalues of $[D]$, which is a nonsymmetric matrix. Several numerical methods have been developed for solving this problem; although the complete mathematical details are beyond the scope of this text,[4] we shall discuss one of them briefly. The method is known as the *QR transformation* and is based on the fact that a general real matrix $[D]$ can be decomposed according to

$$[D] = [Q][R] \tag{10.84}$$

where $[Q]$ is an orthogonal matrix[5] and $[R]$ is an upper triangular matrix having the eigenvalues of $[D]$ as its diagonal elements. The QR method is an iterative technique, and many iterations may be required for convergence of the eigenvalues. The convergence process is hastened if the original matrix is first converted to *upper Hessenberg form*, which is the "almost-triangular" matrix form

$$[D] = \begin{bmatrix} d_{11} & d_{12} & \cdots & d_{1n} \\ d_{12} & d_{22} & \cdots & d_{2n} \\ 0 & d_{23} & \cdots & d_{3n} \\ \cdots & \cdots & \cdots & \cdots \\ 0 & 0 & \cdots & d_{nn} \end{bmatrix} \tag{10.85}$$

For digital computation, subroutines HSBG and ATEIG are used sequentially to obtain eigenvalues by this method. The first of these takes the general matrix $[D]$ as its input, converts it to the form of Eq. (10.85), and returns the transformed matrix as output. Subroutine ATEIG (for almost-triangular eigenvalues) then performs the QR iteration until matrix $[R]$ in Eq. (10.84) is

[4] For a more complete discussion, see Anthony Ralston, *A First Course in Numerical Analysis*, McGraw-Hill, New York, 1965, chap. 10.

[5] An orthogonal matrix is a matrix having the property $[Q]^{-1} = [Q]^T$.

triangular to the desired degree of accuracy. Input to ATEIG is the Hessenberg form of $[D]$, which is destroyed in the computation. The output of the subroutine is composed of two vectors RR(I) and RI(I), I = 1, 2, . . . , N. The vector RR contains the real parts of the eigenvalues, while RI contains the imaginary parts. The eigenvalues are returned in order of decreasing moduli. For undamped-vibration systems the eigenvalues are real. Thus they will be contained in vector RR in descending order of magnitude. As a check, vector RI should be examined for any nonzero values, since their presence will indicate an error condition.

With the eigenvalues known, the natural frequencies are determined by Eq. (10.77) as

$$\omega_r = \sqrt{\frac{1}{\lambda_r}} \qquad r = 1, 2, \ldots, n \tag{10.86}$$

which is accomplished in FORTRAN as

```
      DO 40 I = 1,N
   40 W(I) = SQRT(1/RR(I))
```

Since the eigenvalues are obtained in descending order, the natural frequencies are automatically obtained in ascending order such that the value of computer variable W(1) corresponds to the fundamental frequency. The reader must be cautioned that in using these routines it is necessary to provide duplicate storage of $[D]$ since it is destroyed by ATEIG but must be used again later in obtaining the eigenvectors.

The eigenvectors are calculated by substituting the eigenvalues one at a time into

$$\begin{bmatrix} d_{11}-\lambda_r & d_{12} & \cdots & d_{1n} \\ d_{21} & d_{22}-\lambda_r & \cdots & d_{2n} \\ \cdots & \cdots & \cdots & \cdots \\ d_{n1} & d_{n2} & \cdots & d_{nn}-\lambda_r \end{bmatrix} \begin{Bmatrix} A_1^{(r)} \\ A_2^{(r)} \\ \vdots \\ A_n^{(r)} \end{Bmatrix} = 0 \tag{10.87}$$

Since Eq. (10.87) represents a system of homogeneous linear equations, it is necessary to assign a value to one element of the eigenvector, reduce the system to $n-1$ equations, and solve the resulting inhomogeneous system for $n-1$ amplitude ratios. If we choose $A_1^{(r)} = 1$ for convenience, we then have

$$\begin{bmatrix} d_{22}-\lambda_r & d_{23} & \cdots & d_{2n} \\ d_{32} & d_{33}-\lambda_r & \cdots & d_{3n} \\ \cdots & \cdots & \cdots & \cdots \\ d_{n2} & d_{n3} & \cdots & d_{nn}-\lambda_r \end{bmatrix} \begin{Bmatrix} A_2^{(r)} \\ A_3^{(r)} \\ \vdots \\ A_n^{(r)} \end{Bmatrix} = 0 \tag{10.88}$$

which will yield numerical values for $A_i^{(r)}$, $i = 2, 3, \ldots, n$ for each λ_r, $r = 1, 2, \ldots, n$. Thus, we must solve $n-1$ equations a total of n times, after which we shall have obtained each and every eigenvector (amplitude vector) within a constant multiple.

The system of Eq. (10.88) is readily solved by subroutine GELG, which is based on the method of Gauss elimination. This routine takes the square matrix of coefficients and the right-hand vector of Eq. (10.88) as its inputs and returns the solution vector as its output. Again, this subroutine destroys the input values, so the matrix of coefficients must be stored in duplicate for protection. As the eigenvectors are obtained, they can be used to construct the modal matrix defined by Eq. (10.81) by using a double-subscripted computer variable A(I, J) in which the first subscript corresponds to generalized coordinate q_i, and the second subscript corresponds to principal mode r (natural frequency ω_r). Storing the modal matrix in this manner is extremely useful in solving initial-value and forced-vibration problems.

The procedure discussed above for solving the eigenvalue problem numerically is applicable to any real matrix, gives good accuracy, and is applicable, in theory, to any number of degrees of freedom. The practical limitations involve the computer storage capacity and computing time required as the degrees of freedom increase. Although we have not covered every detail here, a complete sample program for a system having 25 degrees of freedom is given in App. B.

10.9 INITIAL-VALUE PROBLEMS

Let us now return to the problem of free vibration of an undamped system having n degrees of freedom, governed by

$$[m]\{\ddot{q}\} + [k]\{q\} = \{0\} \tag{10.89}$$

where $\{q\}$ is the column matrix, or vector, of generalized coordinates $q_i(t)$, $i = 1, 2, \ldots, n$. We have previously shown that the general solution of Eq. (10.89) can be written as the sum of n solutions of the form

$$\{q^{(r)}\} = \{A^{(r)}\} \sin(\omega_r t + \phi_r) \qquad r = 1, 2, \ldots, n \tag{10.90}$$

or

$$q_i(t) = \sum_{r=1}^{n} A_i^{(r)} \sin(\omega_r t + \phi_r) \qquad i = 1, 2, \ldots, n \tag{10.91}$$

We now consider the case in which vibration of the system results from a prescribed set of initial conditions given by

$$\{q(0)\} = \{q_o\} \qquad \{\dot{q}(0)\} = \{\dot{q}_o\} \tag{10.92}$$

From the solution of the associated eigenvalue problem, the system natural frequencies ω_r are known, and each of the amplitude vectors $\{A^{(r)}\}$ is known within a constant multiple, since we assumed $A_1^{(r)}$ as unity for all r. Eq. (10.91) can thus be written

$$q_i(t) = \sum_{r=1}^{n} \beta_r A_i^{(r)} \sin(\omega_r t + \phi_r) \qquad i = 1, 2, \ldots, n \tag{10.93}$$

where β_r is the aforementioned constant multiple for each mode. By use of Eq.

(10.93), the initial conditions are expressed as

$$q_i(0) = \sum_{r=1}^{n} \beta_r A_i^{(r)} \sin \phi_r$$
$$\dot{q}_i(0) = \sum_{r=1}^{n} \omega_r \beta_r A_i^{(r)} \cos \phi_r \tag{10.94}$$

Since the left-hand sides of Eqs. (10.94) are known, we have a system of $2n$ equations which must be solved for the $2n$ arbitrary constants β_r and ϕ_r ($r = 1, 2, \ldots, n$). This is an apparently formidable task, since the n unknowns ϕ_r appear in the sine and cosine functions. It is simpler to define $2n$ new constants as $B_r = \beta_r \sin \phi_r$ and $C_r = \beta_r \cos \phi_r$ to obtain

$$q_i(0) = \sum_{r=1}^{n} B_r A_i^{(r)}$$
$$\dot{q}_i(0) = \sum_{r=1}^{n} \omega_r C_r A_i^{(r)} \tag{10.95}$$

which are two independent sets of n simultaneous equations. It is easily shown that Eqs. (10.95) are equivalent to the matrix equations

$$[A]\{B\} = \{q_o\} \tag{10.96}$$

and

$$[A][\omega]\{C\} = \{\dot{q}_o\} \tag{10.97}$$

where $[A]$ is the modal matrix, and $\lceil \omega \rfloor$ is the diagonal matrix of natural frequencies. Whether the solution procedure is manual or computerized, the modal matrix and the natural frequencies are known from the solution of the eigenvalue problem. Thus, Eqs. (10.96) and (10.97) can be solved for the constants B_r and C_r. Once these are known, the original constants in Eq. (10.93) are calculated by

$$\phi_r = \tan^{-1}\frac{B_r}{C_r}$$
$$\beta_r = \sqrt{B_r^2 + C_r^2} \tag{10.98}$$

for $r = 1, 2, \ldots, n$. Note that for digital-computer solution, Eqs. (10.96) and (10.97) can be solved using subroutine GELG, and this can be included as part of the computer program for solving the eigenvalue problem. A sample program for the initial-value problem is included in App. B.

Example 10.7 Determine the response of the system of Example 10.4 to the initial conditions $x_1(0) = 2$, $x_2(0) = 1$, $x_3(0) = 1$, $\dot{x}_1(0) = 0$, $\dot{x}_2(0) = 1$, and $\dot{x}_3(0) = -1$. (Let $k/m = 1$.)

SOLUTION The initial conditions are written in matrix form as

$$\{x_0\} = \begin{Bmatrix} 2 \\ 1 \\ 1 \end{Bmatrix} \qquad \{\dot{x}_0\} = \begin{Bmatrix} 0 \\ 1 \\ -1 \end{Bmatrix}$$

Using the natural frequencies and modal vectors determined in Example 10.4, we apply Eq. (10.96) to obtain

$$\begin{bmatrix} 1.0000 & 1.0000 & 1.0000 \\ 1.4235 & 0.8544 & -1.0279 \\ 2.0511 & -0.5399 & 0.1128 \end{bmatrix} \begin{Bmatrix} B_1 \\ B_2 \\ B_3 \end{Bmatrix} = \begin{Bmatrix} 2 \\ 1 \\ 1 \end{Bmatrix}$$

Solving via Cramer's rule gives $B_1 = 0.6577$, $B_2 = 0.7668$, and $B_3 = 0.5754$. Similarly, Eq. (10.97) becomes

$$\begin{bmatrix} 0.3914 & 1.1363 & 2.2485 \\ 0.5572 & 0.9709 & -2.3112 \\ 0.8028 & -0.6135 & 0.2536 \end{bmatrix} \begin{Bmatrix} C_1 \\ C_2 \\ C_3 \end{Bmatrix} = \begin{Bmatrix} 0 \\ 1 \\ -1 \end{Bmatrix}$$

from which we obtain $C_1 = 1.2340$, $C_2 = -0.0861$, and $C_3 = -0.1713$. The phase angles are calculated as

$$\phi_1 = \tan^{-1}\frac{0.6577}{1.2340} = 0.4897 \text{ rad}$$

$$\phi_2 = \tan^{-1}\frac{0.7668}{-0.0861} = 1.6826 \text{ rad}$$

$$\phi_3 = \tan^{-1}\frac{0.5754}{-0.1713} = 1.8601 \text{ rad}$$

while the constants β_r are

$$\beta_1 = \sqrt{(0.6577)^2 + (1.2340)^2} = 1.3983$$

$$\beta_2 = \sqrt{(0.7668)^2 + (-0.0861)^2} = 0.7716$$

$$\beta_3 = \sqrt{(0.5754)^2 + (-0.1713)^2} = 0.6004$$

The complete solution is then

$$x_1(t) = 1.3983 \sin(1.3914t + 0.4897) + 0.7716 \sin(1.1363t + 1.6826) + 0.6004 \sin(2.2485t + 1.8601)$$

$$x_2(t) = 1.9905 \sin(0.3914t + 0.4897) + 0.6593 \sin(1.1363t + 1.6826) - 0.6172 \sin(2.2485t + 1.8601)$$

$$x_3(t) = 2.8681 \sin(0.3914t + 0.4897) - 0.4166 \sin(1.1363t + 1.6826) + 0.0677 \sin(2.2485t + 1.8691)$$

As a check on the accuracy of the solution, it can be verified by direct calculation that the initial conditions are satisfied.

10.10 NORMALIZATION OF THE MODAL MATRIX

Prior to discussing the more general case of forced vibration, we shall find it useful to combine the orthogonality condition for the principal modes with the modal matrix defined in Sec. 10.8, in a process known as normalization. The orthogonality condition is rewritten here for convenience as

$$\{A^{(s)}\}^T[m]\{A^{(r)}\} = 0 \qquad r \neq s \tag{10.99}$$

In our previous discussion of orthogonality, the case in which $r = s$ was not mentioned. For this case, Eq. (10.99) can be written as

$$\{A^{(r)}\}^T[m]\{A^{(r)}\}(\omega_r^2 - \omega_r^2) = 0 \tag{10.100}$$

which is obviously true. The significant consequence of Eq. (10.100) is that we can write

$$\{A^{(r)}\}^T[m]\{A^{(r)}\} = \text{constant} \tag{10.101}$$

and since every amplitude (modal) vector $\{A^{(r)}\}$ is known only within a constant multiple, the modal vectors can be manipulated such that the constant on the right-hand side of Eq. (10.101) can be made to assume any desired value. In particular, if the constant is chosen to be unity such that

$$\{A^{(r)}\}^T[m]\{A^{(r)}\} = 1 \tag{10.102}$$

for all $r = 1, 2, \ldots, n$, then the modal vectors are said to be *orthonormal*, and the process is called *normalization*.

Since we have already discussed both manual and computer methods of calculating the modal vectors, we must find a method of modifying the values obtained so as to satisfy Eq. (10.102). We rewrite Eq. (10.101) in summation form as

$$\sum_{i=1}^{n} \sum_{j=1}^{n} m_{ij} A_i^{(r)} A_j^{(r)} = S_r \qquad r = 1, 2, \ldots, n \tag{10.103}$$

from which it is easily seen that normalization can be accomplished by simply redefining each element of $\{A^{(r)}\}$ $(r = 1, 2, \ldots, n)$ as

$$A_i^{(r)} = \frac{A_i^{(r)}}{\sqrt{S_r}} \qquad i = 1, 2, \ldots, n \tag{10.104}$$

These last two equations are well suited to digital computation, and a FORTRAN subroutine for the normalization process is included in App. B.

If the modal vectors are normalized as outlined above, it can be shown that the modal matrix is transformed such that

$$[A]^T[m][A] = [I] \tag{10.105}$$

where $[I]$ is the $n \times n$ identity matrix. This result has a significant impact in reducing the complexity of solving forced-vibration problems, as will be shown in the following sections.

Example 10.8 Normalize the modal matrix of Example 10.4, and verify that Eq. (10.105) is satisfied.

SOLUTION The mass matrix is

$$[m] = \begin{bmatrix} m & 0 & 0 \\ 0 & m & 0 \\ 0 & 0 & 2m \end{bmatrix}$$

Applying Eq. (10.103) to the first mode gives

$$S_1 = (1)^2 m + (1.4235)^2 m + (2.0511)^2(2m) = 11.4404m$$

so that the first modal vector is normalized by dividing each element by $\sqrt{S_1} = 3.3824\sqrt{m}$. The result is

$$\{A^{(1)}\} = \begin{Bmatrix} 0.2956 \\ 0.4209 \\ 0.6064 \end{Bmatrix} \frac{1}{\sqrt{m}}$$

An identical procedure for the second and third modes gives

$$\{A^{(2)}\} = \begin{Bmatrix} 0.6575 \\ 0.5618 \\ -0.3550 \end{Bmatrix} \frac{1}{\sqrt{m}} \qquad \{A^{(3)}\} = \begin{Bmatrix} 0.6990 \\ -0.7124 \\ 0.0782 \end{Bmatrix} \frac{1}{\sqrt{m}}$$

Forming the modal matrix and its transpose, we have

$$[A]^T[m][A] = \frac{1}{m} \begin{bmatrix} 0.2956 & 0.4209 & 0.6064 \\ 0.6575 & 0.5618 & -0.3550 \\ 0.6930 & -0.7124 & 0.0782 \end{bmatrix} \begin{bmatrix} m & 0 & 0 \\ 0 & m & 0 \\ 0 & 0 & 2m \end{bmatrix}$$

$$\times \begin{bmatrix} 0.2956 & 0.6576 & 0.6930 \\ 0.4209 & 0.5618 & -0.7124 \\ 0.6064 & -0.3550 & 0.0782 \end{bmatrix} = \begin{bmatrix} 1 & 0 & 0 \\ 0 & 1 & 0 \\ 0 & 0 & 1 \end{bmatrix}$$

and Eq. (10.105) holds.

10.11 FORCED VIBRATIONS OF UNDAMPED SYSTEMS

In previous sections of this chapter, we have discussed in detail the solution of free-vibration problems by superposition of the principal-mode solutions. We now consider the response of systems having many degrees of freedom to external forcing functions. It will soon become apparent that the method of modal analysis provides a valuable mathematical tool for solving these problems.

First let us consider the undamped three-degrees-of-freedom system shown in Fig. 10.8, in which each mass is subjected to an external force $F_i(t)$,

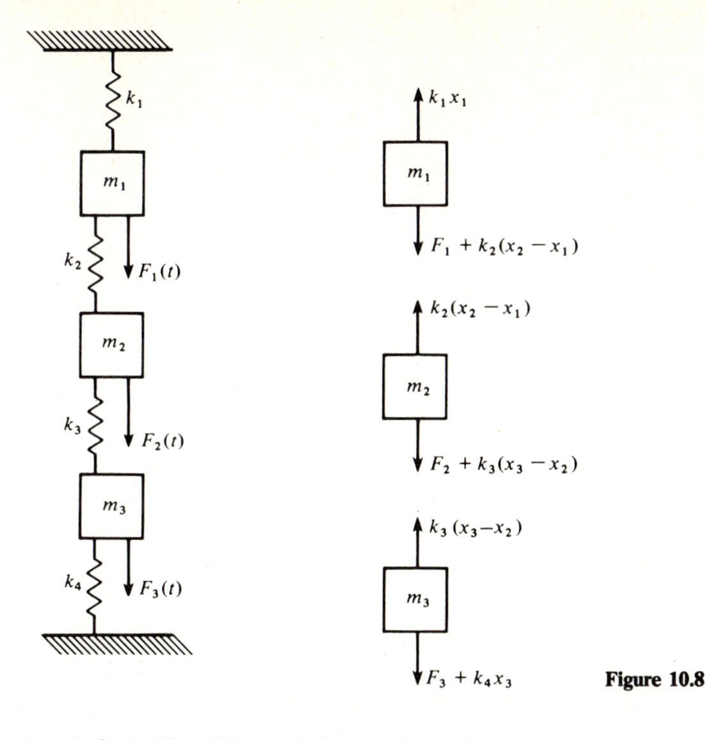

Figure 10.8

$i = 1, 2, 3$. The differential equations of motion are

$$\begin{aligned} m_1\ddot{x}_1 + (k_1 + k_2)x_1 - k_2x_2 &= F_1(t) \\ m_2\ddot{x}_2 + (k_2 + k_3)x_2 - k_2x_1 - k_3x_3 &= F_2(t) \\ m_3\ddot{x}_3 + (k_3 + k_4)x_3 - k_3x_2 &= F_3(t) \end{aligned} \tag{10.106}$$

The system response is described by the displacements $x_1(t)$, $x_2(t)$, and $x_3(t)$, which simultaneously satisfy Eqs. (10.106). Since the equations are inhomogeneous, each displacement is the sum of a homogeneous solution and a particular solution. The homogeneous solutions represent the free response and are assumed to have been determined by modal analysis. Thus, we shall concentrate on determining the particular solutions, which give the forced response of the system.

Equations (10.106) are linear, so the principle of superposition applies. Also, the coupling is such that *each* mass responds to *each* external force. Consequently, one method of solution is to set $F_2(t) = F_3(t) = 0$, solve the resulting equations for the response of the system to $F_1(t)$, and then repeat the process twice to obtain the response to $F_2(t)$ and $F_3(t)$. This method is illustrated by the following example.

Example 10.9 In the system shown in Fig. 10.8, $k_1 = k_2 = k_3 = k$, $k_4 = 2k$, $m_1 = m_2 = m_3 = m$, $F_1(t) = F_1 \sin \omega_f t$, $F_2(t) = F_2 e^{-t}$, and $F_3(t) = 0$. Determine the forced response.

SOLUTION Substituting into Eqs. (10.106) and converting to matrix form, we obtain

$$\begin{bmatrix} m & 0 & 0 \\ 0 & m & 0 \\ 0 & 0 & m \end{bmatrix}\{\ddot{x}\} + \begin{bmatrix} 2k & -k & 0 \\ -k & 2k & -k \\ 0 & -k & 3k \end{bmatrix}\{x\} = \begin{Bmatrix} F_1 \sin \omega_f t \\ F_2 e^{-t} \\ 0 \end{Bmatrix}$$

We first set $F_2(t) = 0$ and assume particular solutions corresponding to $F_1(t)$. As the response to the harmonic force we assume the solutions as

$$x_{i1}(t) = X_{i1} \sin \omega_f t \qquad i = 1, 2, 3$$

where we have introduced a second subscript to indicate response to force 1. Differentiating and substituting into the equations of motion, we obtain the system of algebraic equations

$$\begin{bmatrix} 2k - m\omega_f^2 & -k & 0 \\ -k & 2k - m\omega_f^2 & -k \\ 0 & -k & 3k - m\omega_f^2 \end{bmatrix}\begin{Bmatrix} X_{11} \\ X_{21} \\ X_{31} \end{Bmatrix} = \begin{Bmatrix} F_1 \\ 0 \\ 0 \end{Bmatrix}$$

Simultaneous solution gives

$$X_{11} = \frac{F_1\left[(2k - m\omega_f^2)(3k - m\omega_f^2) - k^2\right]}{|D_1|}$$

$$X_{21} = \frac{-F_1 k(3k - m\omega_f^2)}{|D_1|}$$

$$X_{31} = \frac{F_1 k^2}{|D_1|}$$

where $|D_1|$ is the determinant of the coefficient matrix,

$$|D_1| = (2k - m\omega_f^2)^2(3k - m\omega_f^2) - 5k^3 + k^2 m\omega_f^2$$

Similarly, we set $F_1(t) = 0$ and assume the response to $F_2(t) = F_2 e^{-t}$ to be of the form

$$x_{i2}(t) = X_{i2} e^{-t} \qquad i = 1, 2, 3$$

Substitution of the assumed solutions into the equations of motion gives

$$\begin{bmatrix} m + 2k & -k & 0 \\ -k & m + 2k & -k \\ 0 & -k & m + 3k \end{bmatrix}\begin{Bmatrix} X_{12} \\ X_{22} \\ X_{32} \end{Bmatrix} = \begin{Bmatrix} 0 \\ F_2 \\ 0 \end{Bmatrix}$$

Again the assumed solution form is correct, since algebraic equations are obtained. The solutions to this set of equations are found to be

$$X_{12} = \frac{F_2 k(m + 3k)}{|D_2|}$$

$$X_{22} = \frac{F_2(m + 2k)(m + 3k)}{|D_2|}$$

$$X_{32} = \frac{F_2 k(m + 2k)}{|D_2|}$$

where $$|D_2| = (m + 2k)^2(m + 3k) - 5k^3 - 2k^2 m$$

Since $F_3(t)$ is zero, the complete forced response is

$$x_1(t) = X_{11} \sin \omega_f t + X_{12} e^{-t}$$

$$x_2(t) = X_{21} \sin \omega_f t + X_{22} e^{-t}$$

$$x_3(t) = X_{31} \sin \omega_f t + X_{32} e^{-t}$$

The method used in Example 10.9 to obtain the forced response is straightforward but tedious, since it requires multiple solutions of the equations of motion. This procedure also requires that we be able to accurately assume the functional form of the particular solutions, to reduce the differential equations of motion to an algebraic system. Thus its application is limited to certain forcing functions, and the method is especially cumbersome when applied to systems having many degrees of freedom. One special case is worthy of further discussion, however. Consider a system having n degrees of freedom, governed by

$$[m]\{\ddot{x}\} + [k]\{x\} = \{F\} \qquad (10.107)$$

where $\{F\}$ is the column matrix, or vector, of external forcing functions. Let the external forces be periodic such that

$$F_j(t) = \sum_{p=1}^{\infty} b_{jp} \sin \frac{2\pi p t}{T_j} \qquad j = 1, 2, \ldots, n \qquad (10.108)$$

is the Fourier sine series expansion as in Eq. (7.12). In this manner, each force is expressed as a series of harmonic force components, and the system response to each external force can be expressed as

$$x_{ij} = \sum_{p=1}^{\infty} X_{ij}^{(p)} \sin \frac{2\pi p t}{T_j} \qquad (10.109)$$

where x_{ij} is the displacement of mass i resulting from an external force applied at mass j. The total displacement of each mass can then be obtained as

$$x_i(t) = \sum_{j=1}^{n} x_{ij} = \sum_{j=1}^{n} \left(\sum_{p=1}^{\infty} X_{ij}^{(p)} \sin \frac{2\pi p t}{T_j} \right) \qquad (10.110)$$

For simplicity, but without loss of generality, we now assume that all external forces are zero except, say, $F_1(t)$. Further, $F_1(t)$ is an odd function having period T_1. Under these conditions, the solutions given by Eq. (10.110) can be written

$$x_i(t) = \sum_{p=1}^{\infty} X_{i1}^{(p)} \sin p\omega_f t \tag{10.111}$$

where $\omega_f = 2\pi/T_1$ is the fundamental frequency of the forcing function. The amplitudes $X_{i1}^{(p)}$ can be determined by substituting corresponding harmonic components of Eqs. (10.108) and (10.111) into Eq. (10.107) and solving the resultant algebraic system simultaneously. For our simplified example, Eq. (10.107) will reduce to

$$-p\omega_f^2[m]\begin{Bmatrix} X_{11}^{(p)} \\ X_{21}^{(p)} \\ \vdots \\ X_{n1}^{(p)} \end{Bmatrix} + [k]\begin{Bmatrix} X_{11}^{(p)} \\ X_{21}^{(p)} \\ \vdots \\ X_{n1}^{(p)} \end{Bmatrix} = \begin{Bmatrix} b_{1p} \\ 0 \\ \vdots \\ 0 \end{Bmatrix} \tag{10.112}$$

which can be solved separately for $p = 1, 2, \ldots, n$ to obtain as many terms in the solution series as desired. This method of Fourier analysis is quite amenable to digital-computer solution, since the system represented by Eq. (10.112) is readily solved by a subroutine such as GELG (discussed in Sec. 10.8). Although we have discussed only odd forcing functions, a similar procedure can be applied to any periodic function which can be expanded in a Fourier series according to Eq. (7.2).

In Chap. 8 we discussed the concept of describing the motion of a system in terms of its principal coordinates. The advantage of using the principal coordinates is that the equations of motion are uncoupled. We shall now show how the equations of motion for a system having n degrees of freedom can be uncoupled and thereby obtain a method for determining the forced response to any general excitation.

For the system governed by Eq. (10.107), we introduce the coordinate transformation defined by

$$\{x\} = [A]\{q\} \tag{10.113}$$

where $[A]$ is the normalized modal matrix defined in Sec. 10.10. In terms of the new coordinates $\{q\}$ we obtain

$$[m][A]\{\ddot{q}\} + [k][A]\{q\} = \{F\} \tag{10.114}$$

Premultiplying the last equation by $[A]^T$ gives

$$[A]^T[m][A]\{\ddot{q}\} + [A]^T[k][A]\{q\} = [A]^T\{F\} \tag{10.115}$$

From Eq. (10.105), we have $[A]^T[m][A] = [I]$, and it can also be shown that

$$[A]^T[k][A] = \lceil \omega^2 \rfloor \tag{10.116}$$

where $\lceil \omega^2 \rfloor$ is a diagonal matrix containing the squares of the system natural frequencies. With these observations, we can write Eq. (10.115) as

$$[I]\{\ddot{q}\} + \lceil \omega^2 \rfloor \{q\} = \{Q\} \tag{10.117}$$

where $\{Q\} = [A]^T\{F\}$. Equation (10.117) is equivalent to the n *independent* differential equations

$$\ddot{q}_r + \omega_r^2 q_r = Q_r(t) \qquad r = 1, 2, \ldots, n \tag{10.118}$$

where $q_r = q_r(t)$ are the principal coordinates, and $Q_r(t)$ are the corresponding generalized forces. Thus we have shown that the modal matrix defines the principal coordinates through Eq. (10.113). By first solving the eigenvalue problem associated with the free response of the system, we are able to uncouple the equations of motion and obtain the forced response by solving independent second-order differential equations. In earlier chapters, we discussed many different techniques for solving equations having the form of Eq. (10.118); any of these may be utilized, depending upon the form of the generalized forces $Q_r(t)$. Once the generalized displacements are obtained, the complete response of the system in terms of the original coordinates is found by Eq. (10.113).

10.12 VISCOUSLY DAMPED SYSTEMS

The differential equations of motion for a damped system having n degrees of freedom can be written in matrix form as

$$[m]\{\ddot{x}\} + [c]\{\dot{x}\} + [k]\{x\} = \{F\} \tag{10.119}$$

where $[c]$ is the $n \times n$ damping matrix. In general, the only absolute statement which can be made about the damping matrix is that it is symmetric. To examine the free vibration of the system, we set $\{F\} = 0$ and, as in the analysis of Sec. 8.5, assume solutions of the form

$$x_i(t) C_i e^{st} \qquad i = 1, 2, \ldots, n \tag{10.120}$$

Substitution of the assumed form of the solutions yields the matrix equation

$$s^2[m]\{C\} + s[c]\{C\} + [k]\{C\} = \{0\} \tag{10.121}$$

Equation (10.121) represents the *complex eigenvalue problem*. Its solution requires setting the determinant of coefficients to zero, expanding the determinant into a polynomial of degree $2n$ in s, and solving for all values s which satisfy the polynomial. Generally, this results in n pairs of complex conjugates $s_r = a_r + ib_r$, $r = 1, 2, \ldots, n$. Substitution of these complex eigenvalues into Eq. (10.121) yields $2n$ eigenvectors which also generally occur as n pairs of complex conjugates. These are then used to obtain the general solutions, as was done in Example 8.6 for two degrees of freedom. The general complex eigenvalue problem is beyond the scope of this text, and the interested reader is referred to texts on applied mathematics or numerical analysis for further details. In the following, we concentrate on certain special cases of interest.

Uniform Viscous Damping

When the damping present in a system is an inherent property of the spring material rather than the result of discrete damping elements, the system is often modeled as having uniform viscous damping. In this model, each spring is assumed to act in parallel with a viscous damper which has a damping constant directly proportional to the spring constant. This is expressed as

$$c_i = 2\Gamma k_i \qquad i = 1, 2, \ldots, n \tag{10.122}$$

where 2Γ is the proportionality constant. The damping matrix is then

$$[c] = 2\Gamma[k] \tag{10.123}$$

and Eq. (10.119) becomes

$$[m]\{\ddot{x}\} + 2\Gamma[k]\{\dot{x}\} + [k]\{x\} = \{F\} \tag{10.124}$$

By introducing the coordinate transformation defined by Eq. (10.113), the last equation may be written as

$$[m][A]\{\ddot{q}\} + 2\Gamma[k][A]\{\dot{q}\} + [k][A]\{q\} = \{F\} \tag{10.125}$$

Premultiplying Eq. (10.125) by $[A]^T$ and using Eqs. (10.105) and (10.116), we obtain

$$[I]\{\ddot{q}\} + 2\Gamma[\omega^2]\{\dot{q}\} + [\omega^2]\{q\} = \{Q\} \tag{10.126}$$

which is equivalent to the n independent equations

$$\ddot{q}_r + 2\Gamma\omega_r^2\dot{q}_r + \omega_r^2 q_r = Q_r(t) \qquad r = 1, 2, \ldots, n \tag{10.127}$$

Thus, the equations of motion for a system with uniform viscous damping are uncoupled by the modal matrix. If we let $\zeta_r = \Gamma\omega_r$, Eq. (10.127) can be written

$$\ddot{q}_r + 2\zeta_r\omega_r\dot{q}_r + \omega_r^2 q_r = Q_r(t) \qquad r = 1, 2, \ldots, n \tag{10.128}$$

which is identical in form to the governing equation for damped single-degree-of-freedom systems. The free-vibration response is obtained by setting the generalized forces $Q_r(t) = 0$ for all r and writing the solutions to Eq. (10.128) by analogy with Eq. (5.20) as

$$q_r(t) = e^{-\zeta_r\omega_r t}(a_r \sin\omega_d t + b_r \cos\omega_d t) \qquad r = 1, 2, \ldots, n \tag{10.129}$$

where $\omega_d = \omega_r\sqrt{1 - \zeta_r^2}$. Note, however, that the important distinction here is that the damping is frequency dependent, since $\zeta_r = \Gamma\omega_r$ is a function of the system natural frequency for mode r.

The forced response of a system with uniform viscous damping may be obtained by solving Eq. (10.128) by the methods of Chaps. 5 and 7, depending upon the functional form of the generalized forces $Q_r(t)$.

Uniform Mass Damping

A system is said to possess *uniform mass damping* if the damping which acts on each mass is proportional to the magnitude of the mass. This would correspond physically to a system moving in a viscous medium. In this case, the damping constant for each mass is

$$c_i = \xi m_i \qquad i = 1, 2, \ldots, n \tag{10.130}$$

and the damping matrix becomes

$$[c] = \xi[m] \tag{10.131}$$

The equations of motion are then

$$[m]\{\ddot{x}\} + \xi[m]\{\dot{x}\} + [k]\{x\} = [F] \tag{10.132}$$

They, too, can be uncoupled by the coordinate transformation of Eq. (10.113), since $[A]^T[m][A] = [I]$. The uncoupled equations of motion are

$$\ddot{q}_r + \xi\dot{q}_r + \omega_r^2 q_r = Q_r(t) \qquad r = 1, 2, \ldots, n \tag{10.133}$$

They may be solved by the methods used for Eqs. (10.128) if we set $\xi = 2\zeta_r\omega_r$. Since ξ is constant, we note that uniform mass damping is such that the equivalent damping factor $\zeta_r = \xi/2\omega_r$ decreases with natural frequency.

Light Damping

In many systems, damping is indeed present but is very small. For a general damping matrix, Eq. (10.113) transforms the equations of motion to

$$[I]\{\ddot{q}\} + [A]^T[c][A]\{\dot{q}\} + [\omega^2]\{q\} = \{Q(t)\} \tag{10.134}$$

where $[A]^T[c][A]$ results in a nondiagonal matrix. A frequently used approach for approximating the response of a system with light damping is to ignore all off-diagonal terms of this transformed damping matrix; then Eq. (10.134) may be considered as n independent equations. While not giving exact solutions, this approach will give a better approximation than that obtained by treating the system as undamped.

PROBLEMS

10.1 Obtain the differential equations of motion for the system shown in Fig. 10.9 by direct application of Newton's second law.

10.2 Using Lagrange's equation, determine the differential equations of motion for the system of Fig. 10.9.

10.3 In Fig. 10.9, let $k_1 = k_2 = k_3 = k_4 = k_5 = k$ and $m_1 = m_2 = m_3 = m$, and determine (*a*) the system natural frequencies and (*b*) the amplitude ratios for each of the principal modes.

10.4 Using the data of Prob. 10.3, estimate the fundamental natural frequency using Rayleigh's method. Use the static-equilibrium configuration as an approximation for the fundamental mode shape.

10.5 Figure 10.10 depicts a cantilever beam which supports two concentrated masses rigidly attached to the beam. The beam has a circular cross section with diameter d, and the modulus of elasticity is E. Neglecting the beam's mass, determine the differential equations of motion for elastic free oscillations of the system.

* *Instructor's note:* By providing numerical values for the various physical constants, many of the preceding problems can be assigned for digital-computer solution.

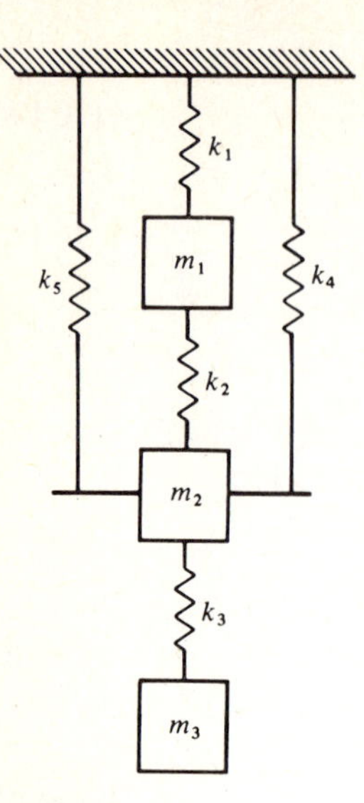

Figure 10.9

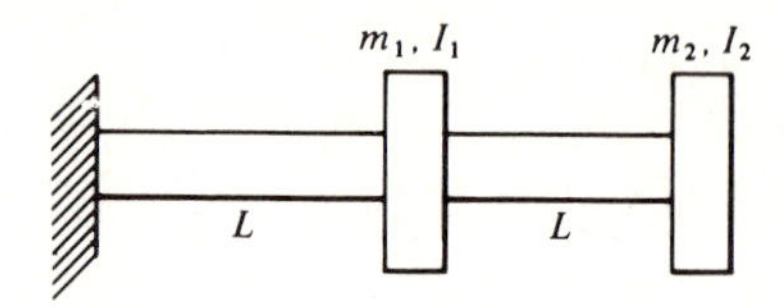

Figure 10.10

10.6 Denoting by I_1 and I_2 the polar moments of inertia of the masses, repeat Prob. 10.5 for torsional oscillations (shear modulus = G).

10.7 Bar AB of Fig. 10.11 is considered to be rigid and massless, and it is free to pivot about A. Derive the differential equations of motion for small oscillations of the system.

10.8 In the system of Fig. 10.11, $m_1 = m_2 = m_3 = m$ and $k_1 = k_2 = k_3 = k$. Determine the natural frequencies and amplitude ratios of the principal-mode solutions for small oscillations.

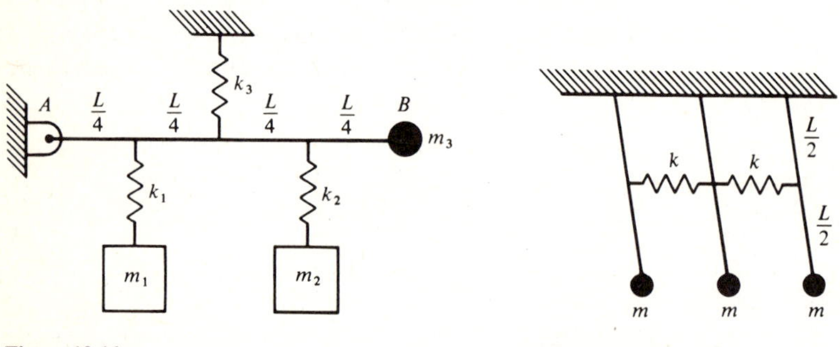

Figure 10.11

Figure 10.12

10.9 In the system shown in Fig. 10.12, the springs are undeformed when the pendulums are in their vertical equilibrium positions.

(*a*) Using Lagrange's equations, derive the differential equations of motion.

(*b*) Linearize the equations of motion by assuming small oscillations, and determine the natural frequencies.

10.10 Three equal masses are held in a vertical plane by a taut string, as shown in Fig. 10.13. The tension T in the string can be considered constant for small oscillations. Determine the differential equations of motion, solve for the principal modes, and plot the corresponding mode shapes.

10.11 Verify the orthogonality condition for the principal modes for the system of Fig. 10.13.

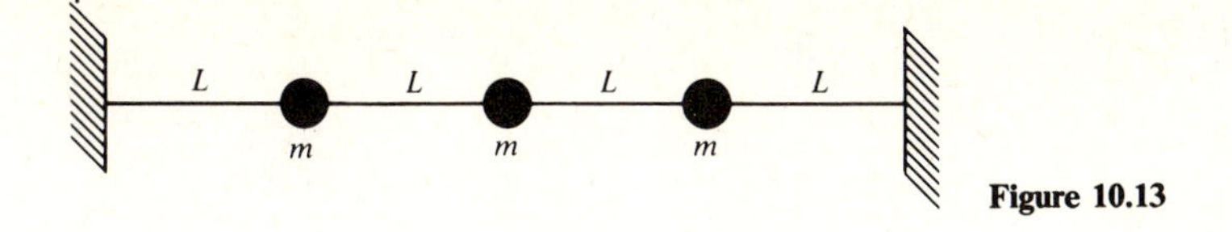

Figure 10.13

10.12 Determine the stiffness matrix for the system of Fig. 10.14. The segments of the beam have uniform moment of inertia I and modulus of elasticity E.

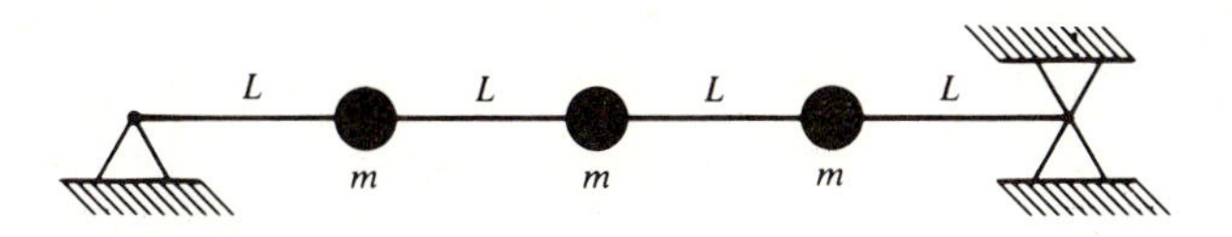

Figure 10.14

10.13 Determine the flexibility matrix for the system shown in Fig. 10.14, and verify that the result is the inverse of the stiffness matrix.

10.14 Three identical masses are connected by springs and rest on a smooth horizontal surface, as shown in Fig. 10.15. If m_3 is displaced to the right through a distance x_o and released, determine the general solution for the resulting motion by superposition of the principal modes. What is the physical significance of the value of the fundamental frequency of this system?

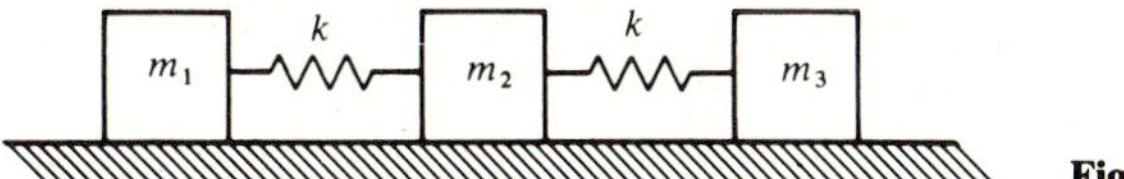

Figure 10.15

10.15 Motion of the system in Fig. 10.16 results from an impulse applied to m_2, giving it an initial velocity v_o to the right. If $m_1 = m_2 = m_3 = 0.5$ kg and $k = 20{,}000$ N/m, determine the displacement of each mass as a function of time.

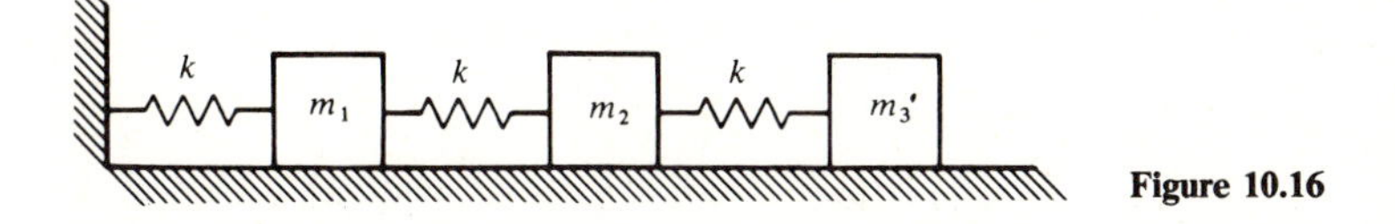

Figure 10.16

10.16 In the system of Fig. 10.9, $m_1 = m_2 = 0.05$ lb · s^2/in, $m_3 = 0.1$ lb · s^2/in, $k_1 = k_2 = k_3 =$ 25 lb/in, and $k_4 = k_5 = 30$ lb/in. Determine the free response of the system if the initial displacements are $x_1(0) = 1$ in, $x_2(0) = 1.5$ in, and $x_3(0) = 0$, and the initial velocities are zero.

10.17 In Fig. 10.10, $L = 10$ in, $d = 1.5$ in, $I_1 = 0.67$ lb · s^2 · in, $I_2 = 1.2$ lb · s^2 · in, and $G = 12 \times 10^6$ lb/in^2. Determine the angular oscillations which result from the initial conditions $\theta_1(0) = 2°$, $\theta_2(0) = 5°$, and $\dot{\theta}_1(0) = \dot{\theta}_2(0) = 0$.

10.18 Verify the orthogonality of the principal modes for Prob. 10.8.

10.19 Show that $[A]^T[k][A] = \lceil \omega_i^2 \rfloor$, where $[A]$ is the normalized modal matrix.

10.20 Discuss the possible motions of the system shown in Fig. 10.17. How many degrees of freedom does the system exhibit?

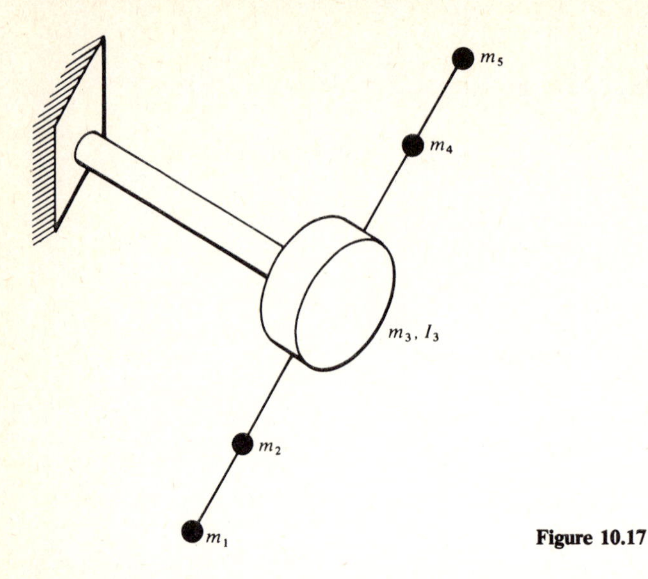

Figure 10.17

10.21 Using Rayleigh's method, estimate the fundamental frequency of the system in Example 8.3. How does this estimate compare with the exact value?

10.22 The system of Fig. 10.11 is excited by a harmonic force $F = F_o \sin \omega_f t$ applied to m_2. Using the direct approach of finding particular solutions for the equations of motion, determine the forced response.

10.23 Solve Prob. 10.22 using modal analysis to uncouple the equations of motion.

10.24 Determine the forced response in Prob. 8.22 by modal analysis.

10.25 For the system and data of Prob. 8.24, obtain the forced response by uncoupling the equations of motion via the modal matrix.

10.26 The system shown in Fig. 10.18 is excited by an external force F which is defined by Fig. 7.3. Using Fourier analysis, obtain the first three terms in the series representations of x_1 and x_2.

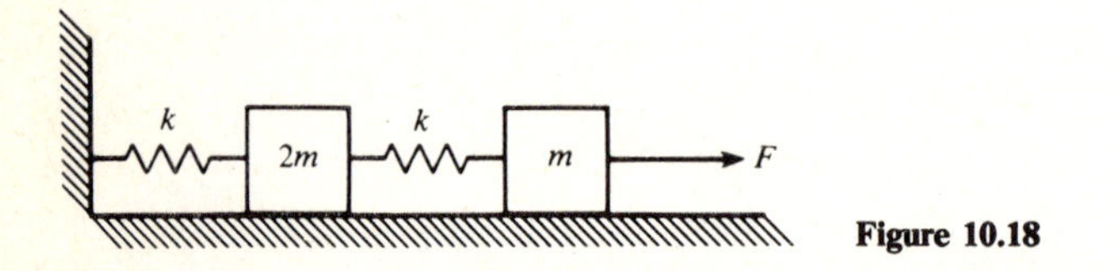

Figure 10.18

10.27 Figure 10.19 depicts an elastic shaft which supports three inertial disks. The shaft is free to rotate in bearings A and B. Obtain the differential equations of motion if the system is excited by the harmonic torque $T = T_o \sin \omega_f t$.

10.28 Two identical masses are attached to the ends of a uniform elastic rod as shown in Fig. 10.20. The rod has flexural rigidity EI and is attached at its midpoint to a vertical support which oscillates harmonically as shown. Derive the differential equations of motion, and obtain the forced response of the system.

10.29 Determine the forced response of the system in Fig. 10.18 if the applied force is a step function having magnitude F_o.

Hint: Solve the uncoupled equations of motion using the convolution integral.

10.30 For the damped system shown in Fig. 10.21, (*a*) derive the equations of motion, (*b*) obtain the natural frequencies and the modal matrix for the corresponding undamped system, and (*c*) show that the equations of motion cannot be uncoupled by the modal-matrix method.

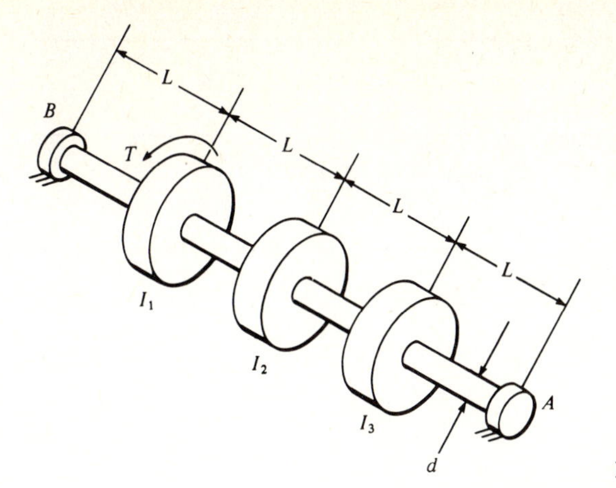

Figure 10.19

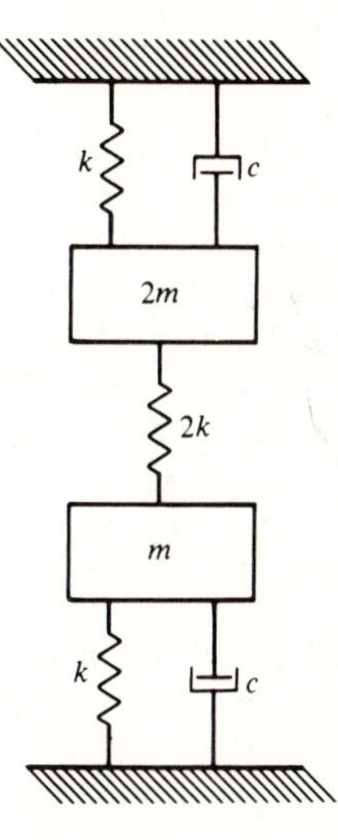

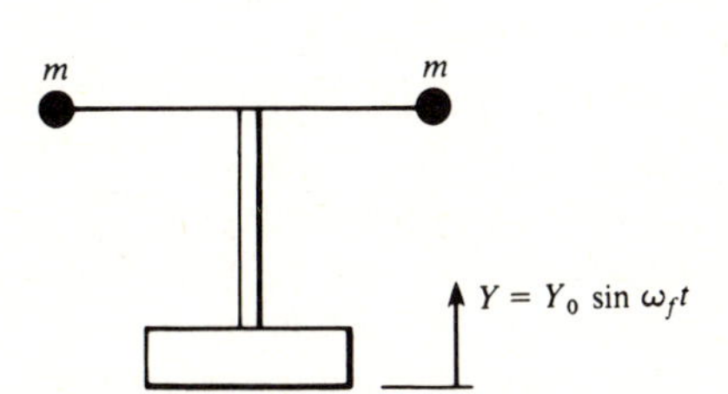

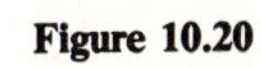

Figure 10.20

Figure 10.21

10.31 Repeat Prob. 10.29 except that the system exhibits uniform viscous damping Γ.

CHAPTER

ELEVEN

VIBRATION OF CONTINUOUS MEDIA

All the systems studied thus far are models of mechanical systems in which the distributions of mass and elastic properties are such that discrete, lumped-mass models suffice to describe system behavior. Many machines and structures are composed of elements which have continuously distributed properties, and discrete-model analysis fails to describe their vibratory characteristics accurately. To adequately describe the dynamic response of such systems requires a distinctly different mathematical approach.

Discrete systems exhibit a finite number of degrees of freedom, since the number of coordinates required to specify system configuration is finite. Continuous systems must be treated as being composed of an infinite number of differential elements, each having mass and elastic properties. The displacement of these elements is described by a continuous function of position and time. Consequently, the governing equations are partial differential equations, and exact solutions can be obtained for only a few special cases.

The study of vibrations of continuous media is a broad subject unto itself. This chapter is intended as an introduction to this area of analysis. We shall discuss a few classical problems for which closed-form solutions can be obtained, as well as certain approximate solution methods.

11.1 THE WAVE EQUATION

Consider a stretched elastic string having its endpoints attached to fixed surfaces and free to vibrate in a vertical plane, as shown in Fig. 11.1*a*. The string is shown in a possible position of motion, and the equilibrium position is shown by the broken line. To aid in the analysis of the motion of the string, the following

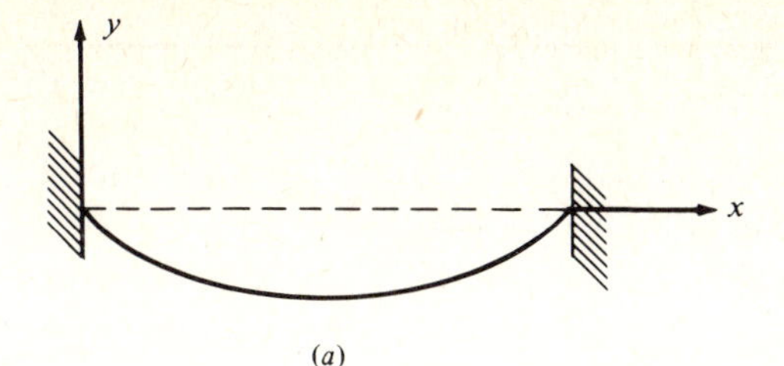

(a)

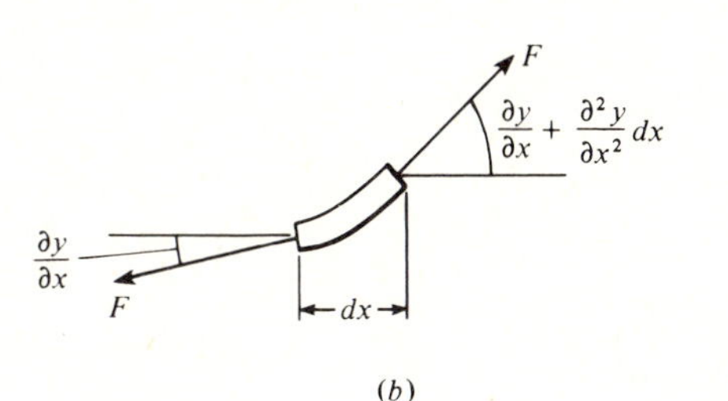

(b)

Figure 11.1

assumptions are made:

1. The displacement of any point on the string is small and occurs only in the vertical direction.
2. Air resistance and internal friction are negligible.
3. The tension in the string is large enough so that gravitational forces may be neglected.

The displacement of any point is a function of its position along the string as well as of time. In the coordinate system shown, the displacement function is

$$y = y(x, t) \tag{11.1}$$

Thus the various derivatives which define slope, velocity, acceleration, and so forth are partial derivatives with respect to the appropriate independent variable x or t.

To obtain the governing equation, we draw a free-body diagram of a differential element of the string in a nonequilibrium position, as in Fig. 11.1b, in which F denotes the tension and $(\partial^2 y/\partial x^2)\,dx$ is the change in slope. If the mass of string per unit length is γ, writing Newton's second law for vertical motion of the differential element gives

$$-F \sin\frac{\partial y}{\partial x} + F \sin\left(\frac{\partial y}{\partial x} + \frac{\partial^2 y}{\partial x^2} dx\right) = \gamma\, dx \frac{\partial^2 y}{\partial t^2} \tag{11.2}$$

From assumption 1, the slopes are also small, and we use the approximation $\sin\theta \simeq \theta$ to obtain

$$-F\frac{\partial y}{\partial x} + F\left(\frac{\partial y}{\partial x} + \frac{\partial^2 y}{\partial x^2} dx\right) = \gamma\, dx \frac{\partial^2 y}{\partial t^2} \tag{11.3}$$

or

$$F\frac{\partial^2 y}{\partial x^2} = \gamma\frac{\partial^2 y}{\partial t^2} \tag{11.4}$$

Equation (11.4) is the governing equation for transverse motion of the string. We note that it is a linear, second-order, partial differential equation with constant coefficients. When written as

$$c^2\frac{\partial^2 y}{\partial x^2} = \frac{\partial^2 y}{\partial t^2} \tag{11.5}$$

where $c = \sqrt{F/\gamma}$, Eq. (11.5) is known as the *one-dimensional wave equation*. The constant c has the units of velocity and is known as the *wave speed*.

Writing Eq. (11.5) as

$$\frac{\partial^2 y/\partial t^2}{\partial^2 y/\partial x^2} = c^2 \tag{11.6}$$

we deduce that the solution $y(x, t)$ must be such that its functional form is maintained by the second partial derivatives with respect to both x and t. To satisfy this requirement, we assume the solution as

$$y(x, t) = f(x)g(t) \tag{11.7}$$

Substituting the assumed solution into Eq. (10.6) gives

$$\frac{f\ddot{g}}{f''g} = c^2 \tag{11.8}$$

which can be written as

$$\frac{\ddot{g}}{g} = c^2\frac{f''}{f} = \eta \tag{11.9}$$

where $\ddot{g} = d^2g/dt^2$, $f'' = d^2f/dx^2$, and η is a constant. The last equation is equivalent to the two ordinary differential equations

$$\ddot{g} - \eta g = 0 \tag{11.10}$$

$$f'' - \frac{\eta}{c^2}f = 0 \tag{11.11}$$

Thus the problem has been reduced to two separate ordinary differential equations. Such a procedure is said to produce a *separable solution*, since the independent variables are separated.

The solutions of Eqs. (11.10) and (11.11) will each contain two arbitrary constants, since we are dealing with second-order equations. Thus, the complete solution to the vibrating-string problem will require the evaluation of four constants of integration. To determine the significance of these constants, note that the solution given by Eq. (11.7) indicates that the string moves in a single spatial configuration described by $f(x)$ with the time history of the motion defined by $g(t)$. Further, since no external forces act on the string, motion will occur only as the result of a disturbance from equilibrium. Therefore, two of the constants are analogous to the initial conditions of discrete systems and can be written as

$$y(x, 0) = f(x)g(0) = f(x)g_o \tag{11.12}$$

$$\frac{\partial y(x, 0)}{\partial t} = f(x)\dot{g}(0) = f(x)\dot{g}_o \tag{11.13}$$

The remaining constants must be chosen such that any physical constraints are satisfied. For the problem at hand, displacement of the ends of the string is prevented by the supports, so the solution must satisfy

$$y(0, t) = f(0)g(t) = 0 \tag{11.14}$$

$$y(L, t) = f(L)g(t) = 0 \tag{11.15}$$

These are known as the *boundary conditions* for the system.

The solutions of the differential equations (11.10) and (11.11) depend on the value of η. If $\eta > 0$, the solutions $g(t)$ and $f(x)$ are exponential functions, and it is easily shown that the boundary conditions cannot be satisfied. If $\eta = 0$, $g(t)$ and $f(x)$ are linear functions of t and x, respectively, and Eqs. (11.14) and (11.15) cannot be satisfied for this case either. Thus we find that $\eta < 0$ is the only possibility for satisfying both the differential equations and the boundary conditions. If $\eta < 0$, we can rewrite Eqs. (11.10) and (11.11) as

$$\ddot{g} + \omega^2 g = 0 \tag{11.16}$$

$$f'' + \frac{\omega^2}{c^2} f = 0 \tag{11.17}$$

which have the solutions

$$g(t) = C_1 \sin \omega t + C_2 \cos \omega t \tag{11.18}$$

$$f(x) = C_3 \sin \frac{\omega}{c} x + C_4 \cos \frac{\omega}{c} x \tag{11.19}$$

The displacements are then given by

$$y(x, t) = (C_1 \sin \omega t + C_2 \cos \omega t)\left(C_3 \sin \frac{\omega}{c} x + C_4 \cos \frac{\omega}{c} x\right) \tag{11.20}$$

Applying the boundary condition given by Eq. (11.14) results in

$$y(0, t) = (C_1 \sin \omega t + C_2 \cos \omega t)C_4 = 0 \tag{11.21}$$

which will be satisfied if $C_4 = 0$. From this result and Eq. (11.15), the second boundary condition becomes

$$y(L, t) = (C_1 \sin \omega t + C_2 \cos \omega t)C_3 \sin \frac{\omega}{c} L = 0 \tag{11.22}$$

This will be satisfied if $C_3 = 0$, but this is the trivial solution since it results in $y(x, t) = 0$ for all x and t. Equation (11.22) is also satisfied by

$$\sin \frac{\omega}{c} L = 0 \tag{11.23}$$

from which

$$\frac{\omega}{c} L = \pi, 2\pi, \ldots, n\pi, \ldots \tag{11.24}$$

Equation (11.24) leads to an infinite number of natural frequencies

$$\omega_n = \frac{n\pi c}{L} = \frac{n\pi}{L}\sqrt{\frac{F}{\gamma}} \qquad n = 1, 2, 3, \ldots \tag{11.25}$$

Each ω_n corresponds to a principal mode having the harmonic mode shape $\sin(n\pi x/L)$. As with discrete systems, any general set of initial conditions results in motion of the string which is described by superposition of the principal modes. The general solution can then be written as

$$y(x, t) = \sum_{n=1}^{\infty} (A_n \sin \omega_n t + B_n \cos \omega_n t) \sin \frac{\omega_n}{c} x \tag{11.26}$$

where $A_n = C_1C_3$ and $B_n = C_2C_3$ for each mode.

Let us now assume that the initial conditions for motion of the string are

$$y(x, 0) = Y_o \sin \frac{\pi}{L} x \tag{11.27}$$

and

$$\frac{\partial y(x, 0)}{\partial t} = 0 \tag{11.28}$$

Note that these conditions correspond to initially deforming the string into the shape of a half sine curve and releasing it from rest. Applying the initial conditions to Eq. (11.26) gives

$$A_n = 0 \qquad \text{for all } n$$
$$B_1 = Y_o$$
$$B_n = 0 \qquad n \neq 1$$

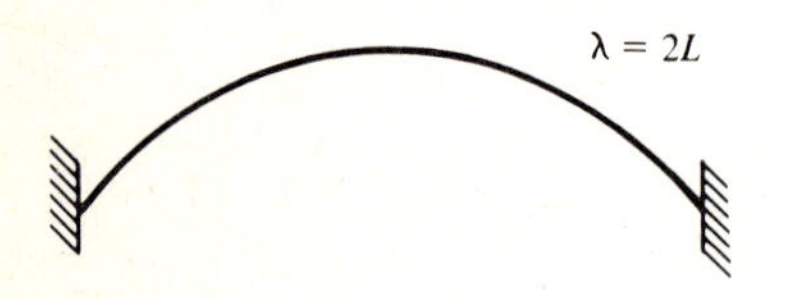

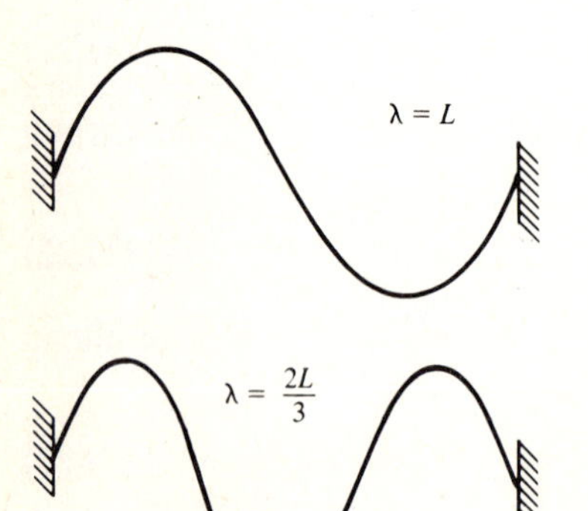

Figure 11.2

and the displacement of every point on the string is given by

$$y(x, t) = Y_o \cos\frac{\pi c}{L} t \sin\frac{\pi}{L} x \tag{11.29}$$

This represents vibration of the string in its fundamental mode. In this mode every point moves harmonically with amplitude $Y_o \sin(\pi x/L)$, and the shape of the string is a half sine wave at all times. Such motion is referred to as a *standing wave*, since all points simply move up and down and the wave does not progress in the axial direction. The *wave length* λ is defined as the length of one complete sine wave, $\sin(n\pi x/L)$. Figure 11.2 shows the standing waves for the first three principal modes.

Example 11.1 Determine the response of a stretched elastic string if it is released from the initial configuration shown in Fig. 11.3.

SOLUTION The initial configuration is described by

$$y(x, 0) = \frac{2Y_o}{L} x \qquad 0 < x < \frac{L}{2}$$

$$y(x, 0) = \frac{2Y_o}{L}(L - x) \qquad \frac{L}{2} < x < L$$

$$\frac{\partial y(x, 0)}{\partial t} = 0$$

This form leads to difficulties in evaluating the arbitrary constants, so we shall expand $y(x, 0)$ into the Fourier sine series

$$y(x, 0) = \sum_{n=1}^{\infty} b_n \sin\frac{n\pi}{L} x$$

where $$b_n = \frac{2}{L}\int_0^L y(x, 0) \sin\frac{n\pi}{L} x \, dx \qquad n = 1, 2, 3, \ldots$$

Evaluating the Fourier coefficients

$$b_n = \frac{2}{L}\int_0^{L/2} \frac{2Y_o}{L} x \sin\frac{n\pi}{L} x \, dx + \frac{2}{L}\int_{L/2}^{L} \frac{2Y_o}{L}(L - x) \sin\frac{n\pi}{L} x \, dx$$

we obtain

$$b_n = \begin{cases} 0 & \text{for } n \text{ even} \\ \dfrac{8Y_o}{n^2\pi^2}(-1)^{(n-1)/2} & \text{for } n \text{ odd} \end{cases}$$

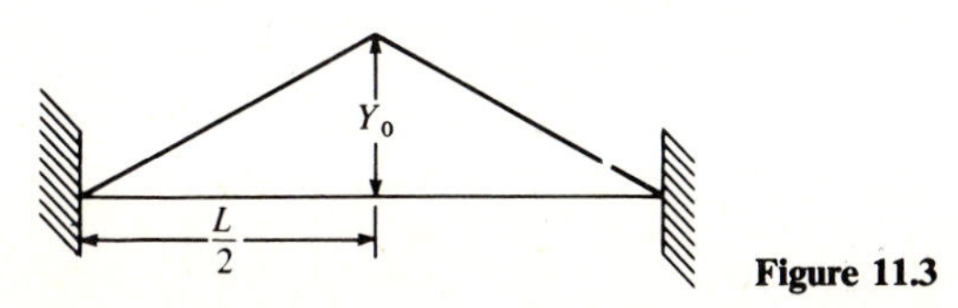

Figure 11.3

The initial configuration is then expressed as

$$y(x, 0) = \sum_{n=1}^{\infty} \frac{8Y_o}{n^2\pi^2}(-1)^{(n-1)/2} \sin\frac{n\pi}{L}x$$

where only odd values of n are included in the summation. The general solution as given by Eq. (11.26) is

$$y(x, t) = \sum_{n=1}^{\infty} (A_n \sin \omega_n t + B_n \cos \omega_n t) \sin\frac{n\pi}{L}x$$

Applying the initial condition on the velocity

$$\frac{\partial y(x, 0)}{\partial t} = 0 = \sum_{n=1}^{\infty} \omega_n A_n \sin\frac{n\pi}{L}x$$

gives $A_n = 0$ for all values of n. The B_n are obtained from

$$y(x, 0) = \sum_{n=1}^{\infty} B_n \sin\frac{n\pi}{L}x$$

Comparing this with the Fourier expansion of $y(x, 0)$, we conclude that the constants B_n in the general solution for motion of the string are the same as the Fourier coefficients b_n. Thus,

$$y(x, t) = \sum_{n=1}^{\infty} b_n \cos \omega_n t \sin\frac{n\pi}{L}x$$

where b_n and ω_n are as previously defined.

11.2 LONGITUDINAL VIBRATIONS

Consider the uniform slender bar shown in Fig. 11.4*a*. Such a bar can execute longitudinal, or axial, vibrations if the equilibrium condition is disturbed axially, as, for example, when the bar is elongated in tension and the load is suddenly removed. Figure 11.4*b* shows the free-body diagram of a differential element of the bar in a deformed position. The equilibrium position of the element is denoted by x, while the deformed position is u. We assume for convenience that the deformed length of the element $(1 + \partial u/\partial x)\, dx$ is greater than the original length. With the cross-sectional area denoted by A, Newton's second law applied to the differential element gives

$$-\sigma A + \left(\sigma + \frac{\partial \sigma}{\partial x} dx\right)A = \rho A\, dx \frac{\partial^2 u}{\partial t^2} \tag{11.30}$$

where ρ is the mass density of the material, and σ is the stress. Equation (11.30) reduces to

$$\frac{\partial \sigma}{\partial x} = \rho \frac{\partial^2 u}{\partial t^2} \tag{11.31}$$

For elastic deformations only, Hooke's law give the stress as

$$\sigma = E\varepsilon = E\frac{\partial u}{\partial x} \tag{11.32}$$

where E is the modulus of elasticity of the material. Combining Eqs. (11.31) and (11.32), we have

$$E\frac{\partial^2 u}{\partial x^2} = \rho\frac{\partial^2 u}{\partial t^2} \tag{11.33}$$

which can be written

$$c^2\frac{\partial^2 u}{\partial x^2} = \frac{\partial^2 u}{\partial t^2} \tag{11.34}$$

where $c = \sqrt{E/\rho}$. Thus the equation of motion is the one-dimensional wave equation as in Eq. (11.5). The general solution can be immediately written as

$$u(x, t) = (C_1 \sin \omega t + C_2 \cos \omega t)\left(C_3 \sin\frac{\omega}{c}x + C_4 \cos\frac{\omega}{c}x\right) \tag{11.35}$$

with ω and the arbitrary constants to be determined by the boundary and initial conditions.

If one end of the bar is fixed and one is free as in Fig. 11.4*a*, we have

$$u(0, t) = 0 \tag{11.36}$$

as the boundary condition for the fixed end, and

$$\frac{\partial u(L, t)}{\partial x} = 0 \tag{11.37}$$

at the free end. The latter condition corresponds to the fact that a free surface is free of stress so the strain must be zero at the end of the bar. Equation (11.36)

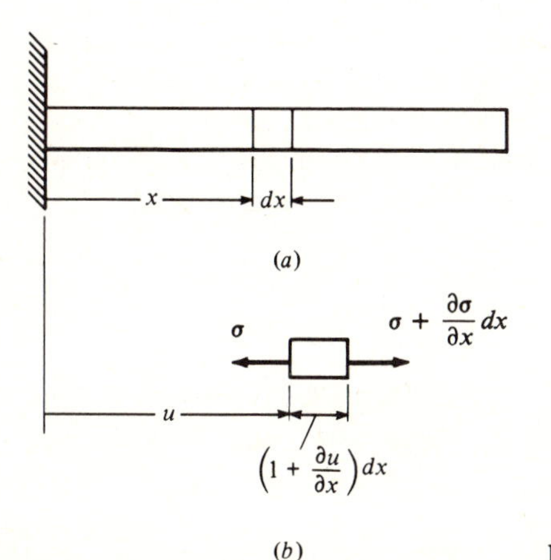

Figure 11.4

gives $C_4 = 0$, while Eq. (11.37) gives

$$(C_1 \sin \omega t + C_2 \cos \omega t) C_3 \frac{\omega}{c} \cos \frac{\omega}{c} L = 0 \tag{11.38}$$

Again we neglect the trivial solution $C_3 = 0$ and obtain

$$\cos \frac{\omega}{c} L = 0 \tag{11.39}$$

leading to the natural frequencies

$$\omega_n = \frac{\pi c}{2L}, \frac{3\pi c}{2L}, \ldots, \frac{(2n-1)\pi c}{2L} \tag{11.40}$$

or

$$\omega_n = \frac{(2n-1)\pi}{2L} \sqrt{\frac{E}{\rho}} \tag{11.41}$$

where $n = 1, 2, 3, \ldots$ identifies the mode of vibration. As with the stretched string, longitudinal vibration of the bar can take place at an infinite number of natural frequencies, and a corresponding infinite number of principal modes exist. Thus, for a fixed-free bar, the general solution is a superposition of the principal modes which can be written as

$$u(x, t) = \sum_{n=1}^{\infty} (A_n \sin \omega_n t + B_n \cos \omega_n t) \sin \frac{\omega_n}{c} x \tag{11.42}$$

with ω_n as defined by Eq. (11.41). The constants A_n and B_n are determined by applying the initial conditions.

The natural frequencies of axial vibrations of a bar are directly dependent upon the boundary conditions. The frequencies given by Eq. (11.41) apply to a fixed-free configuration. As a second example, consider a bar which has both ends fixed. In this case, the boundary conditions are

$$u(0, t) = u(L, t) = 0 \tag{11.43}$$

which, when applied to Eq. (11.35), gives $C_4 = 0$ and

$$\sin \frac{\omega}{c} L = 0 \tag{11.44}$$

The natural frequencies are then

$$\omega = \frac{n\pi c}{L} \qquad n = 1, 2, 3, \ldots \tag{11.45}$$

the same as those for the vibrating string. The following example illustrates yet another form of boundary condition.

Example 11.2 Determine the natural frequencies of the system shown in Fig. 11.5. The system is composed of a rigid 25-lb weight attached to a 0.5-in-diameter steel shaft which is 24 in long. The shaft material has $E = 30 \times 10^6$ lb/in^2 and weighs 0.283 lb/in^3.

SOLUTION The upper end of the shaft is fixed, so one boundary condition is $u(0, t) = 0$. To determine the second boundary condition, we note that the lower end of the shaft is neither fixed nor free. Rather, it is attached to a moving mass which can exert a force on the shaft because of the inertia characteristics of the mass. From the free-body diagram of the mass shown in Fig. 11.5, the equation of motion for the mass is

$$M\frac{\partial^2 u}{\partial t^2}\bigg|_{x=L} = -\sigma_L A = -AE\frac{\partial u}{\partial x}\bigg|_{x=L}$$

which expresses the fact that the mass moves with the end of the shaft as a result of the force due to stress at that point. This is the second boundary condition for the problem. For the general solution given by Eq. (11.35), the first boundary condition gives $C_4 = 0$, while the second condition results in

$$-M\omega^2(C_1 \sin \omega t + C_2 \cos \omega t)\left(C_3 \sin\frac{\omega}{c}L\right)$$
$$= -(C_1 \sin \omega t + C_2 \cos \omega t)\left(C_3 AE\frac{\omega}{c}\cos\frac{\omega}{c}L\right)$$

which is equivalent to

$$M\omega^2 \sin\frac{\omega}{c}L = AE\frac{\omega}{c}\cos\frac{\omega}{c}L$$

This is the frequency equation for the system. Rearranging, we have

$$\frac{AE}{M\omega c} = \tan\frac{\omega}{c}L$$

where ω is the only unknown. The solution procedure is facilitated if we rewrite the frequency equation as

$$\frac{AEL}{Mc^2} = \frac{\omega L}{c}\tan\frac{\omega}{c}L$$

which, after substitution of $c^2 = E/\rho$, becomes

$$\frac{\rho AL}{M} = \frac{\omega L}{c}\tan\frac{\omega}{c}L$$

The left side is the ratio of the total mass of the shaft ρAL to the mass of the attached weight and will be denoted by μ. The frequency equation is then

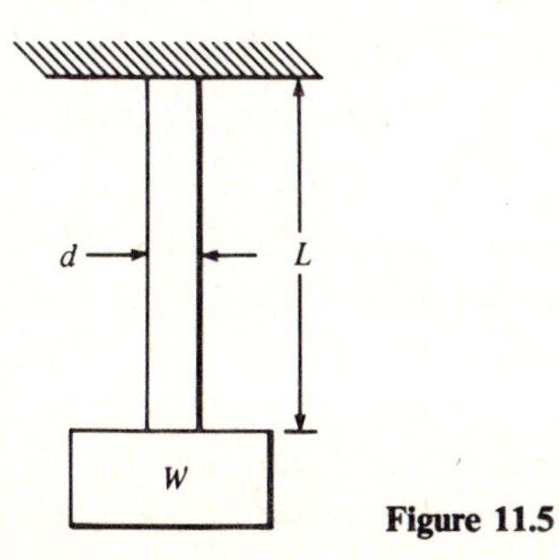

Figure 11.5

$$\mu = \alpha \tan \alpha$$

where $\alpha = \omega L/c$. This form shows us that the natural frequencies depend upon the magnitude of the mass of the shaft relative to the fixed mass.

Substituting known values gives

$$\mu = \frac{\rho AL}{M} = \frac{(0.283/386.4)\pi(0.25)^2(24)}{25/386.4} = 0.053$$

and $$0.053 = \alpha \tan \alpha$$

Solving for the values of α that satisfy this equation leads to an infinite set of natural frequencies given by $\omega_n = \alpha_n c/L$. The first three solutions and corresponding frequencies are

$$\alpha_1 = 0.2281 \qquad \omega_1 = 1924 \quad \text{rad/s}$$

$$\alpha_2 = 3.1584 \qquad \omega_2 = 26{,}634 \text{ rad/s}$$

$$\alpha_3 = 6.3663 \qquad \omega_3 = 53{,}686 \text{ rad/s}$$

These frequencies correspond to 306.2, 4238.9, and 8544.4 Hz, respectively.

An interesting comparison is obtained by neglecting the mass of the shaft and treating the system as a simple spring-mass system. For this assumption the equivalent spring constant for the shaft is $k = AE/L$, and the frequency becomes

$$\omega = \sqrt{\frac{k}{M}} = \sqrt{\frac{\pi(0.25)^2(30 \times 10^6)(386.4)}{(25)(24)}} \simeq 1948 \text{ rad/s}$$

Comparison with the fundamental frequency found above shows that ignoring the mass of the shaft introduces an error of about 1 percent in this case. Of course, for larger mass ratio μ the error increases, since the additional shaft mass reduces the natural frequencies, as we would expect.

11.3 TORSIONAL OSCILLATIONS

In earlier chapters, we considered torsional oscillations of elastic systems by ignoring the inertial properties of the elastic elements and treating all shafts simply as torsional springs. We now undertake a more exact approach, to account for shaft inertia. Consider the elastic circular shaft shown in Fig. 11.6, which is assumed to execute free torsional oscillations because of some initial disturbance. We isolate a differential shaft element of length dx and draw the free-body diagram for the element as shown. Applying the governing equation for pure rotation,

$$\Sigma M_0 = I_0 \ddot{\theta} \tag{11.46}$$

where

$$I_0 = \frac{\pi r^4 \rho}{2} dx \tag{11.47}$$

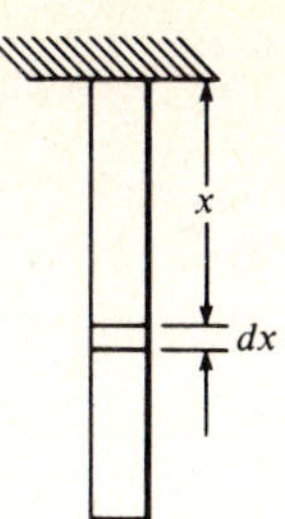

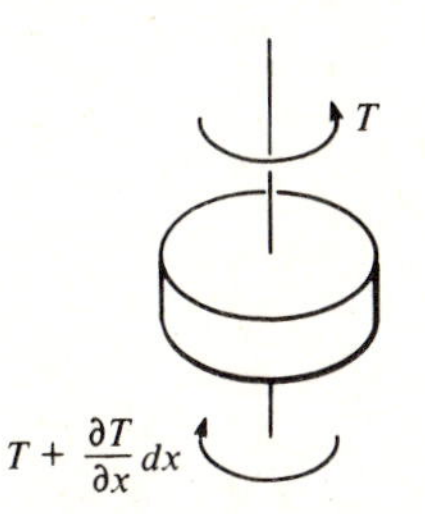

Figure 11.6

is the polar moment of inertia of a shaft with outside radius r and density ρ, we obtain

$$T + \frac{\partial T}{\partial x} dx - T = \frac{\pi r^4 \rho}{2} dx \frac{\partial^2 \theta}{\partial t^2} \tag{11.48}$$

or

$$\frac{\partial T}{\partial x} = \frac{\pi r^4 \rho}{2} \frac{\partial^2 \theta}{\partial t^2} \tag{11.49}$$

Restricting the analysis to elastic deformations only, we use the elementary strength-of-materials relation

$$\tau = G\gamma = Gr\frac{\partial \theta}{\partial x} = \frac{Tr}{J_0} \tag{11.50}$$

to obtain

$$T = GJ_0 \frac{\partial \theta}{\partial x} \tag{11.51}$$

In Eqs. (11.50) and (11.51), τ is maximum shear stress, $\gamma = r(\partial \theta / \partial x)$ is the corresponding shear strain, G is the shear modulus of the material, and $J_0 = \pi r^4/2$ is the polar moment of inertia of the cross-sectional *area* of the shaft. Equation (11.49) can now be written as

$$G\frac{\partial^2 \theta}{\partial x^2} = \rho \frac{\partial^2 \theta}{\partial t^2} \tag{11.52}$$

or, as the one-dimensional wave equation,

$$c^2 \frac{\partial^2 \theta}{\partial x^2} = \frac{\partial^2 \theta}{\partial t^2} \tag{11.53}$$

where $c = \sqrt{G/\rho}$ is the wave speed. By analogy with previous sections, the torsional oscillation is described by

$$\theta(x, t) = (C_1 \sin \omega t + C_2 \cos \omega t)\left(C_3 \sin\frac{\omega}{c} x + C_4 \cos\frac{\omega}{c} x\right) \quad (11.54)$$

The application of a particular set of boundary conditions will result in the determination of the natural frequencies, as illustrated in the following example.

Example 11.3 Determine the fundamental frequency of the torsional system shown in Fig. 3.14 if the mass moment of inertia I_s of the shaft is half that of the disk I_d. (An approximate solution was obtained in Sec. 3.7.)

SOLUTION The boundary condition for the fixed end is

$$\theta(0, t) = 0$$

while the equation of motion for the disk

$$I_d \frac{\partial^2 \theta}{\partial t^2}\bigg|_{x=L} = -GJ_0 \frac{\partial \theta}{\partial x}\bigg|_{x=L}$$

provides the second condition. Once again, the first boundary condition gives $C_4 = 0$ in Eq. (11.54), while the second results in

$$I_d \omega^2 \sin\frac{\omega}{c} L = GJ_0 \frac{\omega}{c} \cos\frac{\omega}{c} L$$

or

$$\frac{GJ_0 L}{I_d c^2} = \frac{\omega L}{c} \tan\frac{\omega}{c} L$$

Substituting $c^2 = G/\rho$ and noting that $\rho J_0 L = I_s$, we have

$$\frac{I_s}{I_d} = \frac{\omega L}{c} \tan\frac{\omega}{c} L$$

For $I_s/I_d = 0.5$,

$$0.5 = \alpha \tan \alpha$$

gives $\alpha_1 = 0.6535$, from which

$$\omega_1 = 0.6535 \frac{c}{L} = \frac{0.6535}{L}\sqrt{\frac{G}{\rho}}$$

is the fundamental frequency.

The approximate value $\omega = 0.961\omega_0$ was obtained in Sec. 3.7, where

$$\omega_0 = \left(\frac{J_0 G}{I_d L}\right)^{1/2}$$

is the frequency for the system with a massless shaft. For comparison, we write

$$\omega_0 = \left(\frac{J_0 L \rho G}{I_d \rho L^2}\right)^{1/2} = \frac{1}{L}\left(\frac{I_s G}{I_d \rho}\right)^{1/2}$$

For $I_s/I_d = 0.5$, the approximate frequency given by the energy approach is

$$\omega = 0.926\omega_0 = \frac{0.6548}{L}\sqrt{\frac{G}{\rho}}$$

Comparing this with ω_1 as obtained from the wave equation reveals an error of 0.2 percent.

11.4 TRANSVERSE VIBRATION OF BEAMS

If an elastic bar such as the simply supported beam shown in Fig. 11.7 is deformed elastically and released, transverse, or lateral, oscillations occur. In the following discussion, we assume that only elastic deflections occur. This being the case, the displacement of any point on the beam is small, and motion occurs only in a direction perpendicular to the axis of the beam. Thus we implicitly ignore the inertial effects of rotation of any section of the beam.

Since the dynamic deflection of any point of the beam is a function of position and time, we shall represent the deflection as

$$y = y(x, t) \tag{11.55}$$

For elastic deflections, the flexure formula becomes

$$\frac{\partial^2 y}{\partial x^2} = -\frac{M}{EI} \tag{11.56}$$

where M is the bending moment at any transverse section, E is the modulus of elasticity of the material, and I is the moment of inertia of the cross-sectional area of the beam about the axis of bending. Except for the use of the partial derivative, this is identical to the elementary strength-of-materials relation.

Before going further, we need to adopt a sign convention for bending moment, deflection, and so forth. Figure 11.8 shows an isolated beam section with bending moments M, shear forces V, and external load per unit length

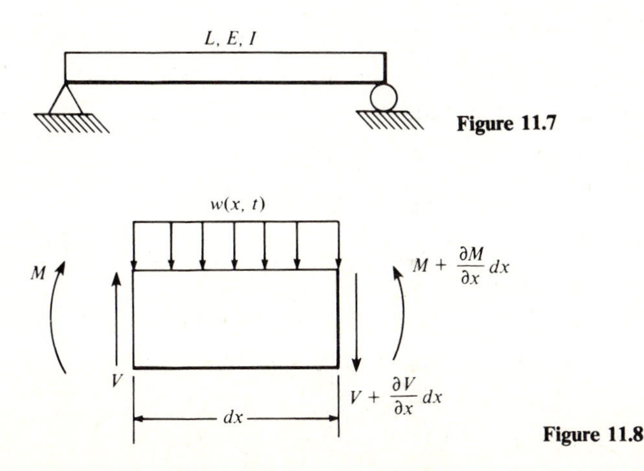

Figure 11.7

Figure 11.8

$w(x, t)$, all shown in the positive sense. By this convention, the negative sign in Eq. (11.56) will give $y(x, t)$ as positive downward.

Neglecting rotary inertia, we sum moments about the left end of the section in Fig. 11.8 to obtain

$$\frac{\partial M}{\partial x}dx - V\,dx - \frac{\partial V}{\partial x}(dx)^2 - w\frac{(dx)^2}{2} = 0 \tag{11.57}$$

Since dx is a differential length, the higher-order terms containing $(dx)^2$ can be neglected. This gives

$$V = \frac{\partial M}{\partial x} \tag{11.58}$$

which is a familiar result from strength of materials, except that the derivative is a partial derivative. With the mass of the beam per unit length denoted as γ, the equation of motion in the vertical direction as given by Newton's second law is

$$\frac{\partial V}{\partial x}dx + w\,dx = \gamma\,dx\frac{\partial^2 y}{\partial t^2} \tag{11.59}$$

By Eqs. (11.56) and (11.58), this becomes

$$\frac{\partial^2}{\partial x^2}\left(-EI\frac{\partial^2 y}{\partial x^2}\right) + w = \gamma\frac{\partial^2 y}{\partial t^2} \tag{11.60}$$

or

$$\frac{\partial^2}{\partial x^2}\left(EI\frac{\partial^2 y}{\partial x^2}\right) + \gamma\frac{\partial^2 y}{\partial t^2} = w \tag{11.61}$$

If the properties of the beam are constant along its length, Eq. (11.61) reduces to

$$EI\frac{\partial^4 y}{\partial x^4} + \gamma\frac{\partial^2 y}{\partial t^2} = w \tag{11.62}$$

where $w = w(x, t)$ is the applied load per unit length.

We consider the case of free vibration by setting $w(x, t) = 0$ to obtain

$$EI\frac{\partial^4 y}{\partial x^4} + \gamma\frac{\partial^2 y}{\partial t^2} = 0 \tag{11.63}$$

Although this is not the one-dimensional wave equation, the form is somewhat similar, and we can write Eq. (11.63) as

$$-c^2\frac{\partial^4 y}{\partial x^4} = \frac{\partial^2 y}{\partial t^2} \tag{11.64}$$

where $c = \sqrt{EI/\gamma}$. We shall seek a separable solution of the form

$$y(x, t) = f(x)g(t) \tag{11.65}$$

from which

$$\frac{\partial^4 y}{\partial x^4} = \frac{d^4 f}{dx^4}g \tag{11.66}$$

and

$$\frac{\partial^2 y}{\partial t^2} = f\frac{d^2 g}{dt^2} \tag{11.67}$$

Substituting into Eq. (11.63) gives

$$-c^2 g \frac{d^4 f}{dx^4} = f \frac{d^2 g}{dt^2} \tag{11.68}$$

or

$$\frac{-c^2 \, d^4 f / dx^4}{f} = \frac{d^2 g / dt^2}{g} \tag{11.69}$$

In principle, the left side of Eq. (11.69) is a function of x alone, while the right side is a function of t only. However, the solution we seek must be valid for all x and t. This will be true only if each side of Eq. (11.69) is constant; thus,

$$\frac{-c^2 \, d^4 f / dx^4}{f} = \frac{d^2 g / dt^2}{g} = \eta \tag{11.70}$$

where η is a constant which must be determined. As in the discussion of the vibrating string, it can be shown that η must be negative to preclude solutions for $g(t)$ corresponding to exponential growth or rigid body motion. With this observation, Eq. (11.70) is equivalent to the two ordinary differential equations

$$\frac{d^2 g}{dt^2} + \omega^2 g = 0 \tag{11.71}$$

and

$$\frac{d^4 f}{dx^4} - \frac{\omega^2}{c^2} f = 0 \tag{11.72}$$

where $\omega^2 = -\eta$ is to be determined.

The solution for Eq. (11.71) can be written by inspection as

$$g(t) = C_1 \sin \omega t + C_2 \cos \omega t \tag{11.73}$$

The solution for Eq. (11.70) is assumed as

$$f(x) = A e^{sx} \tag{11.74}$$

where A and s are constant. Substituting the assumed solution into the governing equation gives

$$\left(s^4 - \frac{\omega^2}{c^2} \right) A e^{st} = 0 \tag{11.75}$$

from which we obtain the four roots

$$s_1 = \sqrt{\frac{\omega}{c}} = \lambda \qquad s_2 = i \sqrt{\frac{\omega}{c}} = i\lambda \tag{11.76}$$

$$s_3 = -\sqrt{\frac{\omega}{c}} = -\lambda \qquad s_4 = -i \sqrt{\frac{\omega}{c}} = -i\lambda$$

The solution is then

$$f(x) = A_1 e^{\lambda x} + A_2 e^{i\lambda x} + A_3 e^{-\lambda x} + A_4 e^{-i\lambda x} \tag{11.77}$$

which is the superposition of the four independent solutions of the fourth-order equation. Equation (11.77) can be written as

$$f(x) = C_3 \frac{e^{\lambda x} - e^{-\lambda x}}{2} + C_4 \frac{e^{\lambda x} + e^{-\lambda x}}{2} + C_5(-i)\frac{e^{i\lambda x} - e^{-i\lambda x}}{2} + C_6 \frac{e^{i\lambda x} + e^{-i\lambda x}}{2} \tag{11.78}$$

or

$$f(x) = C_3 \sinh \lambda x + C_4 \cosh \lambda x + C_5 \sin \lambda x + C_6 \cos \lambda x \tag{11.79}$$

where we have redefined the constants as $A_1 = (C_3 + C_4)/2$, $A_2 = (C_6 - iC_5)/2$, $A_3 = (C_4 - C_3)/2$, and $A_4 = (C_6 + iC_5)/2$.

The solution to Eq. (11.63) is then

$$y(x, t) = (C_1 \sin \omega t + C_2 \cos \omega t)(C_3 \sinh \lambda x + C_4 \cosh \lambda x + C_5 \sin \lambda x + C_6 \cos \lambda x) \tag{11.80}$$

with

$$\omega = \lambda^2 c \tag{11.81}$$

The constants appearing in the solution as well as the natural frequencies are determined by applying the boundary conditions for the beam and the initial conditions of the motion. For a simply supported beam, the boundary conditions are

$$y(0, t) = y(L, t) = 0 \tag{11.82}$$

and

$$\left.\frac{\partial^2 y}{\partial x^2}\right|_{x=0} = \left.\frac{\partial^2 y}{\partial x^2}\right|_{x=L} = 0 \tag{11.83}$$

Equation (11.83) expresses the absence of bending moment at each end of the beam. Applying the boundary conditions to Eq. (11.80) gives

$$\begin{aligned} C_4 + C_6 &= 0 \\ C_3 \sinh \lambda L + C_4 \cosh \lambda L + C_5 \sin \lambda L + C_6 \cos \lambda L &= 0 \\ C_4 - C_6 &= 0 \\ C_3 \sinh \lambda L + C_4 \cosh \lambda L - C_5 \sin \lambda L - C_6 \cos \lambda L &= 0 \end{aligned} \tag{11.84}$$

Equations (11.84) will be satisfied if $C_3 = C_4 = C_6 = 0$ and

$$C_5 \sin \lambda L = 0$$

This is the frequency equation, and it could also be obtained by setting the determinant of coefficients of Eqs. (11.84) to zero. The frequency equation will be satisfied, and nontrivial solutions obtained, if

$$\lambda L = n\pi \qquad n = 1, 2, 3, \ldots \tag{11.85}$$

Combining this result with Eq. (11.81), we obtain the natural frequencies

$$\omega_n = \frac{n^2\pi^2}{L^2} c = n^2\pi^2 \sqrt{\frac{EI}{\gamma L^4}} \tag{11.86}$$

each corresponding to a principal mode of free vibration.

The free response of the simply supported beam is obtained by superposition of the principal modes and can be written in the series form

$$y(x, t) = \sum_{n=1}^{\infty} (A_n \sin \omega_n t + B_n \cos \omega_n t) \sin \frac{n\pi}{L} x \tag{11.87}$$

The constants A_n and B_n are evaluated by applying the initial conditions of motion. From Eq. (11.87), we observe that if vibration occurs as the jth principal mode, the mode shape is a sine curve having $j + 1$ nodes including the end points. The mode shapes are the same as those shown in Fig. 11.2 for the vibrating string.

11.5 FOURIER ANALYSIS

In Example 11.1 we illustrated the use of the Fourier-series expansion to describe the initial configuration of a vibrating string. This approach is convenient and easily applied to many continuous systems, since the general solutions for free motion often involve harmonic functions of the spatial coordinate. Fourier analysis is also quite useful in obtaining the forced response of beams, as is illustrated by the following cases.

Concentrated Loading

Figure 11.9 shows a simply supported beam which is subjected to a harmonic force $F_0 \sin \omega_f t$. The governing equation is

$$EI \frac{\partial^4 y}{\partial x^4} + \gamma \frac{\partial^2 y}{\partial t^2} = w(x, t) \tag{11.88}$$

where $w(x, t)$ expresses the spatial distribution and time dependence of the external load per unit length. The time dependence of the harmonic force is well defined, but it is applied at a single point, and we must express this as a function of x to apply Eq. (11.88). Using a method introduced in Chap. 7, this can be accomplished by writing

$$w(x, t) = F_0 \sin \omega_f t \, \delta(x - a) \tag{11.89}$$

where $\delta(x - a)$ is the Dirac delta function defined in Chap. 7. Considering the Dirac delta function as an odd periodic function having period $2L$ as shown in

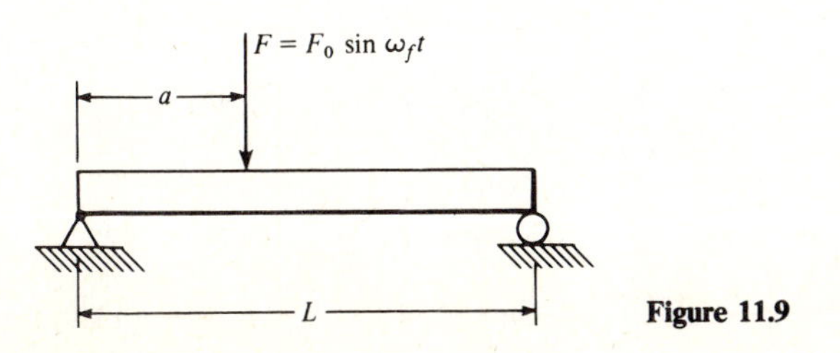

Figure 11.9

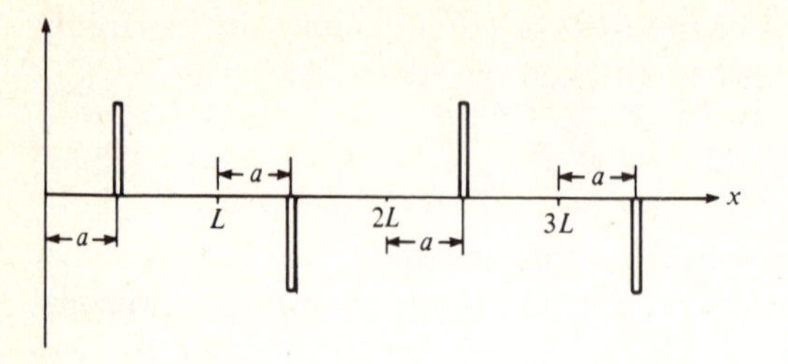

Figure 11.10

Fig. 11.10, we can expand it into a Fourier sine series as

$$\delta(x - a) = \sum_{n=1}^{\infty} b_n \sin\frac{n\pi}{L}x \tag{11.90}$$

where
$$b_n = \frac{2}{L}\int_0^L \delta(x - a)\sin\frac{n\pi}{L}x\,dx = \frac{2}{L}\sin\frac{n\pi a}{L} \tag{11.91}$$

by Eq. (7.21). We now have

$$w(x, t) = \frac{2F_0}{L}\left(\sum_{n=1}^{\infty}\sin\frac{n\pi a}{L}\sin\frac{n\pi}{L}x\right)\sin\omega_f t \tag{11.92}$$

and we note that the expression just obtained has the units of force per unit length as desired.

To determine the forced response, we must now determine the particular solution of Eq. (11.88) for $w(x, t)$ as given above. It is entirely logical to assume that the time response of the beam will have the same frequency as the forcing function, so we shall seek a separable solution

$$y(x, t) = f(x)\sin\omega_f t \tag{11.93}$$

Differentiating as necessary and substituting into the equation of motion give

$$EI\frac{d^4 f}{dx^4} - \gamma\omega_f^2 f = \frac{2F_0}{L}\sum_{n=1}^{\infty}\sin\frac{n\pi a}{L}\sin\frac{n\pi x}{L} \tag{11.94}$$

where we have dropped the common term $\sin\omega_f t$. The particular solution of Eq. (11.94) has the form

$$f(x) = \sum_{n=1}^{\infty} A_n \sin\frac{n\pi x}{L} \tag{11.95}$$

for which Eq. (11.94) becomes

$$\frac{\pi^4 EI}{L^4}\sum_{n=1}^{\infty} n^4 A_n \sin\frac{n\pi x}{L} - \gamma\omega_f^2\sum_{n=1}^{\infty} A_n\sin\frac{n\pi x}{L}$$
$$= \frac{2F_0}{L}\sum_{n=1}^{\infty}\sin\frac{n\pi a}{L}\sin\frac{n\pi x}{L} \tag{11.96}$$

Equating coefficients of $\sin(n\pi x/L)$ gives

$$A_n = \frac{2F_0L^3}{\pi^4 n^4 EI - \gamma L^4\omega_f^2}\sin\frac{n\pi a}{L} \qquad n = 1, 2, 3, \ldots \tag{11.97}$$

and the forced response is obtained as the series

$$y(x, t) = \sum_{n=1}^{\infty} A_n \sin\frac{n\pi x}{L} \sin \omega_f t \tag{11.98}$$

The denominator of Eq. (11.97) shows that a resonant condition will exist if

$$\omega_f = n^2\pi^2\sqrt{\frac{EI}{\gamma L^4}} \tag{11.99}$$

as expected, since the right-hand side represents the natural frequencies of the beam. We also note that the sine series in x automatically satisfies the boundary conditions for the simply supported beam.

Moving Concentrated Load

A classical beam vibration problem concerns the response of a simply supported beam subjected to a force of constant magnitude which moves across the beam with constant speed. Historically this problem was considered in England in the eighteenth century in relation to locomotives crossing railway bridges. If the speed of the moving force is v, the position of the load at any time is $x = vt$, as shown in Fig. 11.11. Using the method of the previous problem, we write

$$w(x, t) = F_0\,\delta(x - vt) \tag{11.100}$$

which will represent load per unit length as desired if we again expand the Dirac delta function as a Fourier sine series. The resulting series representation is

$$w(x, t) = \frac{F_0}{2L}\sum_{n=1}^{\infty} \sin\frac{n\pi vt}{L}\sin\frac{n\pi x}{L} \tag{11.101}$$

which expresses both the spatial and time dependence of the load in series form.

The equation of motion is now written as

$$EI\frac{\partial^4 y}{\partial x^4} + \gamma\frac{\partial^2 y}{\partial t^2} = \frac{F_0}{2L}\sum_{n=1}^{\infty} \sin\frac{n\pi vt}{L}\sin\frac{n\pi x}{L} \tag{11.102}$$

Since this equation involves only even partial derivatives, we assume

$$y(x, t) = \sum_{n=1}^{\infty} A_n \sin\frac{n\pi vt}{L}\sin\frac{n\pi x}{L} \tag{11.103}$$

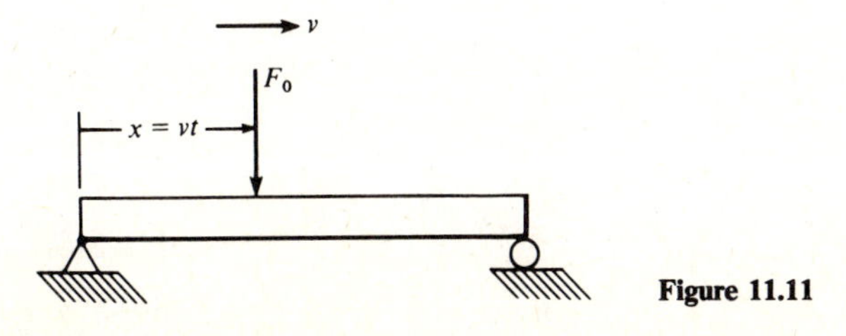

Figure 11.11

Differentiating as required and substituting into Eq. (11.102) yield

$$\sum_{n=1}^{\infty}\left(\frac{EI}{L^4}n^4\pi^4 A_n - \gamma\frac{n^2\pi^2v^2}{L^2}A_n\right)\sin\frac{n\pi vt}{L}\sin\frac{n\pi x}{L}$$

$$= \frac{F_0}{2L}\sum_{n=1}^{\infty}\sin\frac{n\pi vt}{L}\sin\frac{n\pi x}{L} \qquad (11.104)$$

from which we obtain

$$A_n = \frac{F_0 L^3}{2n^2\pi^2(n^2\pi^2 EI - \gamma v^2 L^2)} \qquad n = 1, 2, 3, \ldots \qquad (11.105)$$

The denominator of Eq. (11.105) shows that for each n there exists a speed v which will produce a resonant condition. The lowest speed for which this occurs is

$$v = \frac{\pi}{L}\sqrt{\frac{EI}{\gamma}}$$

This corresponds to the speed for which the lowest equivalent forcing frequency $\pi v/L$ is equal to the fundamental frequency of the beam.

Impulsive Loading

Another case of interest is that of beam vibration occurring as the result of impact. Let a simply supported beam of length L be struck impulsively at position $x = a$. For convenience the impulse is assumed to act at $t = 0$. The loading can be expressed as

$$w(x, t) = \hat{F}_0\,\delta(t)\,\delta(x - a) \qquad (11.106)$$

where the Dirac delta functions express the time and spatial distributions of the loading, respectively. Expanding $\delta(x - a)$ as a Fourier sine series gives

$$w(x, t) = \frac{2\hat{F}_0}{L}\delta(t)\sum_{n=1}^{\infty}\sin\frac{n\pi a}{L}\sin\frac{n\pi x}{L} \qquad (11.107)$$

and the equation of motion can be written

$$EI\frac{\partial^4 y}{\partial x^4} + \gamma\frac{\partial^2 y}{\partial t^2} = \frac{2\hat{F}_0}{L}\delta(t)\sum_{n=1}^{\infty}\sin\frac{n\pi a}{L}\sin\frac{n\pi x}{L} \qquad (11.108)$$

To satisfy the boundary conditions, considering the form of the right-hand side of Eq. (11.108), we assume a series solution of the form

$$y(x, t) = \sum_{n=1}^{\infty} g_n(t)\sin\frac{n\pi x}{L} \qquad (11.109)$$

On substitution of the assumed solution, the equation of motion becomes

$$\frac{\pi^4 EI}{L^4}\sum_{n=1}^{\infty} n^4 g_n(t)\sin\frac{n\pi x}{L} + \gamma\sum_{n=1}^{\infty}\ddot{g}_n(t)\sin\frac{n\pi x}{L} = \frac{2\hat{F}_0}{L}\delta(t)\sum_{n=1}^{\infty}\sin\frac{n\pi a}{L}\sin\frac{n\pi x}{L}$$

$$(11.110)$$

which is equivalent to the n ordinary differential equations

$$\ddot{g}_n(t) + \frac{n^4\pi^4 EI}{\gamma L^4} g(t) = \frac{2\hat{F}_0}{\gamma L} \sin\frac{n\pi a}{L} \delta(t) \qquad n = 1, 2, 3, \ldots \qquad (11.111)$$

These equations can be solved by the method of Sec. 7.3 to obtain

$$g_n(t) = \frac{2\hat{F}_0}{\gamma L \omega_n} \sin\frac{n\pi a}{L} \sin \omega_n t \qquad n = 1, 2, 3, \ldots \qquad (11.112)$$

where $\omega_n = n^2\pi^2\sqrt{EI/\gamma L^4}$ are the natural frequencies. We may note that the presence of ω_n in the denominator ensures convergence of the series solution, since $g_n(t)$ approaches zero as n increases.

In these examples we have implicitly used the fact that the principal modes are orthogonal as for discrete systems. We shall now show this more formally for the beam on simple supports. If we multiply each term of Eq. (11.110) by $\sin(m\pi x/L)$, where m is any integer, and integrate from $x = 0$ to $x = L$, each term will contain the integral

$$\int_0^L \sin\frac{n\pi x}{L} \sin\frac{m\pi x}{L}\, dx = \begin{cases} 0 & \text{for } m \neq n \\ \dfrac{L}{2} & \text{for } m = n \end{cases} \qquad (11.113)$$

This is the statement of the orthogonality condition for the principle modes of the simply supported beam, and it supplies the mathematical justification for reducing Eq. (11.110) to Eqs. (11.111). It can also be shown that the orthogonality condition holds for other types of support as well. In general, the orthogonality condition is expressed as

$$\int_0^L Y_n(x) Y_m(x)\, dx = 0 \qquad \text{for } m \neq n \qquad (11.114)$$

where the $Y(x)$ are the mode shapes and are often called *normal shape functions*.

11.6 APPROXIMATE METHODS

The governing equations and boundary conditions for the elastic bodies discussed so far have been such that exact solutions could be obtained. Although the general solutions are in the form of infinite series of principal modes, each such principal mode is an exact solution. Unfortunately, this is the exception rather than the rule. More often, exact solutions for the oscillation of continuous systems cannot be obtained, and approximate methods must be applied. Approximate solutions usually must be obtained for systems which support discrete masses at nonboundary positions and for many forced-vibration problems. Many techniques are available, but we mention only a few here as examples.

Method of Assumed Modes

In this method, the continuous system is approximated by a discrete system to obtain estimates of the system natural frequencies. The solution for the continuous system is assumed to be described by

$$y(x, t) = \sum_{j=1}^{n} Y_j(x) q_j(t) \tag{11.115}$$

where $q_j(t), j = 1, 2, \ldots, n$, are generalized coordinates, and $Y_j(x), j = 1, 2, \ldots, n$, are any functions which satisfy the boundary conditions for the system. Effectively, this amounts to regarding the continuous system as a discrete system having n degrees of freedom.

The kinetic energy can be written

$$T(t) = \tfrac{1}{2} \sum_{i=1}^{n} \sum_{j=1}^{n} m_{ij} \dot{q}_i(t) \dot{q}_j(t) \tag{11.116}$$

where the m_{ij} are determined by the mass distribution of the system and by functions $Y_i(x)$. Note that the m_{ij} coefficients are symmetric. Similarly, the potential energy is

$$V(t) = \tfrac{1}{2} \sum_{i=1}^{n} \sum_{j=1}^{n} k_{ij} q_i(t) q_j(t) \tag{11.117}$$

where the k_{ij} are also symmetric and depend on the stiffness distribution as well as $Y_i(x)$.

Applying Lagrange's equations

$$\frac{d}{dt}\left(\frac{\partial L}{\partial \dot{q}_r}\right) - \frac{\partial L}{\partial q_r} = 0 \qquad r = 1, 2, \ldots, n \tag{11.118}$$

where $L = T - V$ gives

$$\sum_{i=1}^{n} m_{ir} \ddot{q}_i + \sum_{i=1}^{n} k_{ir} q_i = 0 \qquad r = 1, 2, \ldots, n \tag{11.119}$$

Note that in considering the free vibration of elastic systems, we have utilized the fact that no nonconservative forces enter into the problem. Equation (11.119) can be written in the matrix form

$$[m][\ddot{q}] + [k][q] = 0 \tag{11.120}$$

in which form it can be treated by the methods of Chap. 10 to obtain the system natural frequencies.

Example 11.4 Using the assumed-modes method, estimate the fundamental frequency of free vibration of the beam shown in Fig. 11.12.

SOLUTION The boundary conditions are

$$y(0, t) = y(L, t) = 0 \qquad \left.\frac{\partial y}{\partial x}\right|_{x=0} = 0 \qquad \left.\frac{\partial^2 y}{\partial x^2}\right|_{x=L} = 0$$

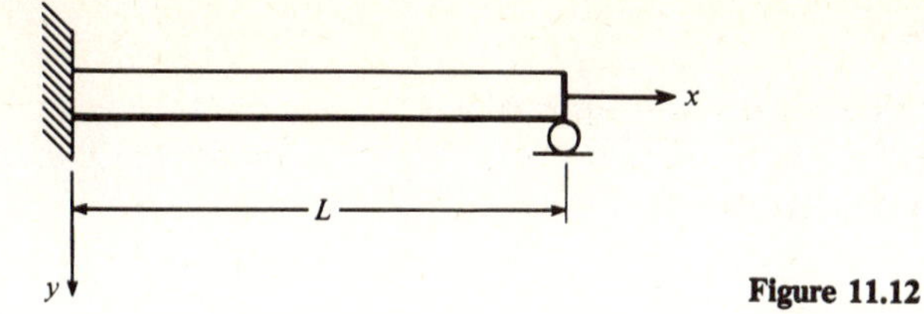

Figure 11.12

Since we desire only the fundamental frequency, we let $n = 1$ and choose

$$q_1(t) = y\left(\frac{L}{2}, t\right)$$

as the generalized coordinate. Next we must choose a function $Y_1(x)$ which satisfies the boundary conditions. Usually the choice of a polynomial in x is appropriate. The function

$$Y_1(x) = 4\left(\frac{x}{L}\right)^3 - 7\left(\frac{x}{L}\right)^4 + 3\left(\frac{x}{L}\right)^5$$

satisfies all boundary conditions and results in

$$y(x, t) = \left[4\left(\frac{x}{L}\right)^3 - 7\left(\frac{x}{L}\right)^4 + 3\left(\frac{x}{L}\right)^5\right]q_1(t)$$

The kinetic energy is

$$\begin{aligned} T &= \frac{1}{2}\int_0^L \left(\frac{\partial y}{\partial t}\right)^2 dm \\ &= \frac{\gamma}{2}\int_0^L \left[4\left(\frac{x}{L}\right)^3 - 7\left(\frac{x}{L}\right)^4 + 3\left(\frac{x}{L}\right)^5\right]^2 \dot{q}_1^2(t)\, dx \\ &= \frac{26\gamma L}{3465}\dot{q}_1^2 \end{aligned}$$

and the elastic potential energy is

$$\begin{aligned} V &= \frac{1}{2}\int_0^L EI\left(\frac{\partial^2 y}{\partial x^2}\right)^2 dx \\ &= \frac{EI}{2}\int_0^L \left(\frac{24}{L^3}x - \frac{84}{L^4}x^2 + \frac{60}{L^5}x^3\right)^2 q_1^2(t)\, dx \\ &= \frac{864EI}{315L^3}q_1^2 \end{aligned}$$

Since the kinetic energy is not a function of $q_1(t)$, Lagrange's equation can be written as

$$\frac{d}{dt}\left(\frac{\partial T}{\partial \dot{q}_1}\right) + \frac{\partial V}{\partial q_1} = 0$$

or

$$\frac{52\gamma L}{3465}\ddot{q}_1 + \frac{1728EI}{315L^3}q_1 = 0$$

Rearranging, we have

$$\ddot{q}_1 + \frac{365.5EI}{\gamma L^4} q_1 = 0$$

from which we obtain the approximate fundamental frequency as

$$\omega_1 = 19.1\sqrt{\frac{EI}{\gamma L^4}}$$

An exact analysis for this case gives $\omega_1 = 15.42\sqrt{EI/\gamma L^4}$, so the error is about 24 percent. A better result could be obtained by letting $Y_1(x)$ correspond to the static-deflection shape of the beam with a load applied at the center.

Lumped-Parameter Method

In this method, a continuous system is divided into a discrete number of segments, and the mass of each segment is treated as if it were concentrated at the mass center of the segment. As an example of the procedure, consider the cantilever beam having variable cross section shown in Fig. 11.13*a*. The beam has variable cross-sectional moment of inertia and mass per unit length denoted by $I(x)$ and $\gamma(x)$, respectively. The modulus of elasticity E is assumed constant over the length of the beam. The beam is subdivided into n discrete masses as shown in Fig. 11.13*b* to obtain an n-degrees-of-freedom system. The mass of each element is

$$m_i = \int_{(i-1)L/n}^{iL/n} \gamma(x)\, dx \qquad i = 1, 2, \ldots, n \tag{11.121}$$

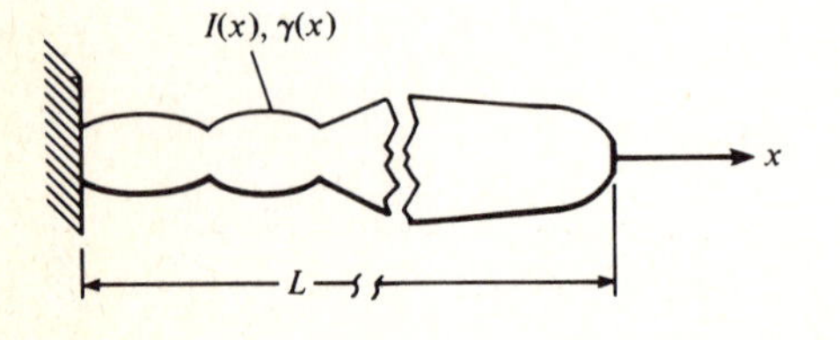

(*a*)

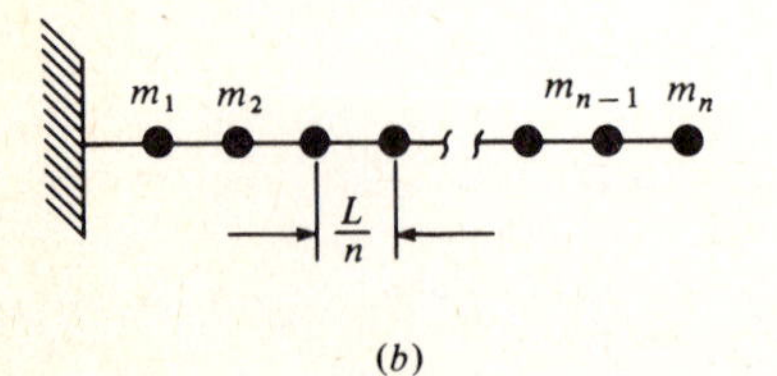

(*b*)

Figure 11.13

For the discrete system, the force on mass i due to displacements y_j, $j = 1, 2, \ldots, n$, can be expressed as

$$F_i = \sum_{j=1}^{n} k_{ij} y_j \qquad i = 1, 2, \ldots, n \tag{11.122}$$

where k_{ij} are the stiffness influence coefficients (elements of the stiffness matrix) as defined in Chap. 10. If the beam executes free vibration, the only forces present are inertial; for harmonic oscillation, they can be written

$$F_i = m_i \ddot{y}_i = \omega^2 m_i y_i \qquad i = 1, 2, \ldots, n \tag{11.123}$$

Combining Eqs. (11.122) and (11.123) gives

$$\omega^2 m_i y_i = \sum_{j=1}^{n} k_{ij} y_j \qquad i = 1, 2, \ldots, n \tag{11.124}$$

which is equivalent to

$$\omega^2 [m][y] = [k][y] \tag{11.125}$$

Equation (11.125) is the eigenvalue problem for the discrete system, which can be solved numerically using the computer techniques discussed in Chap. 10.

PROBLEMS

11.1 Determine the fundamental frequency of a 0.1055-in-diameter copper wire stretched between two supports as in Fig. 11.1. The tension in the wire is 30 lb, and the specific weight of the wire material is 0.297 lb/in^3.

11.2 A stretched wire is set into motion by releasing it from the initial configuration shown in Fig. 11.14. Obtain the solution for the motion of the wire using Fourier analysis.

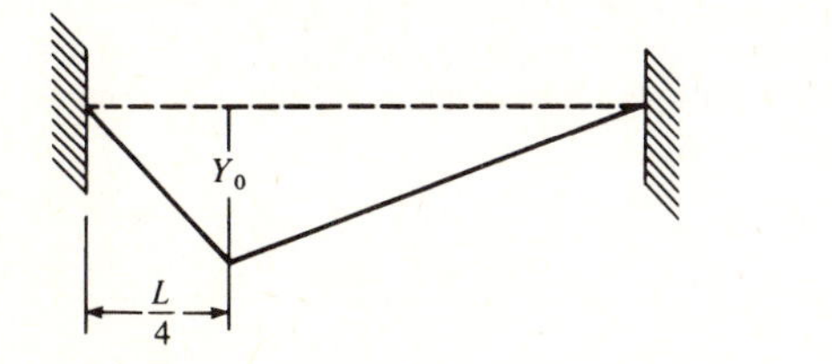

Figure 11.14

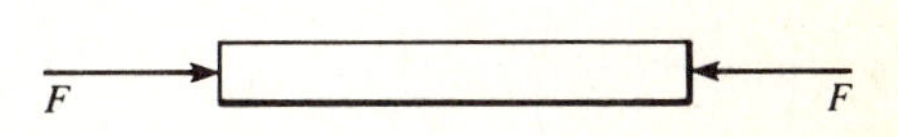

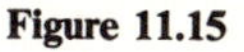

Figure 11.15

11.3 Using the transformation $\eta = x - ct$, $\xi = x + ct$, show that the one-dimensional wave equation can be written as $\partial^2 y/\partial\eta\,\partial\xi = 0$.

11.4 A uniform circular rod of length L is compressed by equal and opposite forces F applied to both ends, as in Fig. 11.15. Determine the vibration resulting from the sudden removal of the forces.

11.5 Determine the general solution for the free longitudinal oscillations of a uniform elastic bar of length L if both ends of the bar are fixed.

11.6 Derive the differential equation governing the free longitudinal oscillations of the tapered circular shaft shown in Fig. 11.16.

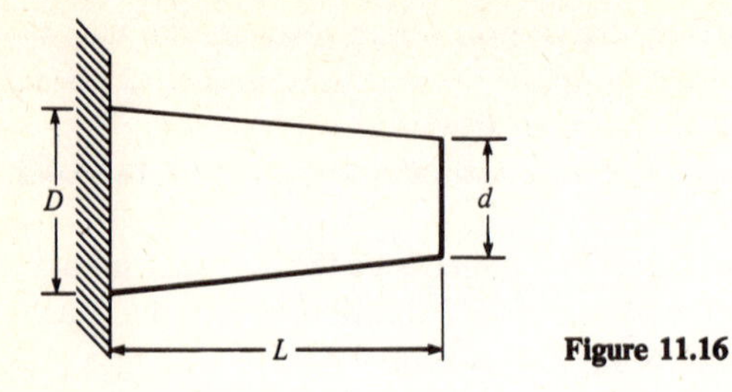

Figure 11.16

11.7 Determine the equation of motion, boundary conditions, and frequency equation for the system shown in Fig. 11.17.

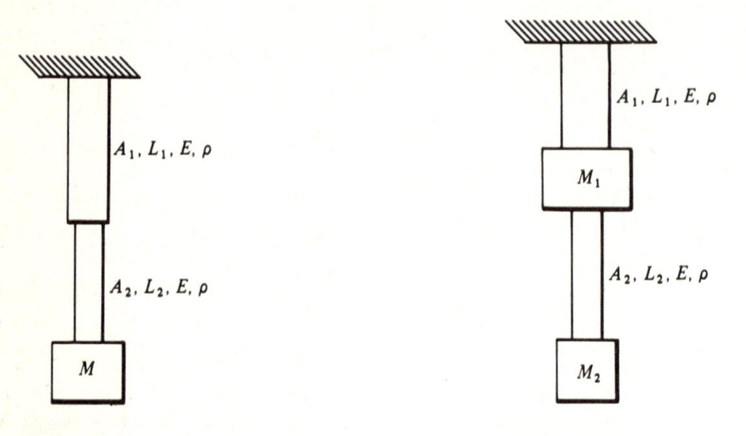

Figure 11.17

Figure 11.18

11.8 Repeat Prob. 11.7 for the system in Fig. 11.18. Assume M_1 is rigid.

11.9 A mass M_1 is dropped from height h onto another mass M_2 as shown in Fig. 11.19. Mass M_2 is attached to a rigid support by a uniform elastic bar having length L. Assuming the impact is perfectly plastic, obtain the differential equation of motion, boundary conditions, initial conditions, and frequency equation for the ensuing longitudinal vibrations.

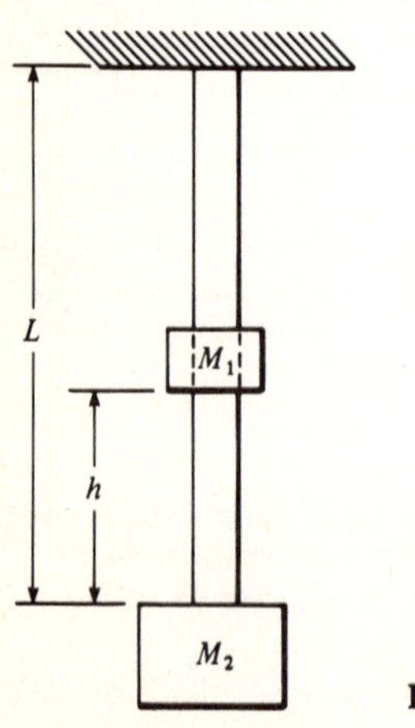

Figure 11.19

11.10 The uniform bar of Fig. 11.4 receives an axial hammer blow at the free end at $t = 0$, resulting in the initial condition $\partial u(L, 0)/\partial t = -V_0$. Obtain the general solution for the resulting vibrations.

11.11 Repeat Prob. 11.10 except that both ends of the bar are free.

11.12 Considering only the fundamental mode, show that the solution of Prob. 11.11 can be written as

$$u(x, t) = \frac{V_0 L}{2\pi c}\left[\sin\frac{\pi}{L}(x + ct) - \sin\frac{\pi}{L}(x - ct)\right]$$

and show that the time required for $u(0, t)$ to reach maximum value is $t = L/c$.

11.13 Determine the response of the system shown in Fig. 11.20 to harmonic motion of the support.

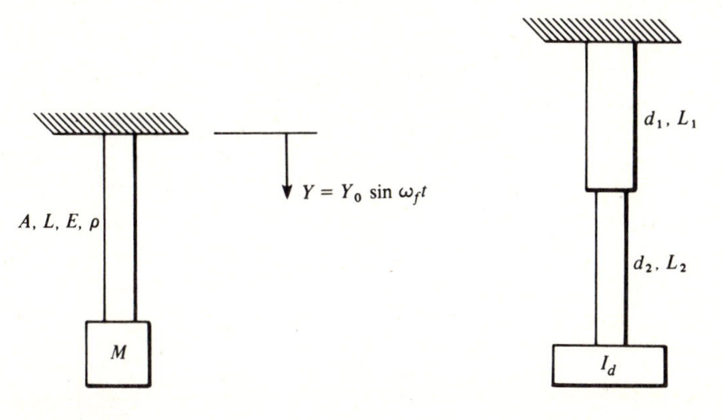

Figure 11.20 **Figure 11.21**

11.14 Obtain the natural frequencies of torsional oscillations of a uniform circular shaft of length L having both ends free.

11.15 Determine the frequency equation for free torsional oscillations of an elastic circular shaft which has both ends fixed.

11.16 In the system of Fig. 11.21, the stepped shaft is steel which weighs 0.283 lb/in^3 and has a modulus of elasticity $E = 30 \times 10^6$ lb/in^2. The shaft dimensions are $d_1 = 2$ in, $L_1 = 20$ in, $d_2 = 1.5$ in, and $L_2 = 30$ in. The attached disk has a polar mass moment of inertia of 0.5 lb · s^2 · in.

(*a*) Determine the fundamental frequency of free torsional oscillations.

(*b*) Ignore the mass of the shaft, and obtain the frequency by the methods of Chap. 3.

(*c*) What error in frequency results in part (*b*)?

11.17 Obtain the differential equation of motion, boundary conditions, and frequency equation for the torsional system of Fig. 11.22.

Figure 11.22 **Figure 11.23**

11.18 A circular shaft having $d = 50$ mm, $L = 1500$ mm, $G = 82 \times 10^3$ N/mm^2, and $\rho = 0.008$ kg/mm^3 is twisted by a constant torque $T = 5000$ N · mm. One end of the shaft is fixed, and the other is free; the torque is applied at the free end. If the torque is suddenly removed, solve for the resulting oscillations.

11.19 Determine the frequency equation for transverse vibrations of the beam shown in Fig. 11.23.

11.20 Repeat Prob. 11.19 using Fig. 11.24.

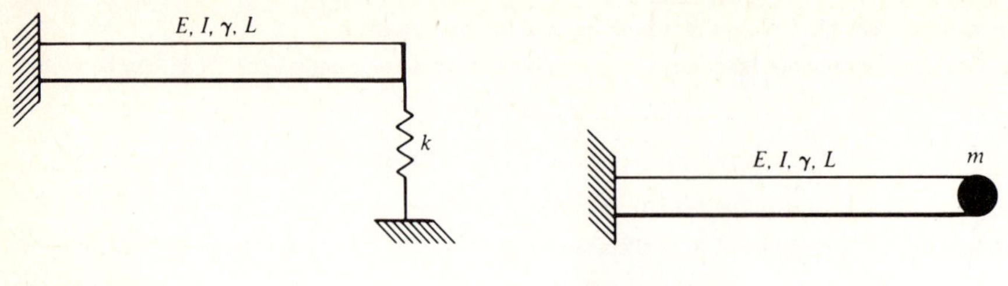

Figure 11.24

Figure 11.25

11.21 Obtain the differential equation of motion, boundary conditions, and frequency equation for transverse oscillations of the system of Fig. 11.25.

11.22 Repeat Prob. 11.21 for the system depicted in Fig. 11.26. Assume the mass m to be in point contact with the beam at all times.

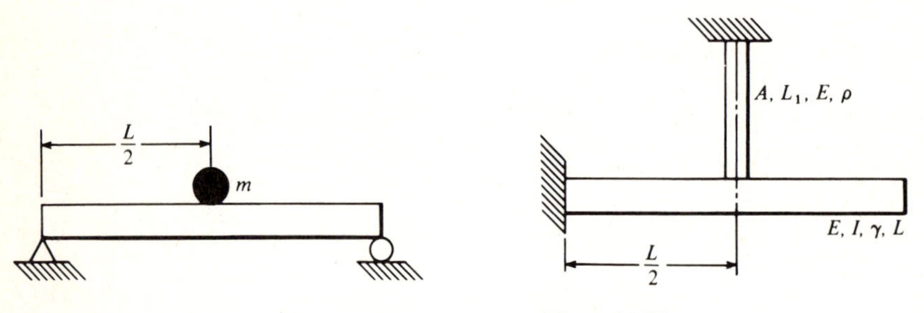

Figure 11.26

Figure 11.27

11.23 The cantilever beam shown in Fig. 11.27 is supported at midspan by a uniform elastic bar. Derive the frequency equation for free oscillations of this system.

11.24 A simply supported beam is subjected to the concentrated force $F = F_0 e^{-\omega_f t}$ as shown in Fig. 11.28. Determine the forced response of the beam.

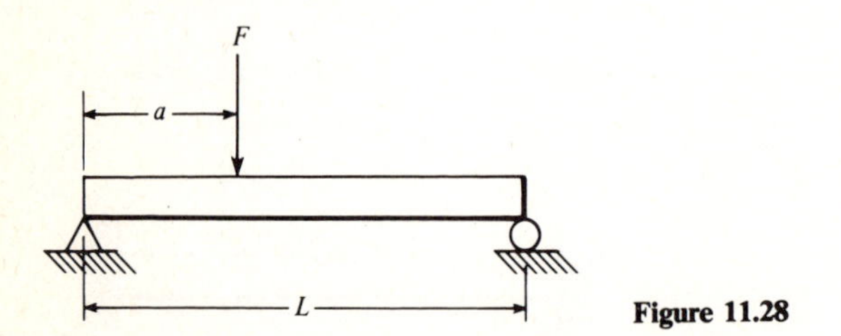

Figure 11.28

11.25 Repeat Prob. 11.24 if the force F is defined by Fig. 11.29.

11.26 A cantilever beam is excited by a harmonic moment $M = M_0 \sin \omega_f t$ applied at the free end, as shown in Fig. 11.30. Determine the complete response of the beam if the initial conditions are zero.

11.27 Derive the differential equation of motion for free transverse oscillations of a uniform, simply supported beam which exhibits uniform viscous damping c per unit length.

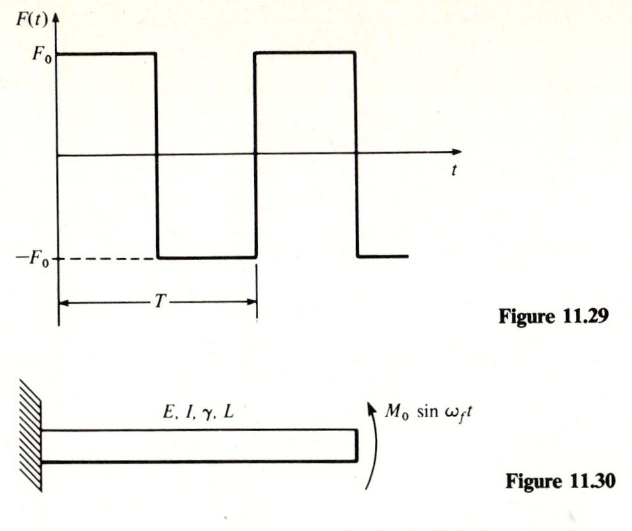

Figure 11.29

Figure 11.30

11.28 Solve the equation of motion for Prob. 11.27 by assuming

$$y(x, t) = \sum_{n=1}^{\infty} f_n(t) \sin \frac{n\pi x}{L}$$

and show that each mode has a critical damping constant given by

$$c_n = 2\gamma \left(\frac{n\pi}{L}\right)^2 \sqrt{\frac{EI}{\gamma}}$$

APPENDIX

A

MATHEMATICAL DETAILS

A.1 MASS MOMENTS OF INERTIA

The mass moment of inertia of a body is a quantity that reflects the inertial resistance of the body to rotation about a specified axis. Consider a particle of mass m attached to a rigid, massless rod which is free to pivot about an axis through point O as in Fig. A.1*a*. If a couple is applied, the system will begin to rotate about the axis (the z axis), and it can be shown that the time required to attain any specified rotational speed is proportional to the product mr^2. This product is known as the *mass moment of inertia* about the axis of rotation. Next consider rotation of a rigid body of finite size about an axis AA as in Fig. A.1*b*. To obtain the mass moment of inertia of the body, we divide the body into mass elements Δm_i and sum the moments of inertia of the individual elements. As the number of mass elements increases, the summation process becomes an integral, and we obtain the basic definition

$$I_A = \int r^2 \, dm \tag{A.1}$$

where I_A denotes the mass moment of inertia about axis A, and the integration is carried out over the entire mass of the body.

Since the mass moments of inertia are referred to a particular axis, it is convenient to express them with respect to coordinate axes. The mass moments

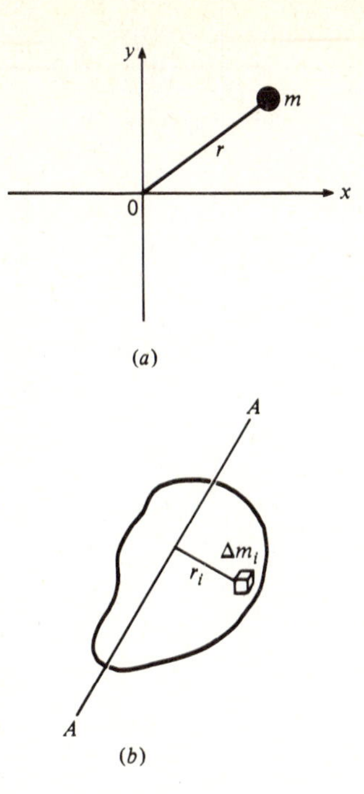

Figure A.1

of inertia of a body with respect to a set of cartesian axes can be written

$$I_x = \int (y^2 + z^2)\, dm$$

$$I_y = \int (x^2 + z^2)\, dm \qquad \text{(A.2)}$$

$$I_z = \int (x^2 + y^2)\, dm$$

Most often the axes are chosen such that the origin coincides with the mass center of the body, in which case the moments of inertia are said to be *centroidal*. Table A.1 gives the centroidal mass moments of inertia of several common bodies which are assumed to have uniform mass distribution.

Table A.1 Centroidal mass moments of inertia of common bodies

Thin circular plate

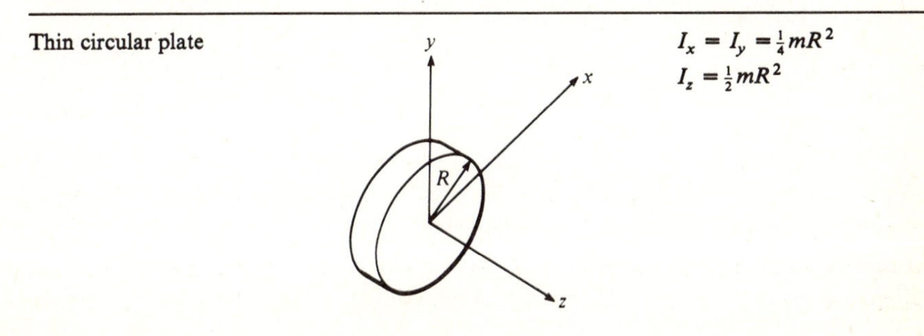

$I_x = I_y = \frac{1}{4}mR^2$
$I_z = \frac{1}{2}mR^2$

Table A.1 Continued

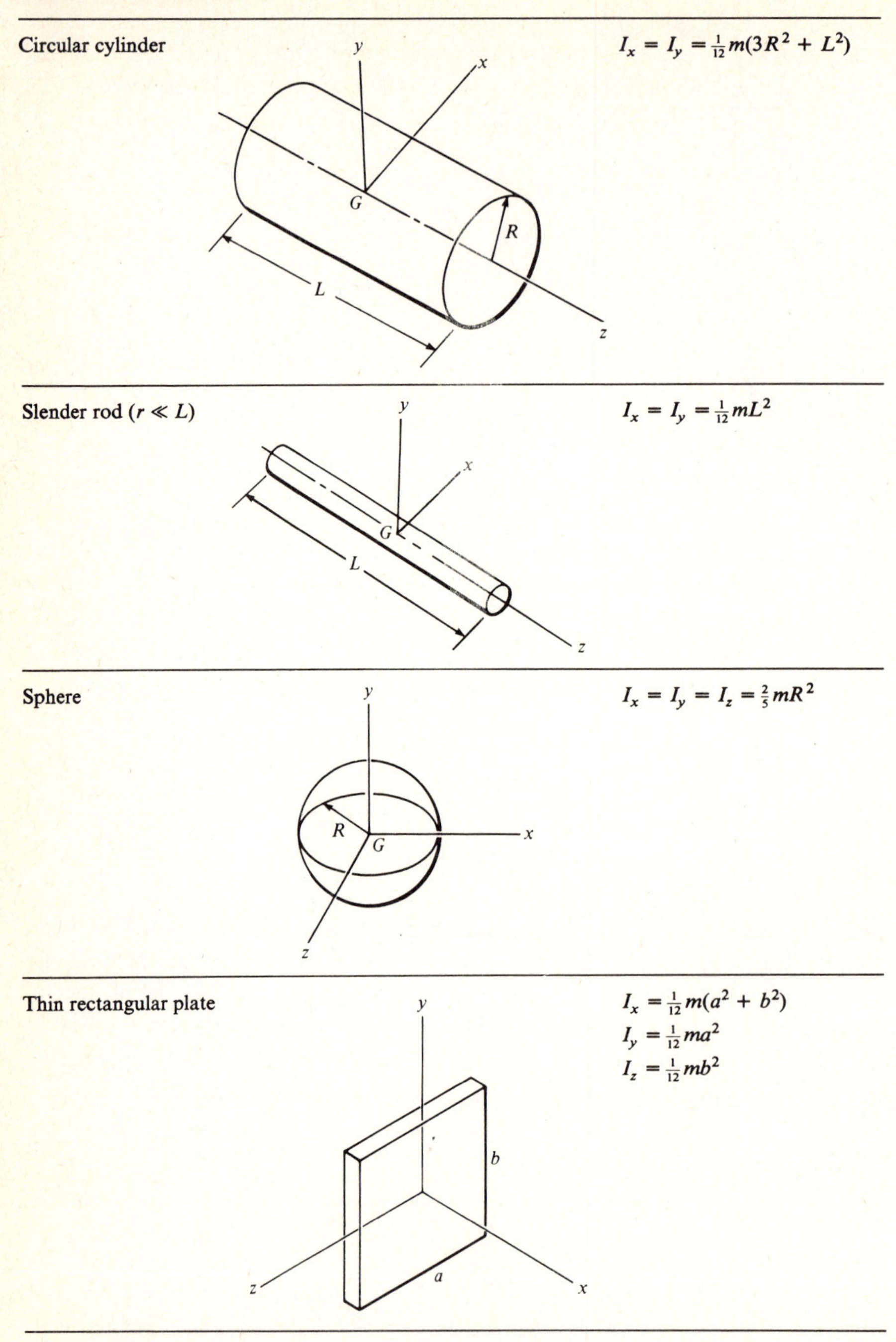

Circular cylinder	$I_x = I_y = \frac{1}{12}m(3R^2 + L^2)$
Slender rod ($r \ll L$)	$I_x = I_y = \frac{1}{12}mL^2$
Sphere	$I_x = I_y = I_z = \frac{2}{5}mR^2$
Thin rectangular plate	$I_x = \frac{1}{12}m(a^2 + b^2)$ $I_y = \frac{1}{12}ma^2$ $I_z = \frac{1}{12}mb^2$

A.2 THE PARALLEL-AXIS THEOREM

Consider a body of mass m and an arbitrary system of rectangular coordinates $0x'y'z'$ as in Fig. A.2. In this system, the mass moments of inertia are

$$I_{x'} = \int (y'^2 + z'^2)\, dm$$

$$I_{y'} = \int (x'^2 + z'^2)\, dm \tag{A.3}$$

$$I_{z'} = \int (x'^2 + y'^2)\, dm$$

Introducing a parallel set of axes $Gxyz$ having its origin at the mass center G as in Fig. A.3, we can write

$$x' = x + \bar{x} \qquad y' = y + \bar{y} \qquad z' = z + \bar{z} \tag{A.4}$$

where $\bar{x}, \bar{y}, \bar{z}$ are the coordinates of G in the system $0x'y'z'$. The first of Eqs.

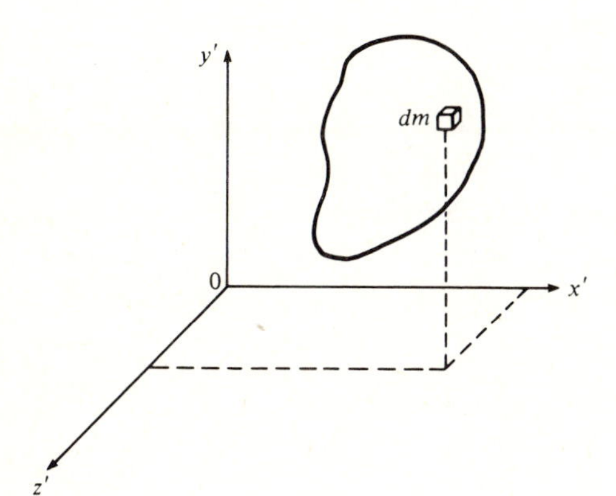

Figure A.2

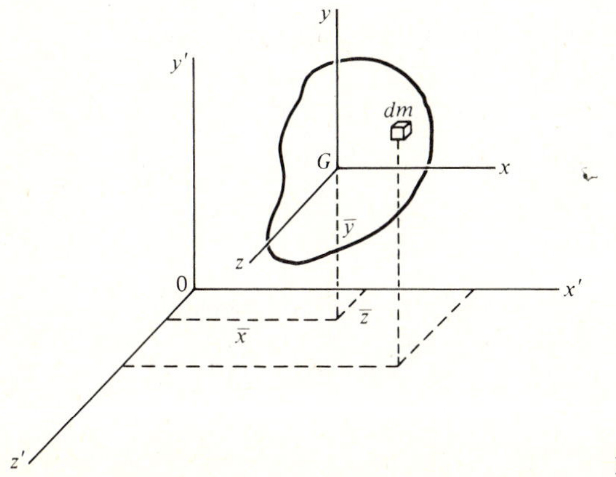

Figure A.3

(A.3) can be written

$$I_{x'} = \int \left[(y + \bar{y})^2 + (z + \bar{z})^2 \right] dm$$

$$= \int (y^2 + z^2)\, dm + 2\bar{y} \int y\, dm + 2\bar{z} \int z\, dm + (\bar{y}^2 + \bar{z}^2) \int dm \quad \text{(A.5)}$$

The first integral in Eq. (A.5) is the mass moment of inertia $\bar{I}_x$ about the centroidal axis x. That is,

$$\bar{I}_x = \int (y^2 + z^2)\, dm \quad \text{(A.6)}$$

The second and third integrals are the first moments of the mass with respect to the xz and xy planes, respectively. Since these planes, by definition, contain the mass center of the body, the two integrals are identically zero. The last integral is simply the total mass of the body, so we have

$$I_{x'} = \bar{I}_x + (\bar{y}^2 + \bar{z}^2)m \quad \text{(A.7)}$$

as the final result. A completely analogous procedure gives

$$I_{y'} = \bar{I}_y + (\bar{x}^2 + \bar{z}^2)m$$
$$\text{(A.8)}$$
$$I_{z'} = \bar{I}_z + (\bar{x}^2 + \bar{y}^2)m$$

These relations are easily generalized by noting that the sum $\bar{y}^2 + \bar{z}^2$ is the square of the distance between the x and x' axes. Similarly, $\bar{x}^2 + \bar{z}^2$ and $\bar{x}^2 + \bar{y}^2$ are the squares of the distances between the y and y' axes and the z and z' axes, respectively. Thus we have the general relation

$$I = \bar{I} + md^2 \quad \text{(A.9)}$$

where I is the moment of inertia relative to an arbitrary axis, $\bar{I}$ is the moment of inertia relative to a *parallel*, *centroidal* axis, m is the total mass of the body, and d is distance between the axes.

A.3 DERIVATION OF THE BEATING EQUATION

For zero initial conditions, the response of an undamped spring-mass system to harmonic excitation can be written as

$$x(t) = \frac{X_0 \omega_f^2}{\omega^2 - \omega_f^2} \left(\sin \omega_f t - \frac{\omega_f}{\omega} \sin \omega t \right) \quad \text{(A.10)}$$

The phenomenon known as *beating* occurs when the system frequency ω is very near the forced frequency ω_f. By using the trigonometric identity

$$\sin \alpha - \sin \beta = 2 \cos \frac{\alpha + \beta}{2} \sin \frac{\alpha - \beta}{2} \quad \text{(A.11)}$$

and noting that $\omega_f/\omega \simeq 1$ for beating, Eq. (A.10) can be written

$$x(t) = \frac{2X_0\omega_f^2}{\omega^2 - \omega_f^2}\cos\frac{\omega_f + \omega}{2}t \sin\frac{\omega_f - \omega}{2}t \tag{A.12}$$

The denominator term can be expressed as

$$\omega^2 - \omega_f^2 = (\omega + \omega_f)(\omega - \omega_f) \simeq -4\omega_f\gamma \tag{A.13}$$

where $\gamma = (\omega_f - \omega)/2$ is a small quantity. Introducing Eq. (A.13) into Eq. (A.12) gives

$$x(t) = -\frac{X_0\omega_f}{2\gamma}\cos \omega_f t \sin \gamma t \tag{A.14}$$

which is the same as Eq. (4.29).

A.4 FOURIER SERIES

For a periodic function of period 2π,

$$f(x) = f(x + 2\pi) \tag{A.15}$$

the series expansion

$$f(x) = \frac{a_0}{2} + \sum_{n=1}^{\infty} a_n \cos nx + \sum_{n=1}^{\infty} b_n \sin nx \tag{A.16}$$

is known as the *Fourier series* (or *full Fourier series*) representation of $f(x)$.

To determine the *Fourier coefficients* a_n, $n = 0, 1, 2, \ldots,$ we multiply both sides of Eq. (A.16) by $\cos mx$, where m is any nonnegative integer, integrate over the interval $-\pi \le x \le \pi$, and interchange the order of integration and summation to obtain

$$\int_{-\pi}^{\pi} f(x) \cos mx \, dx = \int_{-\pi}^{\pi} \frac{a_0}{2} \cos mx \, dx + \sum_{n=1}^{\infty} \int_{-\pi}^{\pi} a_n \cos nx \cos mx \, dx$$
$$+ \sum_{n=1}^{\infty} \int_{-\pi}^{\pi} b_n \sin nx \cos mx \, dx \tag{A.17}$$

Equation (A.17) is more readily interpreted by noting that

$$\begin{aligned}
\int_{-\pi}^{\pi} \cos nx \cos mx \, dx &= \pi\delta_{mn} \\
\int_{-\pi}^{\pi} \sin nx \cos mx \, dx &= 0 \\
\int_{-\pi}^{\pi} \sin nx \sin mx \, dx &= \pi\delta_{mn}
\end{aligned} \tag{A.18}$$

where m and n are nonnegative integers provided *both are not zero*, and δ_{mn} is the *Kronecker delta* defined by

$$\delta_{mn} = \begin{cases} 1 & \text{for } m = n \\ 0 & \text{for } m \neq n \end{cases}$$

For $m = n = 0$,

$$\int_{-\pi}^{\pi} \cos nx \cos mx \, dx = 2\pi$$

$$\int_{-\pi}^{\pi} \sin nx \cos mx \, dx = 0 \tag{A.19}$$

$$\int_{-\pi}^{\pi} \sin nx \cos mx \, dx = 0$$

Setting $m = 0$ in Eq. (A.17) and utilizing Eqs. (A.18) and (A.19) give

$$\int_{-\pi}^{\pi} f(x) \, dx = \int_{-\pi}^{\pi} \frac{a_0}{2} \, dx$$

from which

$$\frac{a_0}{2} = \frac{1}{2\pi} \int_{-\pi}^{\pi} f(x) \, dx \tag{A.20}$$

showing that the term $a_0/2$ is the average value of the function $f(x)$ over the period. For any nonzero m in Eq. (A.17) we obtain

$$a_n = \frac{1}{\pi} \int_{-\pi}^{\pi} f(x) \cos nx \, dx \qquad n = 1, 2, 3, \ldots \tag{A.21}$$

The remaining Fourier coefficients b_n, $n = 1, 2, 3, \ldots$, are obtained by multiplying Eq. (A.16) by sin mx and carrying out an analogous procedure. The results are

$$b_n = \frac{1}{\pi} \int_{-\pi}^{\pi} f(x) \sin nx \, dx \qquad n = 1, 2, 3, \ldots \tag{A.22}$$

Some simplification of the full Fourier series occurs if the function $f(x)$ is an *even* or *odd* function. A function is said to be *even* if $f(x) = f(-x)$, and *odd* if $f(x) = -f(-x)$. In particular, we note that cosine is an even function whereas sine is an odd function. Consider the Fourier-series expansion of an even function $f(x)$. We have

$$f(x) \cos nx = f(-x) \cos(-nx) \qquad \text{(even)}$$

$$f(x) \sin nx = -f(-x) \sin(-nx) \qquad \text{(odd)}$$

so the Fourier coefficients become

$$a_n = \frac{2}{\pi} \int_0^{\pi} f(x) \cos nx \, dx \qquad n = 0, 1, 2, \ldots$$

$$b_n = 0 \qquad n = 1, 2, 3, \ldots$$

The Fourier series for an *even* function is then

$$f(x) = \frac{a_0}{2} + \sum_{n=1}^{\infty} a_n \cos nx \tag{A.23}$$

and is known as a *Fourier cosine series*. Similarly, if $f(x)$ is an odd function, the Fourier coefficients are

$$a_n = 0 \qquad n = 0, 1, 2, \ldots$$

$$b_n = \frac{2}{\pi}\int_0^{\pi} f(x) \sin nx \, dx \qquad n = 1, 2, 3, \ldots$$

and the resulting expansion is the *Fourier sine series*

$$f(x) = \sum_{n=1}^{\infty} b_n \sin nx \tag{A.24}$$

Fourier-series expansions are not limited to functions having period 2π. Suppose $f(x)$ is periodic with period T such that

$$f(x) = f(x + T) \tag{A.25}$$

The change of variable $x = (T/2\pi)t$ is used to transform $f(x)$ into a function $g(t)$ given by

$$g(t) = f\left(\frac{Tt}{2\pi}\right) = f(x) \tag{A.26}$$

which has period 2π since

$$g(t + 2\pi) = f\left(\frac{T}{2\pi}(t + 2\pi)\right) = f\left(\frac{Tt}{2\pi} + T\right) = f\left(\frac{Tt}{2\pi}\right) = g(t) \tag{A.27}$$

The function $g(t)$ can then be expanded in the Fourier series

$$g(t) = \frac{a_0}{2} + \sum_{n=1}^{\infty} a_n \cos nt + \sum_{n=1}^{\infty} b_n \sin nt$$

where

$$a_n = \frac{1}{\pi}\int_{-\pi}^{\pi} g(t) \cos nt \, dt$$

$$b_n = \frac{1}{\pi}\int_{-\pi}^{\pi} g(t) \sin nt \, dt$$

Substituting $f(x) = g(t)$ from Eq. (A.27) and $t = 2\pi x/T$, we obtain

$$f(x) = \frac{a_0}{2} + \sum_{n=1}^{\infty} a_n \cos\frac{2\pi n}{T}x + \sum_{n=1}^{\infty} b_n \sin\frac{2\pi n}{T}x \tag{A.28}$$

and

$$\left.\begin{aligned} a_n &= \frac{1}{\pi}\int_{-\pi}^{\pi} f(x) \cos\frac{2\pi n}{T}x \, dx \\ b_n &= \frac{1}{\pi}\int_{-\pi}^{\pi} f(x) \sin\frac{2\pi n}{T}x \, dx \end{aligned}\right. \tag{A.29}$$

Equations (A.28) and (A.29) can be used to obtain the Fourier-series expansion for *any* periodic function.

A.5 TABLE OF LAPLACE TRANSFORMS

	$\bar{f}(s)$	$f(t)$
1.	1	$\delta(t)$
2.	$\frac{1}{s}$	1
3.	$\frac{1}{s^2}$	t
4.	$\frac{1}{s^n}$ $\quad n = 1, 2, 3, \ldots$	$\frac{t^{n-1}}{(n-1)!}$
5.	$\frac{1}{s+a}$	e^{-at}
6.	$\frac{1}{(s+a)^n}$	$\frac{t^{n-1}}{(n-1)!}e^{-at}$
7.	$\frac{1}{s(s+a)}$	$\frac{1}{a}(1 - e^{-at})$
8.	$\frac{\omega}{s^2+\omega^2}$	$\sin \omega t$
9.	$\frac{s}{s^2+\omega^2}$	$\cos \omega t$
10.	$\frac{1}{s(s^2+\omega^2)}$	$\frac{1}{\omega^2}(1 - \cos \omega t)$
11.	$\frac{s+a}{s^2+\omega^2}$	$\frac{\sqrt{a^2+\omega^2}}{\omega}\sin(\omega t + \phi) \qquad \phi = \tan^{-1}\frac{\omega}{a}$
12.	$\frac{1}{(s+a)^2+b^2}$	$\frac{1}{b}e^{-at}\sin bt$
13.	$\frac{s+a}{(s+a)^2+b^2}$	$e^{-at}\cos bt$
14.	$\frac{1}{s^2+2abs+b^2}$	$\frac{1}{b\sqrt{1-a^2}}e^{-abt}\sin b\sqrt{1-a^2}\,t$
15.	$\frac{s}{(s^2+a^2)^2}$	$\frac{t \sin at}{2a}$
16.	$\frac{s^2-a^2}{(s^2+a^2)}$	$t \cos at$
17.	$\frac{1}{(s+a)(s^2+\omega^2)}$	$\frac{e^{-at}}{a^2+\omega^2} + \frac{1}{\omega\sqrt{a^2+\omega^2}}\sin(\omega t - \phi) \qquad \phi = \tan^{-1}\frac{\omega}{a}$
18.	$\frac{s+c}{(s+a)^2+b^2}$	$\frac{\sqrt{(c-a)^2+b^2}}{b}e^{-at}\sin(bt+\phi) \qquad \phi = \tan^{-1}\frac{b}{c-a}$
19.	$\frac{1}{s^2-a^2}$	$\frac{\sinh at}{a}$
20.	$\frac{s}{s^2-a^2}$	$\cosh at$

A.6 PARTIAL-FRACTION EXPANSION OF LAPLACE TRANSFORMS

Often it is necessary to determine the inverse transform of a rational fraction of the form

$$\bar{f}(s) = \frac{B(s)}{A(s)} = \frac{b_m s^m + b_{m-1} s^{m-1} + \cdots + b_1 s + b_0}{s^n + a_{n-1} s^{n-1} + \cdots + a_1 s + a_0} \tag{A.30}$$

where m and n are real positive integers with $n > m$, and the a_n and b_n are real constants.

Generally, it is not possible to obtain the inverse of a transform having the form of Eq. (A.30) by simple reference to a table of Laplace-transform pairs. Rather, it is necessary to simplify Eq. (A.30) to a more suitable form for table use. To this end, we factor the denominator to obtain

$$\bar{f}(s) = \frac{B(s)}{A(s)} = \frac{b_m s^m + b_{m-1} s^{m-1} + \cdots + b_1 s + b_0}{(s - r_1)(s - r_2) \cdots (s - r_n)} \tag{A.31}$$

where r_i, $i = 1, 2, \ldots, n$ are the roots of $A(s) = 0$. Equation (A.31) can then be written as

$$\frac{B(s)}{A(s)} = \frac{C_1}{s - r_1} + \frac{C_2}{s - r_2} + \cdots + \frac{C_n}{s - r_n} \tag{A.32}$$

which is referred to as a *partial-fraction expansion*. Each term on the right of Eq. (A.32) is a *partial fraction*, and the C_i are constants. In this form, the inverse transform of $B(s)/A(s)$ is found as the sum of the inverse transforms of the individual (simpler) terms on the right-hand side. Before the inverse transforms can be obtained explicitly, it is necessary to evaluate the constants C_i. This can be accomplished by one of two methods, depending on the roots of $A(s)$.

If the roots of $A(s)$ are all distinct, any C_i, $i = 1, 2, \ldots, n$ can be evaluated by multiplying both sides of Eq. (A.32) by $s - r_i$ and then setting $s = r_i$. In this fashion, the left side becomes a constant, and the right side becomes simply C_i.

Example A.1 Determine the partial-fraction expansion of

$$\bar{f}(s) = \frac{s + 1}{s^2 - 5s + 6}$$

SOLUTION The denominator can be factored to obtain

$$\bar{f}(s) = \frac{s + 1}{(s - 2)(s - 3)}$$

and the form of the desired partial fraction is

$$\bar{f}(s) = \frac{s + 1}{(s - 2)(s - 3)} = \frac{C_1}{s - 2} + \frac{C_2}{s - 3}$$

Multiplying both sides by $s - 2$ and letting $s = 2$, we obtain $C_1 = -3$.

Similarly, multiplying by $s - 3$ and letting $s = 3$ give $C_2 = 4$. The partial-fraction expansion is then

$$\bar{f}(s) = \frac{-3}{s-2} + \frac{4}{s-3}$$

If the denominator $A(s)$ has repeated roots, the procedure of the previous example cannot be used to obtain all the constants. Consider the case

$$\frac{B(s)}{A(s)} = \frac{B(s)}{(s-r_1)(s-r_2)^{n-1}} \tag{A.33}$$

in which $A(s)$ has one distinct root and $n - 1$ repeated roots. The partial-fraction expansion is written as

$$\frac{B(s)}{A(s)} = \frac{C_1}{s-r_1} + \frac{C_2}{(s-r_2)^{n-1}} + \frac{C_3}{(s-r_2)^{n-2}} + \cdots + \frac{C_n}{s-r_2} \tag{A.34}$$

The constants C_1 and C_2 can be obtained by the above method as

$$C_1 = (s-r_1)\frac{B(s)}{A(s)}\bigg|_{s=r_1} \qquad C_2 = (s-r_2)^{n-1}\frac{B(s)}{A(s)}\bigg|_{s=r_2}$$

However, an attempt to determine the remaining constants by this method results in an indeterminate form. This dilemma is resolved by noting that

$$(s-r_2)^{n-1}\frac{B(s)}{A(s)} = \frac{(s-r_2)^{n-1}C_1}{s-r_1} + C_2 + (s-r_2)C_3 + (s-r_2)^2C_4$$
$$+ \cdots + (s-r_2)^{n-2}C_n \tag{A.35}$$

Differentiating Eq. (A.35) with respect to s and evaluating at $s = r_2$ gives

$$C_3 = \frac{d}{ds}\left[(s-r_2)^{n-1}\frac{B(s)}{A(s)}\right]_{s=r_2} \tag{A.36}$$

Two differentiations yield

$$C_4 = \frac{1}{2}\frac{d^2}{ds_2}\left[(s-r_2)^{n-1}\frac{B(s)}{A(s)}\right]_{s=r_2} \tag{A.37}$$

Continuing in this manner will yield each constant in turn.

Example A.2 Obtain the partial-fraction expansion of

$$\bar{f}(s) = \frac{1}{s(s-1)^3}$$

SOLUTION We first expand $\bar{f}(s)$ as

$$\bar{f}(s) = \frac{1}{s(s-1)^3} = \frac{C_1}{s} + \frac{C_2}{(s-1)^3} + \frac{C_3}{(s-1)^2} + \frac{C_4}{s-1}$$

Multiplying through by s and evaluating at $s = 0$ give $C_1 = -1$. Next, multiplying by $(s-1)^3$ and letting $s = 1$ result in $C_2 = 1$. To obtain C_3 we apply Eq. (A.36):

$$C_3 = \frac{d}{ds}\left(\frac{1}{s}\right)_{s=1} = \left(-\frac{1}{s^2}\right)_{s=1} = -1$$

C_4 is given by Eq. (A.37) as

$$C_4 = \frac{1}{2}\frac{d^2}{ds^2}\left(\frac{1}{s}\right)_{s=1} = \frac{1}{2}\left(\frac{2}{s^3}\right)_{s=1} = 1$$

The partial-fraction expansion is then

$$\bar{f}(s) = \frac{-1}{s} + \frac{1}{(s-1)^3} - \frac{1}{(s-1)^2} + \frac{1}{s-1}$$

A.7 MATRIX ALGEBRA

The mathematical description of many physical problems is often simplified by the use of rectangular arrays of scalars of the form

$$[A] = \begin{bmatrix} a_{11} & a_{12} & \cdots & a_{1n} \\ a_{21} & a_{22} & \cdots & a_{2n} \\ \cdots & \cdots & \cdots & \cdots \\ a_{m1} & a_{m2} & \cdots & a_{mn} \end{bmatrix} \tag{A.38}$$

Such an array is known as a *matrix*, and the scalar values that make up the array are called the *elements* of the matrix. The position of each element a_{ij} is identified by the *row* subscript i and the *column* subscript j.

The number of rows and columns determines the *order* of a matrix. A matrix having m rows and n columns is said to be of order m by n (or $m \times n$). If the number of rows and columns in a matrix are the same, the matrix is called a *square matrix* and is said to be of order n. A matrix having only one row is called a *row matrix* or *row vector*. Similarly, a matrix with a single column is a *column matrix* or *column vector*.

If the rows and columns of a matrix $[A]$ are interchanged, the resulting matrix is known as the *transpose* of $[A]$ and is denoted by $[A]^T$. Thus, for the matrix defined by Eq. (A.38), we have

$$[A]^T = \begin{bmatrix} a_{11} & a_{21} & \cdots & a_{m1} \\ a_{12} & a_{22} & \cdots & a_{m2} \\ \cdots & \cdots & \cdots & \cdots \\ a_{1n} & a_{2n} & \cdots & a_{mn} \end{bmatrix} \tag{A.39}$$

and we see that if $[A]$ is of order $m \times n$, then $[A]^T$ is of order $n \times m$.

Addition and *subtraction* of matrices can be defined only for matrices of the same order. If $[A]$ and $[B]$ are $m \times n$ matrices, the two are said to be *conformable* for addition or subtraction. The sum of two $m \times n$ matrices is another $m \times n$ matrix whose elements are obtained by summing the corresponding elements of the original two matrices. Symbolically we write

$$[C] = [A] + [B] \tag{A.40}$$

where
$$c_{ij} = a_{ij} + b_{ij} \tag{A.41}$$

for $i = 1, 2, \ldots, m$ and $j = 1, 2, \ldots, n$. The operation of matrix subtraction is similarly defined. Matrix addition or subtraction is both *commutative* and *associative*; that is,

$$[A] + [B] = [B] + [A] \tag{A.42}$$

and
$$[A] + ([B] + [C]) = ([A] + [B]) + [C] \tag{A.43}$$

The *product of a scalar and a matrix* is a matrix in which every element of the original matrix is multiplied by the scalar. If a scalar u multiplies $[A]$, then

$$[B] = u[A] \tag{A.44}$$

where the elements of $[B]$ are given by

$$b_{ij} = ua_{ij} \tag{A.45}$$

for $i = 1, 2, \ldots, m$ and $j = 1, 2, \ldots, n$.

Matrix multiplication is defined in such a way as to facilitate the solution of simultaneous linear equations. The *product of two matrices* $[A]$ and $[B]$

$$[C] = [A][B] \tag{A.46}$$

exists only if the number of columns in $[A]$ is equal to the number of rows in $[B]$. If this condition is satisfied, the matrices are said to be *conformable for multiplication*. If $[A]$ is of order $m \times p$ and $[B]$ is of order $p \times n$, the matrix product $[C] = [A][B]$ is an $m \times n$ matrix having elements defined by

$$c_{ij} = \sum_{k=1}^{p} a_{ik} b_{kj} \tag{A.47}$$

Thus each element c_{ij} is the sum of the products of the elements in the ith row of $[A]$ and the corresponding elements in the jth row of $[B]$. In referring to the matrix product $[A][B]$, the matrix $[A]$ is called the *premultiplier*, and $[B]$ the *postmultiplier*.

In general, matrix multiplication is *not* commutative. That is,

$$[A][B] \neq [B][A] \tag{A.48}$$

Matrix multiplication does satisfy the *associative* and *distributive* laws, and we can therefore write

$$\begin{aligned} ([A][B])[C] &= [A]([B][C]) \\ [A]([B] + [C]) &= [A][B] + [A][C] \\ ([A] + [B])[C] &= [A][C] + [B][C] \end{aligned} \tag{A.49}$$

In addition to being noncommutative, matrix algebra differs from scalar algebra in other ways. For example, the equality

$$[A][B] = [A][C] \tag{A.50}$$

does not necessarily imply $[B] = [C]$ since algebraic summing is involved in forming the matrix products. As another example, if we define the *null matrix* [0] as a matrix having all elements zero, then the product

$$[A][B] = [0] \tag{A.51}$$

does not imply that either $[A]$, $[B]$, or both are null matrices.

The *determinant of a square matrix* is a scalar value which is unique for a given matrix. The determinant of an $n \times n$ matrix $[A]$ is written

$$\det[A] = |A| = \begin{vmatrix} a_{11} & a_{12} & \cdots & a_{1n} \\ a_{21} & a_{22} & \cdots & a_{2n} \\ \cdots & \cdots & \cdots & \cdots \\ a_{n1} & a_{n2} & \cdots & a_{nn} \end{vmatrix} \tag{A.52}$$

and is evaluated by following certain special rules. Consider the 2×2 matrix

$$[A] = \begin{bmatrix} a_{11} & a_{12} \\ a_{21} & a_{22} \end{bmatrix} \tag{A.53}$$

for which the determinant is very simply defined and calculated as

$$|A| = \begin{vmatrix} a_{11} & a_{12} \\ a_{21} & a_{22} \end{vmatrix} = a_{11}a_{22} - a_{12}a_{21} \tag{A.54}$$

With this definition we can obtain the determinants of square matrices of any order.

Next consider the determinant of a 3×3 matrix

$$|A| = \begin{vmatrix} a_{11} & a_{12} & a_{13} \\ a_{21} & a_{22} & a_{23} \\ a_{31} & a_{32} & a_{33} \end{vmatrix} \tag{A.55}$$

which is defined as

$$|A| = a_{11}(a_{22}a_{33} - a_{23}a_{32}) - a_{12}(a_{21}a_{33} - a_{23}a_{31}) + a_{13}(a_{21}a_{32} - a_{22}a_{31}) \tag{A.56}$$

Note that the parenthetical expressions on the right-hand side of Eq. (A.56) are the determinants of the second-order matrices obtained by striking out the first row and the first, second, and third columns, respectively. These are known as *minors*. A *minor* of a determinant is another determinant formed by removing an equal number of rows and columns from the original determinant. We shall denote by $|M_{ij}|$ the minor determinant obtained by striking out row i and column j. With this definition, Eq. (A.56) can be written

$$|A| = a_{11}|M_{11}| - a_{12}|M_{12}| + a_{13}|M_{13}| \tag{A.57}$$

in which case the determinant is said to be expanded in terms of the *cofactors* of the first row. The *cofactor* of an element a_{ij} is found by giving the appropriate sign (plus or minus) to the minor determinant $|M_{ij}|$. If the sum of the row number i and column number j is *even*, the sign of the cofactor is positive; if $i + j$ is *odd*, the sign of the cofactor is negative. Denoting the cofactor by a_{ij}^c, we can write

$$a_{ij}^c = (-1)^{i+j}|M_{ij}| \tag{A.58}$$

Using this definition, the third-order determinant given by Eq. (A.57) can be written

$$|A| = a_{11}a_{11}^c + a_{12}a_{12}^c + a_{13}a_{13}^c \tag{A.59}$$

The determinant of a square matrix of any order can be obtained by expanding the determinant in terms of the cofactors of any row i as

$$|A| = \sum_{j=1}^{n} a_{ij}a_{ij}^c \tag{A.60}$$

or of any column j as

$$|A| = \sum_{i=1}^{n} a_{ij}a_{ij}^c \tag{A.61}$$

The use of Eq. (A.60) or (A.61) requires that the cofactors a_{ij}^c be further expanded to a point such that all minors are of order 2 and can be evaluated by Eq. (A.54).

The *inverse* of a square matrix $[A]$ is denoted by $[A]^{-1}$ and is defined by

$$[A][A]^{-1} = [A]^{-1}[A] = [I] \tag{A.62}$$

where $[I]$ is known as the *identity* matrix. The identity matrix is such that its diagonal elements are unity and all off-diagonal elements are zero.

The concept of the inverse of a matrix is of prime importance in solving simultaneous equations by matrix methods. Consider the algebraic system

$$\begin{aligned} x_1 - 6x_2 + 2x_3 &= 5 \\ 2x_1 - 3x_2 + x_3 &= 4 \\ 3x_1 + 4x_2 - x_3 &= -2 \end{aligned} \tag{A.63}$$

which can be written in matrix form as

$$[A]\{x\} = \{y\} \tag{A.64}$$

where

$$[A] = \begin{bmatrix} 1 & -6 & 2 \\ 2 & -3 & 1 \\ 3 & 4 & -1 \end{bmatrix}$$

$$\{x\} = \begin{Bmatrix} x_1 \\ x_2 \\ x_3 \end{Bmatrix} \qquad \{y\} = \begin{Bmatrix} 5 \\ 4 \\ -2 \end{Bmatrix}$$

If the inverse $[A]^{-1}$ can be found, we can premultiply both sides of Eq. (A.64) to obtain

$$[A]^{-1}[A]\{x\} = [A]^{-1}\{y\} \tag{A.65}$$

or

$$[I]\{x\} = [A]^{-1}\{y\} \tag{A.66}$$

and thus obtain the solution vector $\{x\}$.

We shall illustrate the method of finding the inverse of $[A]$ known as *Gauss-Jordan reduction*. First we write

$$[A][I] = \begin{bmatrix} 1 & -6 & 2 \\ 2 & -3 & 1 \\ 3 & 4 & -1 \end{bmatrix} \begin{bmatrix} 1 & 0 & 0 \\ 0 & 1 & 0 \\ 0 & 0 & 1 \end{bmatrix} \tag{A.67}$$

and seek to transform $[A]$ by simple row and column operations such that matrix $[A]$ in Eq. (A.67) becomes the identity matrix. If this can be done, the identity matrix will be replaced by $[A]^{-1}$. Multiplying the first row by 2 and subtracting from the second row gives

$$\begin{bmatrix} 1 & -6 & 2 \\ 0 & 9 & -3 \\ 3 & 4 & -1 \end{bmatrix} \begin{bmatrix} 1 & 0 & 0 \\ -2 & 1 & 0 \\ 0 & 0 & 1 \end{bmatrix}$$

Similarly, we multiply the first row by 3 and subtract from the third row to obtain

$$\begin{bmatrix} 1 & -6 & 2 \\ 0 & 9 & -3 \\ 0 & 22 & -7 \end{bmatrix} \begin{bmatrix} 1 & 0 & 0 \\ -2 & 1 & 0 \\ -3 & 0 & 1 \end{bmatrix}$$

Next we divide the second row by 9 and the third row by 22 and subtract the resulting second row from the third, obtaining

$$\begin{bmatrix} 1 & -6 & 2 \\ 0 & 1 & -\frac{1}{3} \\ 0 & 0 & \frac{1}{66} \end{bmatrix} \begin{bmatrix} 1 & 0 & 0 \\ -\frac{2}{9} & \frac{1}{9} & 0 \\ \frac{17}{198} & -\frac{1}{9} & \frac{1}{22} \end{bmatrix}$$

We multiply the second row by 6 and add to the first row, and multiply the third row by 66 to obtain

$$\begin{bmatrix} 1 & 0 & 0 \\ 0 & 1 & -\frac{1}{3} \\ 0 & 0 & 1 \end{bmatrix} \begin{bmatrix} -\frac{1}{3} & \frac{2}{3} & 0 \\ -\frac{2}{9} & \frac{1}{9} & 0 \\ \frac{17}{3} & -\frac{22}{3} & 3 \end{bmatrix}$$

Finally, we multiply the third row by $\frac{1}{3}$ and add to the second row to get

$$\begin{bmatrix} 1 & 0 & 0 \\ 0 & 1 & 0 \\ 0 & 0 & 1 \end{bmatrix} \begin{bmatrix} -\frac{1}{3} & \frac{2}{3} & 0 \\ \frac{15}{9} & -\frac{21}{9} & 1 \\ \frac{17}{3} & -\frac{22}{3} & 3 \end{bmatrix}$$

Thus we obtain

$$[A]^{-1} = \begin{bmatrix} -\frac{1}{3} & \frac{2}{3} & 0 \\ \frac{15}{9} & -\frac{21}{9} & 1 \\ \frac{17}{3} & -\frac{22}{3} & 3 \end{bmatrix}$$

and the solution to the system of equations is

$$\{x\} = \begin{bmatrix} -\frac{1}{3} & \frac{2}{3} & 0 \\ \frac{15}{9} & -\frac{21}{9} & 1 \\ \frac{17}{3} & -\frac{22}{3} & 3 \end{bmatrix} \begin{Bmatrix} 5 \\ 4 \\ -2 \end{Bmatrix}$$

or

$$\begin{Bmatrix} x_1 \\ x_2 \\ x_3 \end{Bmatrix} = \begin{Bmatrix} 1 \\ -3 \\ -7 \end{Bmatrix}$$

as may be verified by direct substitution into Eqs. (A.63).

If the determinant of a square matrix [A] is zero, the inverse of [A] is said not to exist, and the matrix is referred to as *singular*.

A.8 ROOTS OF QUARTIC EQUATIONS

The four roots of a quartic equation

$$x^4 + c_3x^3 + c_2x^2 + c_1x + c_0 = 0 \tag{A.68}$$

can be determined by *Brown's method*. To use this method, we write Eq. (A.68) as the difference of two squares:

$$(x^2 + Ax + B)^2 - (Cx + D)^2 = 0 \tag{A.69}$$

Expanding Eq. (A.69) and equating coefficients of similar powers of x in Eqs. (A.68) and (A.69) gives

$$\begin{aligned} A &= \frac{c_3}{2} \\ A^2 + 2B - C^2 &= c_2 \\ AB - CD &= \frac{c_1}{2} \\ B^2 - D^2 &= c_0 \end{aligned} \tag{A.70}$$

Elimination of A, C, and D from Eqs. (A.70) yields a cubic equation for $2B$ of the form

$$y^3 - c_2y^2 + (c_3c_1 - 4c_0)y + c_0(4c_2 - c_3^2) - c_1^2 = 0 \tag{A.71}$$

where $y = 2B$. If we solve Eq. (A.71) and let $y_1 = 2B$ correspond to the *algebraically largest* real root, the roots of Eq. (A.68) will be the same as the roots of the quadratic equations

$$\begin{aligned} x^2 + (A - C)x + B - D = 0 \\ x^2 + (A + C)x + B + D = 0 \end{aligned} \qquad \text{(A.72)}$$

where A, C, and D are determined by substituting $B = y_1/2$ into Eqs. (A.70).

APPENDIX
B

FORTRAN PROGRAMS

The FORTRAN programs included in this appendix can be used in obtaining digital-computer solutions of vibration problems. The majority of these programs are written specifically for modal analysis of systems with multiple degrees of freedom, as discussed in Chap. 10. For uniformity, each such program is written for a system having 25 degrees of freedom. To use the programs for any other number of degrees of freedom requires only that the appropriate DIMENSION statement be changed in each problem. The following list provides a short description of each program or subroutine.

1. QUAD. Calculates the roots of a quadratic equation.
2. CUBIC. Calculates the roots of a cubic equation.
3. QUART. Calculates the roots of a fourth-degree polynomial (quartic equation).
4. MMULT. Calculates the matrix product of two real square matrices.
5. MTSP. Calculates the transpose of a real square matrix.
6. ATBA. Calculates the matrix product $[A]^T[B][A]$, where $[A]$ and $[B]$ are real square matrices.
7. SIMUL. Solves a system of simultaneous linear equations.
8. EIGEN1. Calculates eigenvalues and normalized eigenvectors of a nonsymmetric matrix of the form $[D] = [K]^{-1}[M]$, where $[M]$ and $[K]$ are real symmetric matrices, and $[K]$ is positive-definite.
9. EIGEN2. Calculates eigenvalues and eigenvectors of a real symmetric matrix.
10. NORMAL. Normalizes a matrix of eigenvectors.
11. MPRNT. Subroutine for obtaining a matrix in printed form or on punched cards.

12. MAIN1. Sample main program for modal analysis using other subroutines included in this appendix.
13. MAIN2. Sample main program for modal analysis using IBM subroutines.
14. FREVIB. Calculates the response of a lumped-mass system to initial excitation.

B.1 SUBROUTINE QUAD

This program calculates the roots of a quadratic equation of the form

$$c_1 x^2 + c_2 x + c_3 = 0$$

where c_1, c_2, and c_3 are real constants. The program parameters are

CO. Input vector of coefficients of the quadratic equation, ordered from largest to smallest power of x

R1, R2. Real parts of the roots of the equation

RI. Imaginary part of the roots of the equation

The subroutine provides the values R1, R2, and RI upon return to the calling program.

```
      SUBROUTINE QUAD (CO,R1,R2,RI)
      DIMENSION CO(3)
      B = CO(2)/CO(1)
      C = CO(3)/CO(1)
      D = B**2 - 4.*C
      IF(D) 10,5,5
    5 D = SQRT(D)
      R1 = (-B + D)/2
      R2 = -(B + D)/2
      RI = 0.0
      RETURN
   10 D = SQRT(ABS(D))
      R1 = -B/2
      R2 = R1
      RI = D/2.
      RETURN
      END
```

B.2 SUBROUTINE CUBIC

The roots of a cubic equation of the form

$$c_1 x^3 + c_2 x^2 + c_3 x + c_4 = 0$$

are obtained by determining a real root and using this root to reduce the equation to a quadratic. The quadratic is then solved by subroutine QUAD. The program parameters are:

CC. Input vector containing coefficients of the cubic equation ordered from highest to lowest power of x
RR. Output vector containing the real parts of the roots of the equation
RI. Imaginary part of the roots of the equation
CQ. Vector of coefficients of the quadratic equation generated within the subroutine

Upon return to the calling program, subroutine CUBIC provides the roots of the cubic equation as

$$x_1 = \mathrm{RR}(1)$$
$$x_2 = \mathrm{RR}(2) + i\mathrm{RI}$$
$$x_3 = \mathrm{RR}(3) - i\mathrm{RI}$$

This subroutine requires subroutine QUAD.

```
      SUBROUTINE CUBIC(CC,RR,RI)
      DIMENSION CC(4),RR(3),CQ(3)
      J = 1
      DO 5 I = 2,4
    5 CC(I) = CC(I)/CC(1)
      IF(CC(4))15,10,15
   10 RR(I) = 0.0
      GO TO 75
   15 P = CC(2)
      Q = CC(3)
      R = CC(4)
      A = (3.*Q - P**2)/9.
      B = (2.*P**2 - 9.*P*Q + 27.*R)/54.
      IF (B)25,70,30
   25 B = -B
      J = 2
      X = A**3 + B**2
   30 IF (X)35,40,45
   35 RR(1) = 2.*SQRT(-A)*COS(ATAN(SQRT(X)/B)/3.))
      GO TO 65
   40 RR(1) = -2.*B**(1./3.)
      GO TO 65
   45 IF(A) 50,55,60
   50 RR(1) = -(B + SQRT(X))**(1./3.) - (B - SQRT(X))**(1./3.)
      GO TO 65
   55 RR(1) = -(2.*B)**(1./3.)
      GO TO 65
```

```
60  RR(1) = (SQRT(X) - B)**(1./3.) - (SQRT(X) + B)**(1./3.)
65  IF(J.EQ.2) RR(1) = -RR(1)
    RR(1) = RR(1) - P/3
    GO TO 75
70  RR(1) = -P/3.
75  CQ(1) = 1.0
    CQ(2) = CQ(2) + RR(1)
    CQ(3) = (CC(2) + RR(1))*RR(1) + CC(3)
    CALL QUAD (CQ, RR(2), RR(3), RI)
    RETURN
    END
```

B.3 SUBROUTINE QUART

This program calculates the roots of a quartic equation

$$c_1x^4 + c_2x^3 + c_3x^2 + c_4x + c_5 = 0$$

by Brown's method. The subroutine requires subroutines QUAD and CUBIC, as these are both called during the calculation. The program parameters are:

C. Input vector of coefficients of the equation, ordered from largest to smallest power of x
RR. Vector containing real parts of the roots of the equation
RI. Vector containing imaginary parts of the roots of the equation

Upon return to the calling program, the roots are

$$x_N = \mathrm{RR}(N) + i\mathrm{RI}(N) \qquad N = 1, 2, 3, 4$$

Subroutine QUART is as follows:

```
SUBROUTINE QUART(C, RR, RI)
DIMENSION C(5), RR(4), RI(4), CC(4), CQ(3), RC(3)
C1 = C(2)/C(1)
C2 = C(3)/C(1)
C3 = C(4)/C(1)
C4 = C(5)/C(1)
CC(1) = 1.
CC(2) = -C2
CC(3) = C3*C1 - 4.*C4
CC(4) = C4*(4.*C2 - C1**2) - C3**2
CALL CUBIC(CC, RC, RCI)
IF (RCI.NE.0.0) GO TO 5
Y = AMAX1(RC(1), RC(2), RC(3))
RC(1) = Y
```

```
 5  Y = RC(1)/2.
    IF((Y**2 - C4).GT.0.0) GO TO 10
    X = 0.0
    Z = SQRT((C1/2.)**2 + 2.*Y - C2)
    GO TO 15
10  X = SQRT(Y**2 - C4)
    Z = -(C3/2.-C1*Y/2.)/X
15  CQ(1) = 1.0
    CQ(2) = C1/2.+Z
    CQ(3) = Y + X
    CALL QUAD (CQ,RR(1),RR(2),RI(1))
    CQ(1) = 1.0
    CQ(2) = C1/2 - Z
    CQ(3) = Y - X
    CALL QUAD (CQ,RR(3),RR(4),RI(3))
    RI(2) = -RI(1)
    RI(4) = -RI(3)
    RETURN
    END
```

B.4 SUBROUTINE MMULT

This subroutine calculates the matrix product

$$[C] = [A][B]$$

where $[C]$, $[A]$, and $[B]$ are real square matrices of order N. Program parameters are:

A. Input array containing the elements $[A]$
B. Input array containing the elements of $[B]$
C. Output array containing the elements of $[C]$ on return
N. Order of the matrices

Matrices $[A]$ and $[B]$ are preserved.

```
    SUBROUTINE MMULT(A,B,C,N)
    DIMENSION A(25,25),B(25,25),C(25,25)
    DO 5 I = 1,N
    DO 5 J = 1,N
    S = 0.0
    DO 10 K = 1,N
10  S = S + A(I,K)*B(K,J)
 5  C(I,J) = S
    RETURN
    END
```

stop

B.5 SUBROUTINE MTSP

This subroutine determines the transpose of a real square matrix. The program parameters are:

A. Input array containing the elements of the original matrix
AT. Output array containing the elements of the transpose of the original matrix upon return
N. Order of the matrix

The original matrix is preserved.

```
      SUBROUTINE MTSP(A,AT,N)
      DIMENSION A(25,25),AT(25,25)
      DO 5 I = 1,N
      DO 5 J = 1,N
    5 AT (J,I) = A(I,J)
      RETURN
      END
```

B.6 SUBROUTINE ATBA

This program performs a matrix multiplication of the form

$$[D] = [A]^T[B][A]$$

where $[A]$, $[B]$, and the resultant $[D]$ are real square matrices. The program parameters are:

A. Input array containing the elements of matrix $[A]$
B. Input array containing the elements of matrix $[B]$; upon return, array B contains the elements of $[A]^T[B][A]$
C. Dummy array used for working storage; contains the elements of $[B][A]$ on return
N. Order of the matrices

Matrix $[A]$ is preserved; $[B]$ is destroyed.

```
      SUBROUTINE ATBA(A,B,C,N)
      DIMENSION A(25,25),B(25,25),C(25,25)
      DO 5 I = 1,N
      DO 5 J = 1,N
      S = 0.0
      DO 10 K = 1,N
   10 S = S + B(I,K)*A(K,J)
```

```
    5 C(I,J) = S
      DO 15 I = 1,N
      DO 15 J = 1,N
      S = 0.0
      DO 20 K = 1,N
   20 S = S + A(K,I)*C(K,J)
   15 B (I,J) = S
      RETURN
      END
```

B.7 SUBROUTINE SIMUL

A system of linear algebraic equations of the form

$$[A]\{x\} = \{F\}$$

is solved by the method of Gauss-Jordan elimination. The program parameters are:

A. Input array containing the elements of the real square matrix of coefficients $[A]$

F. One-dimensional input array containing the elements of $\{F\}$; upon return, array F contains the solution vector $\{x\}$

N. Number of equations in the system

IE. Error-code return with solution:

IE = 0	No error
IE = 1	Input error, $N \leq 0$
IE = 2	Coefficient matrix is singular

Input matrix $[A]$ and vector $\{F\}$ are destroyed in the computation.

```
      SUBROUTINE SIMUL(A,F,N,IE)
      DIMENSION A(25,25),F(25)
      IF(N.LE.0) GO TO 99
      IE = 0
      K = 1
    2 P = 0.0
      DO 5 I = K,N
      DO 5 J = K,N
      IF(ABS(A(I,J)).LE.P) GO TO 5
      P = ABS(A(I,J))
      P1 = A(I,J)
    3 I1 = I
      J1 = J
```

```
    5 CONTINUE
      IF (P)100,100,3
      DO 10 J = K,N
      TMP = A(K,J)
      A(K,J) = A(I1,J)
   10 A(I1,J) = TMP
      TMP = F(K)
      F(K) = F(I1)/P1
      F(I1) = TMP
      DO 15 I = K,N
      TMP = A(I,K)
      A(I,K) = A(I,J1)
   15 A(I,J1) = TMP
      DO 20 I = K,N
   20 A(K,I) = A(K,I)/P1
      A(K,K) = J1 - K
      L = K + 1
      DO 25 I = L,N
      DO 30 J = L,N
   30 A(I,J) = A(I,J) - A(I,K)*A(K,J)
   25 F(I) = F(I) - A(I,K)*F(K)
      IF(K.EQ.N) GO TO 35
      K = K + 1
      GO TO 2
   35 J = N - 1
      DO 40 I = 1,J
      SUM = 0.0
      L = N - I
      L1 = L + 1
      DO 45 M = L1,N
   45 SUM = SUM + A(L,M)*F(M)
      F(L) = F(L) - SUM
      M = A(L,L) + 0.5
      IF (M.EQ.0) GO TO 40
      K = M + L
      T = F(K)
      F(K) = F(L)
      F(L) = T
   40 CONTINUE
      RETURN
   99 IE = 1
      RETURN
  100 IE = 2
      RETURN
      END
```

B.8 SUBROUTINE EIGEN1

The eigenvalues and normalized eigenvectors of a nonsymmetric matrix $[S]$ having the special form

$$[S] = [B]^{-1}[A]$$

are calculated. $[A]$ and $[B]$ must both be real symmetric matrices, and $[B]$ is required to be positive-definite. The program parameters are:

A. Input array containing the elements of matrix $[A]$; on return, array A is the matrix of *normalized* eigenvectors
B. Input array containing the elements of matrix $[B]$
C. Array used by the subroutine as working storage
D. One-dimensional array which, upon return, contains the eigenvalues in decreasing order of magnitude
N. Order of the square matrices

Original matrices $[A]$ and $[B]$ are destroyed in the computation. This subroutine also requires subroutines EIGEN2, and ATBA.

```
      SUBROUTINE EIGEN1(A,B,C,D,N)
      DIMENSION A(25,25),B(25,25),C(25,25),D(25)
      CALL EIGEN2(B,C,N)
      DO 5 I = 1,N
      D(I) = 1./SQRT(B(I,I))
    5 C(J,I) = D(I)*C(J,I)
      CALL ATBA (C,A,B,N)
      CALL EIGEN2 (A,B,N)
      DO 10 I = 1,N
   10 D(I) = A(I,I)
      DO 20 I = 1,N
      DO 20 J = 1,N
      S = 0.0
      DO 25 K = 1,N
   25 S = S + C(I,K)*B(K,J)
   20 A(I,J) = S
      NN = N - 1
      DO 30 I = 1,NN
      J = I + 1
      IF (D(I).GT.D(K)) GO TO 30
      X = D(I)
      D(I) = D(K)
      D(K) = X
      DO 35 L = 1,N
      X = A(L,I)
      A(L,I) = A(L,K)
```

```
35  A(L,K) = X
30  CONTINUE
    DO 40 I = 1,N
    X = 0.0
    DO 45 J = 1,N
45  X = X + A(J,I)**2
    DO 40 J = 1,N
40  A(J,I) = A(J,I)/SQRT(X)
    RETURN
    END
```

B.9 SUBROUTINE EIGEN2

This subroutine calculates the eigenvalues and eigenvectors of a real symmetric matrix $[B]$. The program parameters are:

B. Input array containing the elements of matrix $[B]$; on return, the eigenvalues are stored as the diagonal elements of array B
R. On return, this array contains the eigenvectors of matrix $[B]$; the eigenvectors are stored in column order
N. Order of matrix $[B]$

The original matrix $[B]$ is destroyed in the computation, and the eigenvalues are in no particular order on return. The calculated eigenvectors are *not* normalized.

```
      SUBROUTINE EIGEN2(B,R,N)
      DIMENSION B(25,25),R(25,25)
      REAL NORM1,NORM2
      E = 1.0E-6
      DO 5 I = 1,N
      DO 5 J = 1,N
      R(I,J) = 0.0
    5 R(I,I) = 1.0
      NORM1 = 0.0
      DO 10 I = 1,N
      DO 10 J = 1,N
      IF(I.EQ.J) GO TO 10
      NORM1 = NORM1 + B(I,J)**2
   10 CONTINUE
      IF (NORM1.EQ.0.0) GO TO 40
      NORM1 = SQRT(2.0*NORM1)
      NORM2 = NORM1*E/N
      NDK = 0
      TH = NORM1
```

```
15  TH = TH/N
20  DO 30 I = 2,N
    I1 = I - 1
    DO 30 J = 1,I1
    IF(ABS(B(J,I)).LT.TH) GO TO 30
    NDK = 1
    U = 0.5*(B(I,I) - B(J,J))
    S = 1.0
    IF(U.LT.0.0) S = -1.0
    V = -B(J,I)
    W = S*V/SQRT(V**2 + U**2)
    SINT = W/SQRT(2.*(1.+SQRT(1.-W**2)))
    COST = SQRT(1.-SINT**2)
    W = 2.*V*SINT*COST
    U = B(J,J)
    V = B(I,I)
    DO 25 L = 1,N
    X = B(L,J)
    Y = B(L,I)
    B(L,J) = X*COST - Y*SINT
    B(J,L) = B(L,J)
    B(L,I) = X*SINT + Y*COST
    B(I,L) = B(L,I)
    X = R(L,J)
    Y = R(L,I)
    R(L,J) = X*COST - Y*SINT
25  R(L,I) = X*SINT + Y*COST
    B(J,J) = U*COST**2 + V*SINT**2 + W
    B(I,I) = U*COST**2 + V*SINT**2 - W
30  CONTINUE
    IF (NDK.EQ.0) GO TO 35
    NDK = 0
    GO TO 20
35  IF(TH.GE.NORM2) GO TO 15
40  CONTINUE
    RETURN
    END
```

B.10 SUBROUTINE NORMAL

This routine normalizes the column vectors of a real square matrix of eigenvectors $[A]$ with respect to a second real square symmetric matrix $[B]$ such that

$$[A]^T[B][A] = [I]$$

where $[I]$ is the identity matrix. The program parameters are:

A. Input array containing the elements of original matrix $[A]$; on return, this array contains the normalized elements
B. Input array containing the elements of matrix $[B]$
N. Order of matrices $[A]$ and $[B]$

Original matrix $[A]$ is destroyed in the computations; $[B]$ is preserved.

```
      SUBROUTINE NORMAL(A,B,N)
      DIMENSION A(25,25),B(25,25)
      DO 5 K = 1,N
      S = 0.0
      DO 10 I = 1,N
      DO 10 J = 1,N
   10 S = S + B(I,J)*A(I,K)*A(J,K)
      DO 5 I = 1,N
      A(I,K) = A(I,K)/SQRT(S)
      RETURN
      END
```

B.11 SUBROUTINE MPRNT

This routine is used to print on line or punch on cards the elements of a real square matrix $[A]$. The complete matrix can be printed or punched in row order, or only the diagonal elements can be obtained. Row-column location is included as part of the punched output. The program parameters are:

A. Input array containing elements of matrix $[A]$
N. Order of matrix $[A]$
IP. Input integer used to specify output mode:

IP = 0	For printed output
IP = 1	For punched cards
IP = 2	For both

IO. Input integer used to specify output values desired:

IO = 0	For complete matrix
IO = 1	For diagonal elements only

The subroutine is as follows:

```
      SUBROUTINE MPRNT(A,N,IP,IO)
      DIMENSION A(25,25)
      IF (IP.GE.1) GO TO 20
```

```
  2  WRITE (6,101)
     IF (IO)5,5,15
  5  DO 10 I = 1,N
     WRITE (6,102)I
 10  WRITE (6,103)(A(I,J),J = 1,N)
     GO TO 40
 15  WRITE (6,104)
     WRITE (6,103)(A(I,I),I = 1,N)
     GO TO 40
 20  IF (IO)25,25,30
 25  WRITE (5,105)((I,J,A(I,J),J = 1,N),I = 1,N)
     GO TO 35
 30  WRITE (5,105)(I,I,A(I,I),I = 1,N)
 35  IF (IP.EQ.2) GO TO 2
 40  CONTINUE
101  FORMAT (1H1)
102  FORMAT ('0','ROW',I3)
103  FORMAT ('0',10E13.7)
104  FORMAT ('0','DIAGONAL ELEMENTS')
105  FORMAT(I3,I3,E14.7)
     RETURN
     END
```

B.12 MAIN1 MODAL-ANALYSIS PROGRAM

This sample main program calculates the natural frequencies and the normalized modal matrix of a lumped-mass system having n degrees of freedom. The required input values are as follows:

N. Number of degrees of freedom
NM. Number of *nonzero* elements in the mass matrix
NK. Number of *nonzero* elements in the stiffness matrix
MM. The mass matrix; initialized at zero so only nonzero elements are required as input
KM. The stiffness matrix; initialized at zero so only nonzero elements are required as input.

The program calculates eigenvalues and normalized eigenvectors using subroutine EIGEN1. The system natural frequencies are obtained from the eigenvalues by direct calculation. The program output includes: (1) a printout of the natural frequencies in ascending order of magnitude, (2) the natural frequencies on punched cards in ascending order, (3) a printout of the normalized modal matrix, and (4) the elements of the modal matrix on punched cards with row and column identification. The punched-card output is convenient as input for other programs which calculate system response. Program MAIN1 requires subroutines EIGEN1, EIGEN2, ATBA, and MPRNT.

```
      REAL MM,KM
      DIMENSION MM(25,25),KM(25,25)
      DIMENSION A(25,25),B(25,25),C(25,25),D(25)
      DIMENSION W(25)
      DATA MM, KM/1250*0.0/
      READ (5,100)N,NM,NK
      DO 5 I = 1,NM
    5 READ (5,101)J,K,MM(J,K)
      DO 10 I = 1,NK
   10 READ (5,101)J,K,KM(J,K)
      DO 15 I = 1,N
      DO 15 J = 1,N
      A(I,J) = MM(I,J)
   15 B(I,J) = KM(I,J)
      CALL EIGEN1(A,B,C,D,N)
      DO 20 I = 1,N
   20 W(I) = 1.0/SQRT(D(I))
      WRITE (6,102)
      WRITE (6,103)(I,W(I),I = 1,N)
      WRITE (5,104) (I,W(I),I = 1,N)
      WRITE (6,105)
      CALL MPRNT(A,N,2,0)
  100 FORMAT (3I3)
  101 FORMAT (2I3,E14.7)
  102 FORMAT ('1 NATURAL FREQUENCIES')
  103 FORMAT ('0W('I3') = 'E14.7)
  104 FORMAT (I3, E14.7,I3,E14.7,I3,E14.7,I3,E14.7)
  105 FORMAT (' MODAL MATRIX')
      STOP
      END
```

B.13 MAIN2 MODAL-ANALYSIS PROGRAM

The input requirements and output data of this program are identical to those of program MAIN1. The only difference in the two is that MAIN2 solves the eigenvalue problem by calling subroutines from the IBM Scientific Subroutine Package as discussed in Chap. 10. MAIN2 also calls subroutines MPRNT and NORMAL.

```
      REAL MM,KM
      DIMENSION MM(25,25),KM(25,25)
```

```
      DIMENSION AK(325),D(25,25),RR(25),RI(25)
      DIMENSION IANA(25)
      DATA MM,KM/1250*0.0/
      READ (5,100)N,NM,NK
      DO 5 I = 1,NM
    5 READ (5,101)J,K,MM(J,K)
      DO 10 I = 1,NK
   10 READ (5,101)J,K,KM(J,K)
      L = 1
      DO 15 J = 1,N
      DO 15 I = 1,J
      AK(L) = KM(I,J)
   15 L = L + 1
      E = 1.0E - 6
      CALL SINV(AK,N,E,IER)
      CALL MPRD(AK,MM,D,N,N,1,0)
      CALL HSBG(N,D,1)
      CALL ATEIG(N,D,RR,RI,IANA,1)
      DO 20 I = 1,N
   20 RR(I) = 1./SQRT(RR(I))
      DO 25 I = 1,N
      IF (RI(I).NE.0.0) GO TO 30
   25 CONTINUE
      WRITE (6,102)
      WRITE (6,103)(I,RR(I),I = 1,N)
      WRITE (5,104)(I,RR(I),I = 1,N)
      CALL NORMAL(D,MM,N)
      WRITE (6,105)
      CALL MPRNT(D,N,2,0)
      GO TO 35
   30 WRITE (6,106)
   35 CONTINUE
  100 FORMAT (3I3)
  101 FORMAT (2I3,E14.7)
  102 FORMAT ('1 NATURAL FREQUENCIES')
  103 FORMAT ('0W('I3,') = ',E14.6)
  104 FORMAT (I3,E14.7,I3,E14.7,I3,E14.7,I3,E14.7)
  105 FORMAT ('1 MODAL MATRIX')
  106 FORMAT ('1 IMAGINARY EIGENVALUES OBTAINED')
      STOP
      END
```

B.14 PROGRAM FREVIB

This is a sample main program for calculating the free response of a system with n degrees of freedom to some initial excitation. The program requires that the system natural frequencies and the corresponding normalized modal matrix be available as input. These values are obtained as output from programs MAIN1 or MAIN2. Required input values are:

N. Number of degrees of freedom
N1. Number of modes to be included in displacement calculations ($N1 \leq N$)
W. One-dimensional array of natural frequencies
A. Two-dimensional array containing the elements of the normalized modal matrix
TF. Total time for which system response is to be calculated
DT. Incremental time between displacement calculations
XO. One-dimensional array of initial displacements
XDOTO. One-dimensional array of initial velocities

The program calls subroutine SIMUL twice to evaluate the arbitrary constants. Thus, SIMUL is also required. Program output includes the displacement, velocity, and acceleration for each coordinate, as well as the corresponding time.

```
      DIMENSION W(25),A(25,25),XO(25),XDOTO(25)
      DIMENSION C(25,25),B(25),PHI(25)
      DATA T/0.0/
      READ (5,100)N,N1
      READ (5,101)TF,DT
      READ (5,102)(I,W(I),I = 1,N)
      READ (5,103)((I,J,A(I,J),J = 1,N),I = 1,N)
      READ (5,104)(XO(I),XDOTO(I),I = 1,N)
      DO 5 I = 1,N
      DO 5 J = 1,N
    5 C(I,J) = A(I,J)
      CALL SIMUL(C,XO,N,IE)
      DO 10 I = 1,N
      DO 10 J = 1,N
   10 C(I,J) = W(J)*A(I,J)
      CALL SIMUL(C,XDOTO,N,IE)
      DO 15 I = 1,N
      B(I) = SQRT(XO(I)**2 + XDOTO(I)**2)
   15 PHI(I) = ATAN(XO(I)/XDOTO(I))
      WRITE (6,200)
   35 T = T + DT
      WRITE (6,201)T
      WRITE (6,202)
      DO 20 I = 1,N
      X = 0.0
      XD = 0.0
      XDD = 0.0
      DO 25 J = 1,N
      ANG = W(J)*T + PHI(J)
```

```
      X = X + B(J)*A(I,J)*SIN(ANG)
      XD = XD + W(J)*B(J)*A(I,J)*COS(ANG)
   25 XDD = XDD - B(J)*A(I,J)*SIN(ANG)*W(J)**2
   20 WRITE (6,203) I,X,XD,XDD
      IF(T.GT.TF) GO TO 30
      GO TO 35
  100 FORMAT (2I3)
  101 FORMAT (2F6.2)
  102 FORMAT (I3,E14.7,I3,E14.7,I3,E14.7,I3,E14.7)
  103 FORMAT (2I3,E14.7)
  104 FORMAT (2F10.2)
  200 FORMAT ('1')
  201 FORMAT (1X,'T = ',F8.2)
  202 FORMAT ('0','COORD',5X,'DISPL',10X,'VEL',12X,'ACCEL'/)
  203 FORMAT (1X,I3,1X,3E15.7)
   30 STOP
      END
```

BIBLIOGRAPHY

The following list is but a representative sample of the many excellent references on vibrations and related subjects.

Church, A. H.: *Mechanical Vibrations*, 2d ed., Wiley, New York, 1963.

Den Hartog, J. P.: *Mechanical Vibrations*, 4th ed., McGraw-Hill, New York, 1956.

Hamming, R. W.: *Numerical Methods for Scientists and Engineers*, McGraw-Hill, New York, 1962.

Harris, C. M., and C. E. Crede: *Shock and Vibration Handbook*, McGraw-Hill, New York, 1976.

James, M. L., G. M. Smith, and J. C. Wolford: *Analog Computer Simulation of Engineering Systems*, International Textbook, Scranton, Pa., 1966.

Kreider, D. L., R. G. Kuller, D. R. Ostberg, and F. W. Perkins: *An Introduction to Linear Analysis*, Addison-Wesley, Reading, Mass., 1966.

Kunz, K. S.: *Numerical Analysis*, McGraw-Hill, New York, 1957.

Meirovitch, L.: *Analytical Methods in Vibrations*, Macmillan, New York, 1967.

———: *Elements of Vibration Analysis*, McGraw-Hill, New York, 1975.

Myklestad, N. O.: *Fundamentals of Vibration Analysis*, McGraw-Hill, New York, 1956.

Ralston, A.: *A First Course in Numerical Analysis*, McGraw-Hill, New York, 1965.

Timoshenko, S., and D. H. Young: *Vibration Problems in Engineering*, 3d ed., Van Nostrand, Princeton, N. J., 1955.

INDEX